In 1970 Bernd Fischer proved his beautiful theorem classifying the almost simple groups generated by 3-transpositions and, in the process, discovered three new sporadic groups, now known as the Fischer groups. Since then, the theory of 3-transposition groups has become an important part of finite simple group theory, but Fischer's work has remained unpublished. This is particularly unfortunate since, despite the depth of Fischer's result, the number and complexity of the examples it covers, and its numerous consequences, the proof can be understood knowing only some elementary group theory and finite geometry.

Part I of *3-Transposition Groups* contains the first published proof of Fischer's Theorem written out completely in one place. It can serve as a text for an intermediate level graduate course on finite groups. Prerequisites consist only of that part of an introductory undergraduate course in abstract algebra covering groups and linear algebra, plus some material from the author's earlier book, *Finite Group Theory*, on the elementary theory of finite groups.

Parts II and III are aimed at specialists in finite groups. They establish the existence, uniqueness, and structural results for the Fischer groups, necessary for the classification of the finite simple groups. Parts II and III are a step in the author's program (begun in *Sporadic Groups*) to supply a strong foundation for the theory of sporadic groups.

T0291593

CAMBRIDGE TRACTS IN MATHEMATICS

General Editors

B. BOLLOBAS, P. SARNAK, C. T. C. WALL

124   3-Transposition groups

124   5-Transposition groups

MICHAEL ASCHBACHER
California Institute of Technology

3-Transposition groups

CAMBRIDGE UNIVERSITY PRESS
Cambridge, New York, Melbourne, Madrid, Cape Town, Singapore, São Paulo, Delhi

Cambridge University Press
The Edinburgh Building, Cambridge CB2 8RU, UK

Published in the United States of America by Cambridge University Press, New York

www.cambridge.org
Information on this title: www.cambridge.org/9780521571968

First published 1997
This digitally printed version 2008

A catalogue record for this publication is available from the British Library

Library of Congress Cataloguing in Publication data
Aschbacher, Michael, 1944–
3-transposition groups / Michael Aschbacher.
p.   cm. – (Cambridge tracts in mathematics; 124)
Includes bibliographical references (pp. 253–4) and index.
ISBN 0-521-57196-0 (hardback)
1. Finite groups.   I. Title.   II. Series.
QA177.A79   1997
512'.2–dc20   96–26497
                                               CIP

ISBN 978-0-521-57196-8 hardback
ISBN 978-0-521-10102-8 paperback

Contents

PART I
FISCHER'S THEORY

Introduction

Around 1970 Bernd Fischer proved his beautiful theorem classifying almost simple finite groups generated by 3-transpositions, and in the process discovered three new sporadic simple groups, now termed *Fischer groups*. Fischer's Theorem was deep; the list of groups appearing in its conclusion included not only the three sporadic Fischer groups but also the symmetric groups and several families of classical groups. Nevertheless Fischer used none of the sophisticated machinery of the day. His proof required little more than some elementary group theory and finite geometry, combined in a new and original way.

The hypotheses of Fischer's Theorem were unusual for the time too. Fischer considered the following setup. Let G be a finite group. A *set of* 3-*transpositions* of G is a set D of involutions of G (i.e., elements of order 2) such that D is the union of conjugacy classes of G, D generates G, and for all a, b in D, the order of the product ab is 1, 2, or 3.

One example comes to mind immediately; namely the class D of transpositions forms a conjugacy class of 3-transpositions of any symmetric group G. For if $t = (x, y)$ and $s = (u, v)$ are distinct transpositions then ts has order 2 if $\{x, y\} \cap \{u, v\}$ is empty, whereas ts has order 3 if the intersection is nonempty.

Fischer then imposed some extra constraints to single out the most interesting examples, which are almost simple. There are various equivalent hypotheses; one choice is the following:

Fischer's Theorem. *Let D be a conjugacy class of 3-transpositions of the finite group G. Assume the center of G is trivial and the derived subgroup of G is simple. Then one of the following holds:*

(1) $G \cong S_n$ is the symmetric group of degree n and D is the set of transpositions of G.

1

(2) $G \cong Sp_n(2)$ is the symplectic group of dimension n over the field of order 2 and D is the set of transvections.

(3) $G \cong U_n(2)$ is the projective unitary group of dimension n over the field of order 4 and D is the set of transvections.

(4) $G \cong O_n^\epsilon(2)$ is an orthogonal group of dimension n over the field of order 2 and D is the set of transvections.

(5) $G \cong PO_n^{\mu,\pi}(3)$ is the subgroup of an n-dimensional projective orthogonal group over the field of order 3 generated by a conjugacy class D of reflections.

(6) G is a Fischer group of type $M(22)$, $M(23)$, or $M(24)$, determined up to isomorphism, and D is a uniquely determined class of involutions in G.

To be more precise, our proof of Fischer's Theorem shows in cases (1)–(5) only that D is determined up to conjugation in $\mathrm{Aut}(G)$. Thus statements like "*the* set of transpositions" should really read "the set of transpositions with respect to some representation of G." The definition of groups of type $M(22)$, $M(23)$, and $M(24)$ will be given later in this Introduction after we introduce more notation. The notation $PO_n^{\mu,\pi}(3)$ is defined precisely in Section 11, which contains a discussion of the classical groups appearing in cases (2)–(5) of Fischer's Theorem.

The general outline of our proof of Fischer's Theorem is close to that of Fischer in his Warwick Lecture notes [F1]. (Unfortunately these notes remain largely unpublished. Only the first few sections appear in the *Inventiones* article [F2].) However many of the details are different. For example a number of modifications due to Richard Weiss in [W1] are incorporated. Most particularly the proof of the uniqueness result on triple graphs in Section 12 is taken from [W1]. Other references on 3-transpositions appear in the References. Some are discussed in Chapter 6.

The theory of 3-transpositions and Fischer's Theorem in particular are of importance to specialists in finite group theory. Fischer's Theorem is the best means of establishing (in conjunction with the involvement of $M(24)$ in the normalizer in the Monster of a subgroup of order 3) the existence and uniqueness of the Fischer groups for purposes of the classification of the finite simple groups. The theory of 3-transpositions is the basic tool for studying the subgroup structure of the Fischer groups. It is also useful in establishing certain sporadic embeddings of 3-transposition groups in other simple groups and is used in establishing the uniqueness and studying the structure of other sporadics. Finally the theory of 3-transpositions is part of a larger theory that describes root elements in groups of Lie type from an abstract group theoretic point of view. Thus the book serves as a basic reference in finite simple group theory.

In Part I of *3-Transposition Groups* we develop the theory of 3-transpositions and use that theory to give a proof of Fischer's Theorem. In Part II we use 3-transposition theory to prove the existence and uniqueness of the three Fischer groups as groups with certain involution centralizers. Finally in Part III we use 3-transposition theory to study the subgroup structure of the Fischer groups. In Chapter 6 there is also a brief discussion of Fischer spaces, including Marshall Hall's result describing the largest 4-generator 3-transposition group of width 1, and there is an overview of some more recent literature on infinite 3-transposition groups and 3-transposition groups that are not almost simple. At the end of the chapter there is a very brief indication of the role of 3-transpositions in the study of root elements in groups of Lie type.

The treatment in Part I assumes only the part of an introductory course in algebra covering groups and vector spaces, plus some material from the text *Finite Group Theory* [FGT] on the elementary theory of finite groups. The majority of the results needed from [FGT] are summarized in Chapter 1. Section 10 and part of Section 11 also summarize some of the material from Chapter 7 of [FGT] on classical groups. Thus the first half of the book is accessible to someone with a minimal background. It offers an unusual opportunity for a student to see the proof of a deep, complex result at an early stage in his or her mathematical development. In the process, the student is exposed to Coxeter systems, the classical groups and their geometries, sporadic groups, rank 3 permutation groups, the generalized Fitting subgroup, and extraspecial groups, among other useful group theoretic topics.

Parts II and III are more difficult going and require more background, although even here, with the exception of an appeal to Glauberman's Z^*-Theorem, all necessary background can be found either in [FGT] or in the author's text [SG] on sporadic groups. This last half of the book is aimed at simple group theorists or those with some knowledge of finite groups who wish to learn more about the classification of the finite simple groups. It is part of the author's program to supply a strong foundation for the theory of the sporadic groups.

In the remainder of this Introduction we define the Fischer groups and the centralizer of involution hypotheses with which we will characterize those groups, and discuss briefly what goes on in *3-Transposition Groups*. The Introductions to Parts II and III provide a more detailed overview of those portions of the book. The introduction to Chapter 5 gives an outline of the proof of Fischer's Theorem.

Let D be a conjugacy class of 3-transpositions of a finite group G. For $d \in D$ define D_d, A_d, to be the set of elements $b, c \in D$ such that db, dc has order 2, 3, respectively. Let $d \in D$, $H = \langle D_d \rangle$, $c \in A_d$, and $L = \langle D_d \cap D_c \rangle$. Define G to be of *type* $M(22)$ if $Z(G) = 1$, $H/Z(H) \cong U_6(2)$, and $L/Z(L) \cong PO_6^{+,\pi}(3)$.

Define G to be of *type* $M(23)$ if $Z(G) = 1$, $H/Z(H)$ is of type $M(22)$, and $L/Z(L) \cong PO_7^{-\pi,\pi}(3)$. Finally define G to be of *type* $M(24)$ if $Z(G) = 1$, $H/Z(H)$ is of type $M(23)$, and $L/Z(L) \cong S_3/P\Omega_8^+(3)$. Our proof of Fischer's Theorem shows that up to isomorphism there is at most one group of type M for $M = M(22)$, $M = M(23)$, and $M = M(24)$.

In Part II we characterize the Fischer groups as the unique finite groups with certain involution centralizers. A finite group G is of *type* $\tilde{M}(22)$ if G possesses an involution d such that $C_G(d)$ is quasisimple with $C_G(d)/\langle d \rangle \cong U_6(2)$ and d is not weakly closed in $C_G(d)$ with respect to G. A finite group G is of *type* $\tilde{M}(23)$ if G possesses an involution d such that $C_G(d)$ is quasisimple with $C_G(d)/\langle d \rangle$ of type $M(22)$ *and* of type $\tilde{M}(22)$, and d is not weakly closed in $C_G(d)$ with respect to G.

We eventually prove in Part II that G is of type M if and only if G is of type $\tilde{M}$ for $M = M(22)$ and $M(23)$. By Fischer's Theorem there is at most one group of type M. Thus if we show $\tilde{M}$ implies M and show groups of type $\tilde{M}$ exist, then we have also shown that M implies $\tilde{M}$, that groups of type $\tilde{M}$ are unique, and that groups of type M exist. In Chapter 11 we show $\tilde{M}$ implies M and groups of type $\tilde{M}$ exist.

Our proof of the uniqueness of groups of type $\tilde{M}$ provides the necessary centralizer of involution characterization of the two smaller Fischer groups. In Chapter 11 we also characterize the largest Fischer group $M(24)$ and its derived subgroup $M(24)' = F_{24}$ by the centralizer of a 2-central involution. Namely we prove that groups of type F_{24} and $\text{Aut}(F_{24})$ are unique and that the automorphism group of a group of type F_{24} is of type $M(24)$. A finite group G is of *type* F_{24} if it possesses an involution z such that $F^*(C_G(z)) = Q$ is extraspecial of order 2^{13}, $C_G(z)/Q \cong \mathbb{Z}_2/U_4(3)/\mathbb{Z}_3$, and z is not weakly closed in Q with respect to G. Define G to be of *type* $\text{Aut}(F_{24})$ if it has a subgroup G_0 of index 2 of type F_{24} such that $O_2(C_G(z)) = O_2(C_{G_0}(z))$ for z a 2-central involution of G_0.

In the language introduced in [SG], G is of type F_{24} if it satisfies Hypothesis $\mathcal{H}(6, \mathbb{Z}_2/U_4(3)/\mathbb{Z}_3)$. The majority of the sporadics satisfy Hypothesis $\mathcal{H}(w, L)$ for suitable w and L, and this is the preferred hypothesis to characterize the sporadic group in the author's program to solidify the foundations of the theory of the sporadics.

The interplay among questions involved in the study of the existence, uniqueness, and structure of the Fischer groups is delicate. Often results in one of the three areas make possible an easy proof of a result in another area. Thus to obtain an attractive and complete treatment of the Fischer groups, one needs to take a path that sometimes wanders from one area to another, and one must be careful to keep track of what one knows about the Fischer groups and when.

For example the three Fischer groups are shown to exist in Chapter 11 of *3-Transposition Groups* at the same time that the uniqueness of these groups as groups of type $\tilde{M}(22)$, $\tilde{M}(23)$, and F_{24} is established. Namely the largest Fischer group $M(24)$ is the section $N_M(X)/X$ for a suitable subgroup X of order 3 in the Monster M. However the construction in [SG] does not tell us this section is generated by 3-transpositions; it only says the section is of type $\text{Aut}(F_{24})$. Thus to establish the existence of $M(24)$, we need to show each group of type $\text{Aut}(F_{24})$ is a 3-transposition group, and hence of type $M(24)$. The existence of a group of type $M(24)$ is then used in the proof of the uniqueness of groups of type F_{24}.

After the existence and uniqueness results in Part II, we can write $M(22)$ for the unique group of type $M(22)$ (*and* the unique group of type $\tilde{M}(22)$) and similarly write $M(23)$ and $M(24)$ for the remaining Fischer groups, and $M(24)'$ or F_{24} for the derived group of $M(24)$. In Part III we investigate the structure of the Fischer groups. In Chapter 12 we determine the automorphism groups of $M(22)$, $M(23)$, and $M(24)'$. We then go on to determine the conjugacy classes of subgroups of prime order and their normalizers in each of the Fischer groups.

The author would like to thank Jon Hall and Franz Timmesfeld for carefully reading portions of early drafts of this manuscript, catching numerous small errors, and suggesting improvements. I also take this opportunity to acknowledge the influence of Bernd Fischer on my early development as a group theorist. Learning the beautiful theory of 3-transpositions from his Warwick Lecture notes led to my first substantial result on finite groups, and the point of view pioneered by Fischer in those notes has continued to influence not only my thinking about groups, but the thinking of many others of my generation.

1

Preliminaries

We assume familiarity with that portion of an elementary course in abstract algebra covering groups and vector spaces. In addition we assume some slightly less elementary results on groups that can be found in the text *Finite Group Theory* [FGT].

In Chapter 1 we record much of the basic terminology and notation used in the book. We also list most of the results from [FGT] that we will require, although Chapter 4 contains the results from [FGT] on bilinear forms and classical groups. Section 3 contains results on permutation groups, Section 4 contains results on linear groups, and Section 2 contains the rest of the group theoretic results from [FGT]. Section 5 consists of a brief discussion of certain Coxeter groups.

1. Categories

For a slightly more expansive discussion of the topics in this section see Sections 2 and 4 in [FGT].

A *category of sets with structure* $\mathcal{A}$ consists of a collection $\mathrm{Ob}(\mathcal{A})$ of *objects*, and for each $A, B \in \mathrm{Ob}(\mathcal{A})$ a set $\mathrm{Mor}(A, B)$ of *morphisms* of A to B. Each object A consists of a set $\bar{A}$ together with some "structure" on $\bar{A}$. $\mathrm{Mor}(A, B)$ consists of all functions from $\bar{A}$ to $\bar{B}$ that "preserve structure." We require that the identity function 1_A on $\bar{A}$ preserves structure. Thus $1_A \in \mathrm{Mor}(A, A)$. We also require that if $f : \bar{A} \to \bar{B}$ and $g : \bar{B} \to \bar{C}$ preserve structure then so does the composition $fg : \bar{A} \to \bar{C}$. That is if $f \in \mathrm{Mor}(A, B)$ and $g \in \mathrm{Mor}(B, C)$ then $fg \in \mathrm{Mor}(A, C)$.

Examples:

(1) *Category of sets and functions.* Objects are sets and $\mathrm{Mor}(A, B)$ consists of all functions from A into B. Thus in this case we have no extra structure.

(2) *Category of groups and group homomorphisms.* Objects are groups and
Mor(A, B) is the set of all group homomorphisms of A into B. The structure is the group structure. Morphisms must preserve the group multiplication.

(3) *Category of graphs.* Objects are graphs. A *graph* is an ordered pair $\Gamma =
(V, E)$ where V is a set and E is a symmetric relation on V. The relation
E supplies the graph structure on V. For $v \in V$ we write $\Gamma(v)$ for the set
of $u \in V$ distinct from v such that $(v, u) \in E$ and set $v^\perp = \{v\} \cup \Gamma(v)$.
A morphism $\alpha : (A, R) \to (B, S)$ is a function from A to B such that
$a^\perp \alpha \subseteq a\alpha^\perp$ for each $a \in A$.

An *inverse* for $\alpha \in \text{Mor}(A, B)$ is a morphism $\beta \in \text{Mor}(B, A)$ such that
$\alpha\beta = 1_A$ and $\beta\alpha = 1_B$. If α possesses an inverse we say α is an *isomorphism*
and say A and B are isomorphic.

REMARK: α has at most one inverse, which we write as α^{-1}. Indeed α^{-1} is the
inverse of α as a function from $\bar{A}$ to $\bar{B}$. It is not in general sufficient to know α
has an inverse as a function to conclude it has an inverse in a category $\mathcal{A}$ of sets
with structure, although this is true in the category of groups, rings, modules,
fields (i.e., in most algebraic categories). For example it is easy to construct a
bijective morphism in the category of graphs whose inverse function does not
preserve structure, and hence is not a morphism. So in general to check that a
bijective morphism is an isomorphism, one must check that the inverse function
preserves structure.

Given an object A in a category $\mathcal{A}$, define an *automorphism* of A to be an
isomorphism of A with itself. Denote by $\text{Aut}(A)$ the group of all automorphisms
of A under the operation of composition of functions. One can check that $\text{Aut}(A)$
is indeed a group.

Let G be a group. A *representation* of G in the category $\mathcal{A}$ is a group homomorphism $\pi : G \to \text{Aut}(A)$ for some object A in $\mathcal{A}$. π is *faithful* if π is
an injection. Two representations $\pi : G \to \text{Aut}(A)$ and $\sigma : G \to \text{Aut}(B)$ are
equivalent if there exists an isomorphism $\alpha : A \to B$ such that for all $g \in G$
and all $a \in A$, $(ag\pi)\alpha = (a\alpha)(g\sigma)$. Representations $\pi_i : G_i \to \text{Aut}(A_i)$,
$i = 1, 2$, in the category $\mathcal{A}$ are *quasiequivalent* if there exist isomorphisms
$\alpha : A_1 \to A_2$ and $\beta : G_1 \to G_2$ such that for all $g \in G_1$ and $a \in A_1$,
$(ag\pi_1)\alpha = (a\alpha)(g\beta\pi_2)$. Equivalently the following diagram commutes for all
$g \in G_1$:

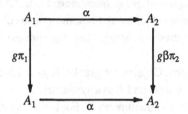

Notice equivalence and quasiequivalence are equivalence relations.

Examples:

(4) A *permutation representation* of G is a representation in the category of sets. That is a permutation representation is a group homomorphism $\pi : G \to \text{Sym}(A)$, since the symmetric group $\text{Sym}(A)$ is the group of automorphisms of A in the category of sets.

(5) Let F be a field. A *linear representation* or *FG-representation* is a representation of G in the category of vector spaces over F. Thus $\pi : G \to GL(A)$, where $GL(A)$ is the *general linear group* on A; that is $GL(A)$ is the group of all invertible linear maps on the vector space A.

(6) Notice that if $\pi : G \to \text{Aut}(A)$ is a representation of G in the category $\mathcal{A}$ then π is also a permutation representation of G on $\bar{A}$. For $\text{Aut}(A)$ is a subgroup of $\text{Sym}(\bar{A})$.

(1.1) Let $\pi, \sigma : G \to \text{Aut}(A)$ *be faithful representations of a group* G *in a category* $\mathcal{A}$. *Then* π *is quasiequivalent to* σ *if and only if* $G\pi$ *is conjugate to* $G\sigma$ *in* $\text{Aut}(A)$.

Proof: This is Exercise 1.7 in [FGT].

2. Elementary group theory

We take as our basic reference in finite group theory the text *Finite Group Theory* [FGT]. In particular we adopt the notation and terminology of [FGT], although we list some of the most frequently used notation and terminology in this section. We assume the elementary results from Section 1 in [FGT] plus scattered results from later sections, many of which are also recorded later. Results on permutation representations are collected in Section 3 and results on linear representations are collected in Section 4.

Let G be a group. We write $H \leq G$ to indicate that H is a subgroup of G and write $H \trianglelefteq G$ to indicate H is a normal subgroup of G. The order of a set S

is denoted by $|S|$ and the order of an element $g \in G$ is denoted by $|g|$. If $S \subseteq G$ then $\langle S \rangle$ denotes the subgroup of G generated by S.

For $x, g \in G$, set $x^g = g^{-1}xg \in G$, and, for $X \subseteq G$, $X^g = \{x^g : x \in X\}$ and $X^G = \{X^g : g \in G\}$. Thus X^g is the *conjugate* of X under g and X^G the *conjugacy class* of X in G. Define

$$N_G(X) = \{g \in G : X^g = X\}$$

to be the *normalizer* in G of X, and

$$C_G(X) = \{g \in G : x^g = x \text{ for all } x \in X\}$$

to be the *centralizer* in G of X. For $X, Y \subseteq G$, define $XY = \{xy : x \in X, y \in Y\}$. Define $Z(G) = C_G(G)$ to be the *center* of G.

If p is a prime then $O_p(G)$ denotes the largest normal p-subgroup of G and $O^p(G)$ denotes the subgroup of G generated by all p'-*elements* of G, that is by the elements of order prime to p. Write $\mathrm{Syl}_p(G)$ for the set of Sylow p-subgroups of G.

We write $\mathbf{Z}_n$ for the cyclic group of order n and $GF(q)$ for the field of order q. Given a prime p, we write E_{p^n} for the direct product of n copies of Z_p, and call this group the *elementary abelian p-group of rank n*. An *involution* is an element of G of order 2.

For $x, y \in G$ write $[x, y]$ for $x^{-1}y^{-1}xy \in G$. Thus $[x, y]$ is the *commutator* of x with y. For $X, Y \leq G$, write

$$[X, Y] = \langle [x, y] : x \in X, \ y \in Y \rangle.$$

The *commutator subgroup* or *derived subgroup* of G is $G' = [G, G]$. Recall G' is the smallest normal subgroup of G with an abelian factor group.

Let $H \leq G$. A *complement* to H in G is a subgroup K of G such that $H = HK$ and $H \cap K = 1$. Our group G is said to be an *extension* of a group X by a group Y if there exists $H \trianglelefteq G$ with $H \cong X$ and $G/H \cong Y$. The extension is said to *split* if H has a complement in G. In that event the complement is isomorphic to Y.

Let $\pi : A \rightarrow \mathrm{Aut}(G)$ be a group homomorphism. Thus in the terminology of Section 1, π is a representation of A on G in the category of groups. Form the set product $S = A \times G$ of A and G and define a group structure on S via

$$(a, g)(b, h) = (ab, g^{b\pi}h),$$

where $g^{b\pi}$ denotes the image of g under the automorphism $b\pi$. The group S is the *semidirect product* of G by A with respect to π, and is denoted by $S(A, G, \pi)$. We recall that $\{(1, g) : g \in G\}$ is a normal subgroup of S isomorphic to G and $\{(a, 1) : a \in A\}$ is a complement to this normal subgroup isomorphic to A, so that S is a split extension of G by A. Conversely each split extension GA of G by A is isomorphic to a semidirect product $S(A, G, \pi)$, where $\pi : A \to \text{Aut}(G)$ is the conjugation map defined by $a\pi : g \mapsto g^a$ for $a \in A$ and $g \in G$. We will need the following elementary fact, which is Lemma 10.3 in [FGT]:

(2.1) Let $S_i = S(A_i, G_i, \pi_i)$, $i = 1, 2$ be semidirect products such that π_1 is quasiequivalent to π_2. Then $S_1 \cong S_2$.

The *exponent* of G is the least positive integer n such that $g^n = 1$ for all $g \in G$. The *Frattini subgroup* is the intersection of all maximal subgroups of G and is denoted by $\Phi(G)$. By 23.2 in [FGT]:

(2.2) If G is a p-group then $\Phi(G)$ is the smallest normal subgroup of G with an elementary abelian factor group.

Our group G is *nilpotent* if G is the direct product of its Sylow groups. The largest normal nilpotent subgroup of G is called the *Fitting subgroup* of G and denoted by $F(G)$.

A subgroup H of G is *subnormal* in G if there is a series

$$H = G_1 \trianglelefteq \cdots \trianglelefteq G_n = G.$$

Define G to be *simple* if 1 and G are the only normal subgroups of G, and *quasisimple* if $G = G'$ and $G/Z(G)$ is simple. The *components* of G are its quasisimple subnormal subgroups. Write $E(G)$ for the subgroup generated by all components of G and set $F^*(G) = F(G)E(G)$, the *generalized Fitting subgroup* of G.

(2.3) Let G be a finite group. Then

(1) If L and K are distinct components of G then

$$[L, K] = [E(G), F(G)] = 1.$$

(2) $C_G(F^*(G)) = Z(F^*(G))$.

Proof: See Section 31 in [FGT].

(2.4) Thompson $A \times B$ Lemma.   Let $A \times B \le N_G(P)$ with B and P p-groups, and A a group of order prime to p such that $[C_P(B), A] = 1$. Then $[A, P] = 1$.

Proof: See 24.2 in [FGT].

(2.5) Baer–Suzuki Theorem. *Let X be a p-subgroup of G. Then either $X \leq O_p(G)$ or there exists $g \in G$ with $\langle X, X^g \rangle$ not a p-group.*

Proof: See 39.6 in [FGT].

(2.6) Gaschütz's Theorem. *Let p be prime, V a normal abelian p-subgroup of G, and $P \in \mathrm{Syl}_p(G)$. Then G splits over V if and only if P splits over V.*

Proof: See 10.4 in [FGT].

(2.7) Burnside's $p^a q^b$-Theorem. *Let p, q be distinct primes and $1 \neq G$ a simple group of order $p^a q^b$. Then $|G| = p$ or q.*

Proof: See 35.13 in [FGT].

Let p be a prime. A p-group G is said to be *extraspecial* if $G' = \Phi(G) = Z(G)$ is of order p. For example it turns out that if p is odd then there is a unique nonabelian group of order p^3 and exponent p. This group is extraspecial and is denoted by p^{1+2}. There are two extraspecial 2-groups of order 2^{2n+1}; they are D_8^n and $D_8^{n-1} Q_8$, where $D_8^m Q_8^k$ denotes the central product (i.e., the direct product modulo the relation identifying the centers of the factors) of m copies of the dihedral group D_8 of order 8 with k copies of the quaternion group Q_8 of order 8. Sometimes we write 2^{1+2n} to denote an extraspecial group of order 2^{2n+1}. See Section 23 in [FGT] or Section 8 in [SG] for much more on extraspecial groups.

We write

$$H_n / H_{n-1} / \cdots / H_1$$

for a group G with a normal series $1 = G_0 \trianglelefteq G_1 \trianglelefteq \cdots \trianglelefteq G_n = G$ such that $G_i / G_{i-1} \cong H_i$ for each i.

Let $W \leq H \leq G$. We say that W is *weakly closed* in H with respect to G if W is the unique G-conjugate of W contained in H. In particular if W is weakly closed in H with respect to G then W is normal in H and indeed $N_G(H) \leq N_G(W)$. We say H *controls fusion* in W if whenever n is a positive integer and $x = (x_1, \ldots, x_n)$ and $y = (y_1, \ldots, y_n)$ are n-tuples from W conjugate in G, then x and y are conjugate in H.

(2.8) *Let p be a prime, T be a p-subgroup of G, and W an abelian subgroup of T. Then*

(1) *If $T \in \mathrm{Syl}_p(G)$ and W is weakly closed in T with respect to G, then $N_G(W)$ controls fusion in W.*

(2) *If* $x \in W$, $T \in \mathrm{Syl}_p(C_G(x))$, *and* W *is weakly closed in* T *with respect to* G, *then* $x^G \cap W = x^{N_G(W)}$.

Proof: Part (1) is 7.7.1 in [SG]; its proof is essentially the same as that of part (2), which we now give. Suppose $g \in G$ with $x^g \in W$. Then W, $W^{g^{-1}} \leq C_G(x)$ and by Sylow's Theorem there is $h \in C_G(x)$ with $W^{g^{-1}h^{-1}} \leq T$. Then as W is weakly closed in T, $W = W^{g^{-1}h^{-1}}$, so $hg \in N_G(W)$ and $x^{hg} = x^g$.

3. Permutation representations

For a more detailed discussion of permutation representations, see Chapters 2 and 5 in [FGT].

In this section X is a set and $\mathrm{Sym}(X)$ is the *symmetric group* on X; that is $\mathrm{Sym}(X)$ is the group of all permutations on X under composition of functions. Assume further that G is a group and $\pi : G \to \mathrm{Sym}(X)$ is a permutation representation of G (i.e., π is a group homomorphism).

If X and Y have the same cardinality then $\mathrm{Sym}(X) \cong \mathrm{Sym}(Y)$. Thus we write S_n for the symmetric group on a set of order n.

For $x \in X$ and $\alpha \in \mathrm{Sym}(X)$ write $x\alpha$ for the image of x under α. Thus by definition of multiplication in $\mathrm{Sym}(X)$,

$$x(\alpha\beta) = (x\alpha)\beta, \qquad x \in X, \ \alpha, \beta \in \mathrm{Sym}(X).$$

Usually we suppress the representation π and write xg for $x(g\pi)$, when $x \in X$ and $g \in G$. One feature of this notation is that

$$x(gh) = (xg)h, \ x \in X, \ g, h \in G.$$

The relation $\sim$ on X defined by $x \sim y$ if and only if there exists $g \in G$ with $xg = y$ is an equivalence relation on X. The equivalence class of x under this relation is $xG = \{xg : g \in G\}$ and is called the *orbit* of x under G. As the equivalence classes of an equivalence relation partition a set, X is partitioned by the orbits of G on X.

Let Y be a subset of X. G is said to *act* on Y if Y is a union of orbits of G. Notice G acts on Y precisely when $yg \in Y$ for each $y \in Y$, and each $g \in G$. For $x \in X$, define

$$G_x = \{g \in G : xg = x\}.$$

G_x is a subgroup of G called the *stabilizer* of x in G. For $S \subseteq G$ define

$$\mathrm{Fix}(S) = \{x \in X : xs = x \text{ for all } s \in S\}.$$

Thus $\mathrm{Fix}(S)$ is the set of *fixed points* of S. We have

(3.1)   *If* $H \trianglelefteq G$ *then* G *acts on* $\mathrm{Fix}(H)$.

Our representation π is *transitive* if G has just one orbit on X; equivalently for each $x, y \in X$ there exists $g \in G$ with $xg = y$. G will also be said to be transitive on X.

Let $H \leq G$. Then $\alpha : G \to \mathrm{Sym}(G/H)$ is a transitive permutation representation of G on the coset space G/H, where $\alpha : Hx \mapsto Hxg$. We call α the *representation of G on the cosets of H by right multiplication*. Notice H is the stabilizer of the coset H in this representation.

(3.2)   *Let G be transitive on X, $x \in X$, and $H = G_x$. Then*
  (1) The map $xg \mapsto Hg$ is an equivalence of the permutation representation of G on X with the representation of G by right multiplication on G/H.
  (2) X has cardinality $|G : G_x|$ for each $x \in X$.
  (3) $G_{xg} = (G_x)^g$ for each $g \in G$.
  (4) If $\beta : G \to \mathrm{Sym}(Y)$ is a transitive permutation representation and $y \in Y$, then π is equivalent to β if and only if G_x is conjugate to G_y in G.

Proof: See Section 5 of [FGT]. $\quad\blacksquare$

From now on assume X is finite and set $S = \mathrm{Sym}(X)$. We assume familiarity with the cycle notation describing permutations in S. In particular each $g \in S$ can be written uniquely as a product of disjoint cycles and the order of g is the least common multiple of the lengths of these cycles. Two permutations are conjugate in S if and only if they have the same cycle structure. A *transposition* is an element of S moving exactly two points of X. That is a transposition is a cycle of length 2.

(3.3)   *S is generated by its transpositions.*

A permutation is said to be an *even permutation* if it can be written as the product of an even number of transpositions, and to be an *odd permutation* otherwise. Denote by $\mathrm{Alt}(X)$ the set of all even permutations of X. Then $\mathrm{Alt}(X)$ is a normal subgroup of S of index 2 and a permutation is even if and only if it has an even number of cycles of even length. The group $\mathrm{Alt}(X)$ is the *alternating group* on X. The isomorphism type of $\mathrm{Alt}(X)$ depends only on the cardinality of X, so we may write A_n for $\mathrm{Alt}(X)$ when $|X| = n$.

Given a positive integer m, G is said to be *m-transitive* on X if G acts transitively on ordered m-tuples of distinct points of X.

Let G be transitive on X. A partition Q of X is *G-invariant* if G permutes the members of Q. Q is *nontrivial* if $Q \neq \{X\}$ or $\{\{x\} : x \in X\}$. G is said to be *imprimitive* on X if there exists a nontrivial G-invariant partition Q of X. In that event Q is said to be a *system of imprimitivity* for G on X. G is *primitive* if G is transitive and not imprimitive. By 5.19 in [FGT]:

(3.4)   *Let G be transitive on X and $x \in X$. Then G is primitive on X if and only if G_x is maximal in G.*

We recall that 2-transitive representations are primitive.

4. Linear representations

See Chapter 4 in [FGT] for more on linear representations.

In this section F is a field, n is a positive integer, and V is an n-dimensional vector space over F. The group of vector space automorphisms of V is the *general linear group* $GL(V)$. As the isomorphism type of V depends only on n and F, the same is true for $GL(V)$, so we can also write $GL_n(F)$ for $GL(V)$.

(4.1)   *Let $X = (x_1, \ldots, x_n)$ be an ordered basis for V, and for $g \in \mathrm{End}_F(V)$ let $M_X(g) = (g_{ij})$ be the matrix defined by $x_i g = \sum_j g_{ij} x_j$, $g_{ij} \in F$. Then*

   (1) The map $M_X : g \mapsto M_X(g)$ is an isomorphism of $GL(V)$ with the group G of all nonsingular n by n matrices over F.

   (2) Let $Y = (y_1, \ldots, y_n)$ be a second ordered basis of V, let h be the unique element of $GL(V)$ with $x_i h = y_i$, $1 \leq i \leq n$, and $B = M_Y(h)$. Then $M_X = h^* M_Y = M_Y B^*$ is the composition of h^* with M_Y and of M_Y with B^*, where h^* and B^* are the conjugation automorphisms induced by h and B on $GL(V)$ and G, respectively.

Because of 4.1, we can think of subgroups of $GL(V)$ as groups of matrices if we choose.

Let G be a group. An *FG-representation* is a representation of G in the category of F-spaces. That is an FG-representation is a group homomorphism $\pi : G \to GL(V)$ of G into the general linear group on some F-space V. An FG-representation π on V can be thought of as a homomorphism from G into the group of all n by n nonsingular matrices over G, by composing π with the isomorphism M_X.

For $y \in \mathrm{End}_F(V)$ define the *determinant* of y to be $\det(y) = \det(M_X(y))$. That is the determinant of y is the determinant of its associated matrix. If A is a matrix and B is a nonsingular matrix then $\det(A^B) = \det(A)$, so $\det(y)$ is independent of the choice of basis X by 4.1.2.

Define the *special linear group* $SL(V)$ to be the set of elements of $GL(V)$ of determinant 1. The determinant map is a group homomorphism of $GL(V)$ onto the multiplicative group of F with $SL(V)$ the kernel of this homomorphism, so

$SL(V)$ is a normal subgroup of $GL(V)$ and $GL(V)/SL(V) \cong F^{\#}$. Also write $SL_n(F)$ for $SL(V)$.

Associated to V is its *projective geometry* $PG(V)$. By definition $PG(V)$ is the graph whose vertices are the nonzero proper subspaces of V, with two subspaces A and B adjacent if $A \leq B$ or $B \leq A$. We refer to this relation as *incidence*; thus incidence is just set theoretic inclusion. We only consider automorphisms of our graph that preserve dimension. We call the subspaces of dimension $1, 2$, and $n - 1$, *points, lines,* and *hyperplanes*, respectively.

For $g \in GL(V)$ define $gP : PG(V) \to PG(V)$ by $gP : U \mapsto Ug$, for $U \in PG(V)$. Evidently $P : GL(V) \to \text{Aut}(PG(V))$ is a representation of $GL(V)$ on $PG(V)$ preserving dimension. Denote the image of $GL(V)$ under P by $PGL(V)$. $PGL(V)$ is the *projective general linear group*. The notation $PGL_n(F)$ is also used for $PGL(V)$. The image of $SL(V)$ under P is denoted by $PSL(V)$ or $PSL_n(F)$. The group $PSL(V)$ is the *projective special linear group*. Sometimes $PSL_n(F)$ is denoted by $L_n(F)$.

A *scalar transformation* of V is a member g of $\text{End}_F(V)$ such that $vg = av$ for all v in V and some a in F independent of v. A *scalar matrix* is a matrix of the form aI, $a \in F$, where I is the identity matrix. It turns out that the nonzero scalar transformations form the center of $GL(V)$ and that $Z(GL(V)) = \ker(P)$. Hence the projective general linear group $PGL(V)$ is isomorphic to the group of all n by n nonsingular matrices modulo the subgroup of scalar matrices.

Given any FG-representation $\pi : G \to GL(V)$, π can be composed with P to obtain a homomorphism $\pi P : G \to GL(V)$. Observe that πP is a representation of G on the projective geometry $PG(V)$.

For v in V and α in $\text{End}_F(V)$, $[v, \alpha] = v\alpha - v$ is the *commutator* of v with α. This corresponds with the notion of commutator in Section 2. Indeed we can form the semidirect product of V by $GL(V)$ with respect to the natural representation, and in this group the two notions agree. Similarly for $G \leq GL(V)$,

$$[V, G] = \langle [v, g] : v \in V, \ g \in G \rangle,$$

and for $g \in G$, $[V, g] = [V, \langle g \rangle]$.

A *transvection* is an element t of $GL(V)$ such that $[V, t]$ is a point of $PG(V)$, $C_V(t)$ is a hyperplane of $PG(V)$, and $[V, t] \leq C_V(t)$. $[V, t]$ and $C_V(t)$ are called the *center* and *axis* of t, respectively. One can check that transvections are of determinant 1.

Let $G \leq GL(V)$. An *FG-submodule* or *G-submodule* of V is a subspace U of V such that G acts on U. Define G to be *irreducible* on V if 0 and V are the only FG-submodules of V.

Suppose p is a prime and $E \cong E_{p^n}$ is an elementary abelian p-group of rank n. Let $F = GF(p)$ and regard F as the integers modulo p. Then E is a vector space over F where scalar multiplication is defined by $((p) + m)x = mx$ for $x \in E$, and mx denotes the mth power of x. Hence the group $\mathrm{Aut}(E)$ of automorphisms of E in the category of groups is just the group $GL(E)$ of automorphisms of E regarded as an F-space. This observation is used implicitly throughout the book.

5. Coxeter systems

In discussing 3-transpositions it will be convenient to have at our disposal a few elementary results on Coxeter systems. A more complete discussion can be found in Chapter 10 of [FGT].

A *Coxeter matrix* is a symmetric n by n matrix $M = (m_{ij})$ with 1's on the main diagonal and such that $1 < m_{ij}$ is an integer for each $i \neq j$. Associated to each Coxeter matrix M is a Coxeter diagram $\Delta = \Delta(M)$. This diagram consists of n nodes indexed by the integers $1 \leq i \leq n$, with an edge of weight $m_{ij} - 2$ joining distinct nodes i and j. We will be concerned only with the case when $m_{ij} = 2$ or 3 for all $i \neq j$, and we will be most interested in the following Coxeter diagrams:

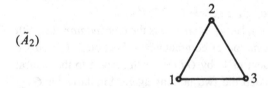

(A_n)

$(\tilde{A}_2)$

A *Coxeter system* with Coxeter matrix $M = (m_{ij})$ is a pair (G, S) where $S = (s_i : 1 \leq i \leq n)$ is a family of elements of G and

$$G = \mathrm{Grp}(S : (s_i s_j)^{m_{ij}} = 1, \ 1 \leq i, \ j \leq n)$$

where the notation $G = \mathrm{Grp}(S : w \in W)$ indicates that G is the largest group generated by a family $S = (s_i : 1 \leq i \leq n)$ of elements satisfying the relations $w \in W$, or more precisely:

(5.1) *Suppose (G, S) is a Coxeter system with Coxeter matrix $M = (m_{ij})$, H is a group, and $T = (t_i : 1 \leq i \leq n)$ is a family of elements of H such that*

$(t_i t_j)^{m_{ij}} = 1$ *for all* $1 \le i, j \le n$. *Then there exists a unique homomorphism* $\pi : G \to H$ *with* $s_i \pi = t_i$ *for all* i.

Standard arguments (cf. Section 28 in [FGT]) show that Coxeter systems exist for each Coxeter matrix M and that if (G, S) and $(\bar{G}, \bar{S})$ are Coxeter systems with matrix M then the map $s_i \mapsto \bar{s}_i$ extends to an isomorphism of G with $\bar{G}$. That is each Coxeter matrix M determines a unique Coxeter system.

Let $\bar{W}(n, m)$ be the wreath product of $\mathbf{Z}_m$ by S_n. (The case $m = \infty$, where we set $Z_\infty = \mathbf{Z}$, is allowed.) Thus $\bar{W}(n, m) = \bar{H}\bar{L}$ is the split extension of $\bar{L} = L_1 \times \ldots \times L_n$ by $H = S_n$, where $L_i = \langle x_i \rangle \cong \mathbf{Z}_m$ and $x_i^h = x_{ih}$ for $h \in H$. (This split extension can be constructed using the semidirect product construction of Section 2.)

Define a set $S(n, m)$ of involutions of $\bar{W}(n, m)$ as follows. Let $s_i = (i, i+1) \in H$ for $1 \le i < n$, and let $s_n = s_2^{s_1 x_3}$. Then let $S(n, m) = \{s_1, \ldots, s_n\}$ and $W(n, m) = \langle S(n, m) \rangle$. Notice

(5.2) $W(n, m) = HL$ *where*

$$L = \{x_1^{k_1} \cdots x_n^{k_n} : \sum_i k_i \text{ is even}\}$$

is the direct product of $n - 1$ *copies of* $\mathbf{Z}_m$.

The theory of root systems and reflection groups establishes the following lemma; see for example 30.19 in [FGT]. Alternatively a direct elementary proof is possible as in Exercise 1.2.

(5.3) *Let* $G = S_{n+1}$ *be the symmetric group on* $\{1, \ldots, n\}$, $s_i = (i, i + 1)$, *and* $S = (s_i : 1 \le i \le n)$. *Then* (G, S) *is a Coxeter system of type* A_n.

(5.4) $(W(3, \infty), S(3, \infty))$ *is a Coxeter system of type* $\tilde{A}_2$.

Proof: This follows from the theory of affine root systems (cf. Bourbaki [B]) but we give an elementary proof here.

Let $M = (m_{ij})$ be a Coxeter matrix of type $\tilde{A}_2$. Thus $m_{ij} = 3$ for all $i \ne j$. Let (X, T) be a Coxeter system of type $\tilde{A}_2$. Then

(i) $(t_i t_j)^3 = 1$ for all $i \ne j$, and hence $t_i^{t_j} = t_j^{t_i}$.

Let $a = t_1$, $b = t_2$, and $c = t_3$. Observe next that

(ii) $[ac^b, a^b c] = 1$.

For using (i) repeatedly we get

$$[ac^b, a^bc] = c^b aca^b ac^b a^b c = c^b ac(b^a a)(ca)^b c = c^b ac(ab)(ca)^b c$$
$$= c^b (acaca)bc = c^b(c)bc = c^b b^c = (c^b)^2 = 1.$$

Let $Y = \langle a^b c, ac^b \rangle$. As $(a^b c)^b = ac^b$, b acts on Y. Similarly

$$(a^b c)^a = bc^a = bc^a = bc(baa^b)c = c^b aa^b c = (ac^b)^{-1}(a^b c)$$

and

$$(ac^b)^a = c^b a = (ac^b)^{-1},$$

so $K = \langle a, b \rangle$ acts on Y. Then as $c \in KY$, $X = \langle a, b, c \rangle = KY$.

Next we let $G = W(3, \infty) = HL$, where $H = S_3$ and L are as in Lemma 5.2 and $S = S(3, \infty)$. Check that for each $i \neq j$, s_i is an involution and $|s_i s_j| = 3$. So by 5.1, there is a surjective homomorphism $\pi : X \to G$ with $t_i \pi = s_i$ for each i. Observe also that $s_1^{s_2} = s_2^{s_1} = (1, 3)$, so $s_1^{s_2} s_3 = (1, 3)(1, 3)^{x_3} = x_3^{-(1,3)} x_3 = x_1^{-1} x_3$. A similar calculation shows

(*) $\qquad\qquad s_1^{s_2} s_3 = x_1^{-1} x_3 \quad$ and $\quad s_1 s_3^{s_2} = x_1 x_2^{-1}.$

By (*), $\pi : a^b c \mapsto u = x_1^{-1} x_3$ and $ac^b \mapsto v = x_1 x_2^{-1}$. Thus as $L = \langle u \rangle \times \langle v \rangle$ and $Y = \langle a^b c \rangle \langle ac^b \rangle$, we have $K\pi = H$ and $Y\pi = L$.

Next $\pi : K \to H$ is a bijection and $X = KY$ with $K\pi \cap Y\pi = H \cap L = 1$, so $\ker(\pi) \leq Y$. Similarly as $L = \langle u \rangle \times \langle v \rangle$ and $Y = \langle a^b c \rangle \langle ac^b \rangle$,

$$\ker(\pi) = (\ker(\pi) \cap \langle a^b c \rangle)(\ker(\pi) \cap \langle ac^b \rangle).$$

Then as $\langle u \rangle \cong \langle v \rangle \cong \mathbf{Z}$, $\ker(\pi) = 1$, so indeed π is an isomorphism.

(5.5) *Let $G = W(3, 3)$, $S = S(3, 3)$, $s = s_1$, $y_1 = ss_2$, and $y_2 = ss_3$. Then*

(1) *G is the split extension of $R = \langle y_1, y_2 \rangle \cong 3^{1+2}$ by $\langle s \rangle$, with s inverting y_1 and y_2 and $Z(G) = \langle [y_1, y_2] \rangle$.*

(2) *If $M = \langle d_1, d_2, d_3 \rangle$ is a group with $d_i^2 = (d_i d_j)^3 = (d_2 d_1 d_2 d_3)^3 = 1$ then there exists a surjective group homomorphism $\pi : G \to M$ with $s_i \pi = d_i$ for all i.*

(3) *If $M = \langle t, x, y \rangle$ with $t^2 = x^3 = y^3 = (xy)^3 = (x^{-1}y)^3 = 1$ and t inverting x and y, then there is a surjective homomorphism $\pi : G \to M$ with $s\pi = t$, $y_1\pi = x$, and $y_2\pi = y$.*

(4) *If D is a set of 3-transpositions in M, $d_i \in D$ for $i = 1, 2, 3$, with $M = \langle d_1, d_2, d_3 \rangle$ and $|d_i d_j| = 3 = |d_1^{d_2} d_3|$ for all $i \neq j$, then there*

exists a surjective group homomorphism $\pi : G \to M$ *with* $s_i\pi = d_i$ *for all* i.

Proof: Assume the hypotheses of (2). By 5.4 there is a surjection $\alpha : W(3, \infty) \to M$ with $s_i\alpha = d_i$. Let $K = \ker(\alpha)$. By (*) of 5.4 and our hypothesis, $(x_1^{-1}x_3)\alpha = d_1^{d_2}d_3$ has order dividing 3, so $x_1^3 x_3^{-3} \in K$. Then $x_1^3 x_2^{-3} = (x_1^3 x_3^{-3})^{s_2} \in K$. Hence $\bar{K} \cap L \leq K$, where $\bar{K} = \langle x_1^3, x_2^3, x_3^3 \rangle \leq \bar{W}(3, \infty)$. Then as $W(3, 3) \cong W(3, \infty)\bar{K}/\bar{K} \leq \bar{W}(3, \infty)/\bar{K}$ we can factor α through G. Hence (2) is established. Notice (2) implies (4).

By (2) or direct calculation, $y_1 = (1, 3, 2)$ and $y_2 = (1, 2, 3)x_1^{-1}x_3$ are of order 3 and inverted by s. By direct calculation, $[y_1, y_2] = (x_1x_2x_3)^{-1} \in Z(G)$. Thus (1) holds by Exercise 1.3.

Finally assume the hypothesis of (3). Let $d_1 = t$, $d_2 = tx$, and $d_3 = ty$. Check that the relations of (2) are satisfied and appeal to (2) to establish (3).

Exercises

1. Let V be an n-dimensional vector space over $GF(2)$, $S_3 \cong G \leq GL(V)$, and s, t distinct involutions in G. Let $m = \dim(C_V(t))$ and assume $V = [V, t] \oplus [V, s]$. Then
   (1) $n = 2m$.
   (2) $C_V(t) = [V, t]$.
   (3) $C_V(G) = 0$.
   (4) V is the direct sum of 2-dimensional irreducible G-submodules.
   (5) If $\pi : G \to GL(U)$ is a representation of G on a $GF(2)$-space U with $\dim(C_U(t)) = m$ and $U = [U, t] \oplus [U, s]$, then π is equivalent to the representation of G on V.
   [*Hint:* Use Exercise 4.2.3 in [FGT].]
2. Prove Lemma 5.3.
   [*Hint:* Let R be a group generated by elements $r_1, \ldots, r_n$ such that $(r_ir_j)^{m_{ij}} = 1$. Prove $|R| \leq (n + 1)!$ by induction on n by showing $R = H \cup Hr_n H$, where $H = \langle r_1, \ldots, r_{n-1} \rangle$.]
3. Let G be a group, $x, y \in G$, and $z = [x, y]$. Assume $|x| = |y| = |z| = 3$, $z \in Z(G)$, and $G = \langle x, y \rangle$. Prove $G \cong 3^{1+2}$.
4. Let V be a 2-group and $g \in G \leq \mathrm{Aut}(V)$ such that g is of odd order and
   (a) $\Phi(C_V(g)) = 1$ and $C_G(g)$ is irreducible on $C_V(g)$.
   (b) $|V : C_V(g)| = |[C_V(g), h]|$ with $C_G(g) \cap N_G(\langle h \rangle)$ irreducible on $[C_V(g), h]$ for some $h \in g^G \cap C_G(g)$.
   Prove $\Phi(V) = 1$ and G is irreducible on V.
   [*Hint:* Use Exercise 4.1 in [FGT].]

5. Let $G = A_7$ be the alternating group on a set X of order 7, $x \in X$, $H = G_x \cong A_6$, and $P \in \text{Syl}_3(H)$. Prove H is the unique maximal subgroup of G containing $N_H(P)$.

   [*Hint:* Use Exercise 5.6.2 in [FGT].]

6. Let t be an involution acting on a solvable group G of odd order with $C_G(t) \cong E_9$, $Z_3 \cong Y \leq Z(G\langle t \rangle)$, and $H = [H, t]$ a 3-subgroup of G with $C_G(H)$ a 3-group. Prove G is a 3-group.

2
Commuting graphs of groups

In this chapter we collect the results on graphs we will need to study groups generated by 3-transpositions. One such result consists of the list of standard numerical constraints on strongly regular graphs, which appears as Theorem 6.3. In the context of the graph of a rank 3 permutation group, this was first proved by D. Higman in [Hi]. A proof of Theorem 6.3 is contained in Section 16 of [FGT], so no proof is included here. There is also a brief discussion of the *lines* defined by a graph.

Finally most effort is devoted to the notion of a *contraction* of a graph, particularly as applied to the commuting graph of a locally conjugate conjugacy class of a finite group. This material comes from [A2] and [AH]. It is an abstraction of ideas introduced by Fischer in Section 1 and 3 of [F1] and [F2], most particularly Theorem 3.3.5 of those references.

6. Graphs

A *graph* is a pair $\Gamma = (V, E)$ where V is a set of *vertices* (or points or objects) and E is a symmetric relation on V called *adjacency* (or incidence or something else). The ordered pairs in the relation E are called the *edges* of the graph. We say u is *adjacent* to v if $(u, v) \in E$ is an edge in Γ.

Example: Let G be a group and D a collection of subsets of G. Define $\mathcal{D}(D)$ to be the graph with vertex set D such that $A, B \in D$ are adjacent in $\mathcal{D}(D)$ if and only if $[A, B] = 1$, that is when $ab = ba$ for all $a \in A$ and $b \in B$. We call $\mathcal{D}(D)$ the *commuting graph* of D.

In the remainder of this section we assume Γ is a graph. In addition in each of the lemmas, assume Γ is finite.

For $v \in V$, write $\Gamma(v)$ for the set of u in V distinct from v and adjacent to v. Denote by Γ^c the *complementary graph* to Γ that has vertex set V and with u adjacent to v in Γ^c if and only if u is not adjacent to v in Γ. Define $v^\perp = \{v\} \cup \Gamma(v)$. Notice that $\Gamma^c(v) = V - v^\perp$.

A *morphism* $\alpha : \Gamma \to \Gamma'$ of graphs is a function $\alpha : V \to V'$ that preserves adjacency; that is for each $v \in V$, $v^\perp \alpha \subseteq v\alpha^\perp$.

Example: Let G be a group and D a union of conjugacy classes of subsets of G. Form the commuting graph $\Gamma = \mathcal{D}(D)$. Then G is represented as a group of automorphisms of $\mathcal{D}(D)$ via its action on D by conjugation. Further for $A \in D$, $\Gamma(A)$ consists of those $B \in D$ distinct from A such that $B \subseteq C_G(A)$.

Given distinct vertices u and v in Γ, the *line* through u and v in Γ is defined to be

$$u * v = \bigcap_{x \in u^\perp \cap v^\perp} x^\perp.$$

The line $u * v$ is said to be *singular* if u is adjacent to v and *hyperbolic* otherwise.

Example: In Chapter 4 we consider nondegenerate symmetric or alternating bilinear forms f and unitary forms f on a vector space U. There a subspace W of U is defined to be "singular" if the restriction of f to U is trivial. Let S be the set of singular points of the projective space $PG(U)$ and define a graph Γ on S by decreeing that $A, B \in S$ are adjacent if $f(A, B) = 0$. Then for distinct $A, B \in S$, the line $A * B$ in Γ through A and B consists of the points on the line $A + B$ of $PG(U)$. The line $A + B$ is "singular" in the sense just defined if and only if $A * B$ is singular in Γ. Moreover in Chapter 4 we find there are hyperbolic lines in $PG(U)$ defined by f and $A + B$ is hyperbolic in $PG(U)$ if and only if $A * B$ is hyperbolic in Γ.

(6.1) Let $v \in V$ and $u \in \Gamma(v)$. Assume $G \leq \operatorname{Aut}(\Gamma)$ is transitive on V and G_v is transitive on $\Gamma(v)$. Then

(1) *Each pair of distinct adjacent vertices lies on a unique singular line.*

(2) *G is transitive on singular lines.*

(3) *$N_G(u * v)$ is 2-transitive on $u * v$.*

(4) *$u * v = \{v\} \cup V_{u,v}$, where $V_{u,v}$ consists of those $x \in \Gamma(v)$ with $x^\perp \cap \Gamma(v) = u^\perp \cap \Gamma(v)$.*

(5) *$u^\perp \cap v^\perp = x^\perp \cap y^\perp$ for each $x, y \in u * v$ with $x \neq y$.*

Proof: As $u \in \Gamma(v)$, $v \in v^\perp \cap u^\perp$, and therefore $u * v \subseteq v^\perp$. Indeed if $x \in u * v - \{v\}$ then $v \in u^\perp \cap v^\perp \subseteq x^\perp \cap v^\perp$, so $x \in \Gamma(v)$. Then as G_v is transitive on $\Gamma(v)$, $|u^\perp \cap v^\perp| = |x^\perp \cap v^\perp|$, so as $u^\perp \cap v^\perp \subseteq x^\perp \cap v^\perp$, we conclude $u^\perp \cap v^\perp = x^\perp \cap v^\perp$. Therefore $u * v = x * v$. This implies (1), (4), and (5), and then (1) and transitivity of G on V and G_v on $\Gamma(v)$ imply (2) and (3).

(6.2) Let $v \in V$ and $w \in \Gamma^c(v)$. Assume $G \leq \mathrm{Aut}(\Gamma)$ is transitive on V and G_v is transitive on $\Gamma^c(v)$. Assume $v^\perp \cap w^\perp$ is not contained in $u^\perp$ for any $u \in \Gamma(v)$. Then

(1) Each pair of distinct nonadjacent vertices lies on a unique hyperbolic line.
(2) G is transitive on hyperbolic lines.
(3) $N_G(v * w)$ is 2-transitive on $v * w$.
(4) $\Gamma(v, w) = \Gamma(x, y)$ for each $x, y \in v * w$ with $x \neq y$.

Proof: Argue as in the previous lemma.

Theorem 6.3 (D. Higman). *Assume G is a group of automorphisms of Γ primitive on V, $v \in V$, $u \in \Gamma(v)$, and $w \in \Gamma^c(v)$. Assume further that G_v is transitive on $\Gamma(v)$ and $\Gamma^c(v)$. Let $n = |V|$, $k = |\Gamma(v)|$, $l = |\Gamma^c(v)|$, $\lambda = |\Gamma(v) \cap \Gamma(u)|$, and $\mu = |\Gamma(v) \cap \Gamma(w)|$. Then*

(1) $n = 1 + k + l$.
(2) $\mu l = k(k - \lambda - 1)$.
(3) *Either*
  (a) $k = l$ and $\mu = \lambda + 1 = k/2$, or
  (b) $\delta^2 = (\lambda - \mu)^2 + 4(k - \mu)$ is a square and δ divides

$$2k + (\lambda - \mu)(k + l).$$

Proof: See 16.3 and 16.7 in [FGT].

A transitive group of permutations G on a set X is said to be of *rank* 3 if the stabilizer G_x of a point x of X has exactly three orbits on X. Thus the group G of Lemma 6.3 is rank 3 on V. The graph of a rank 3 group was first discussed by D. Higman in [Hi]. These graphs are important in simple group theory and very important in the study of 3-transposition groups as we will see in Lemma 9.5.4.

Given graphs Γ and $\bar{\Gamma}$ define a *bimorphism* of Γ to $\bar{\Gamma}$ to be a map $\alpha : V \to \bar{V}$ such that $\alpha : \Gamma \to \bar{\Gamma}$ and $\alpha : \Gamma^c \to \bar{\Gamma}^c$ are both morphisms. Define Γ to be *simple* if whenever $\alpha : \Gamma \to \bar{\Gamma}$ is a surjective bimorphism then either α is an isomorphism or $|\bar{V}| = 1$.

Given a partition P of V denote by Γ/P the graph with vertex set P and A and B in P adjacent if and only if some vertex in A is adjacent to some vertex in B in Γ. Observe the inclusion map $i : \Gamma \to \Gamma/P$ is a morphism. Further if $\alpha : \Gamma \to \bar{\Gamma}$ is a surjective morphism then the set P of fibers of α forms a partition of V and the map $\alpha^{-1}(x) \mapsto x$ is a bijective morphism $\Gamma/P \to \bar{\Gamma}$ of graphs. Further if α is a bimorphism then this map is an isomorphism. That is each bimorphic image of Γ is a factor graph.

Define a *contraction* of Γ to be a partition $P \neq \{V\}$ of V such that the inclusion map $i : \Gamma \to \Gamma/P$ is a bimorphism. Thus the nontrivial bimorphic images of Γ correspond to the contractions of Γ, and Γ is simple if and only if it has no nontrivial contractions.

Define a *clump* of Γ to be a proper subset C of V such that $c^\perp \subseteq C \cup \Gamma(C)$ for each $c \in C$, where $\Gamma(C) = \bigcap_{x \in C} \Gamma(x)$. Notice

(6.4) *A partition $P \neq \{V\}$ of V is a contraction if and only if each member of P is a clump.*

Partially order the set of contractions of Γ via $P < Q$ if each member of P is contained in a member of Q. Observe that if $P < Q$ then Q corresponds to a contraction of Γ/P, so there is a natural bimorphism of Γ/P onto Γ/Q. The maximal contractions are of greatest interest; they correspond to the nontrivial simple bimorphic images of Γ.

Example: Once again let f be a nondegenerate symmetric or alternating bilinear form or a unitary form on a vector space U. This time let Γ be the graph with vertex set the set of singular vectors in $U^{\#}$ such that u is adjacent to v if and only if $f(u, v) = 0$. Then $\langle u \rangle^{\#}$ is a clump of Γ and indeed the set of singular points of $PG(U)$ is the unique maximal contraction of Γ.

A *path* from u to v is a sequence of vertices $u = u_0, u_1, \ldots, u_n = v$ such that $u_{i+1} \in u_i^\perp$ for each i. The relation $\sim$ on V, defined by $u \sim v$ if and only if there is a path from u to v in Γ, is an equivalence relation on V. The equivalence classes of this relation are the *connected components* of Γ, and Γ is said to be *connected* if it has just one connected component.

Theorem 6.5. *Assume $G \leq \mathrm{Aut}(\Gamma)$ is transitive on V and Γ and Γ^c are connected. Let $v \in V$, U a connected component of $\Gamma(v)^c$ of maximal order, and $C = \Gamma(U)$. Then*

(1) *$P = C^G$ is the unique maximal contraction of Γ, so Γ/P is the unique nontrivial simple bimorphic image of Γ.*

(2) *P is a system of imprimitivity for G on V, so there is an induced representation of G on Γ/P.*

(3) $\Gamma(C)^c$ *and* $\Gamma^c(C)$ *are connected subgraphs of* Γ^c *and* Γ, *respectively.*

Proof: First observe

(a) C is a clump and $\Gamma(C) = U$.

For if $c \in C$ then by maximality of $|U|$, U is a connected component of $\Gamma(c)^c$. Hence $U \subseteq \Gamma(x)$ for each $x \in \Gamma(c) - U$. That is $\Gamma(c) - U \subseteq \Gamma(U) = C$, so $\Gamma(c) \subseteq U \cup C \subseteq \Gamma(C) \cup C$, and hence C is a clump and $\Gamma(C) = U$.
We next claim

(b) If A, B are clumps with $A \cap B \neq \emptyset$ then $A \cup B$ is a clump.

Namely let $x \in A \cap B$ and $a \in A$. Then $a^{\perp} \subseteq A \cup \Gamma(A)$ and $a^{\perp} \cap \Gamma(A) \subseteq \Gamma(x) \subseteq B \cup \Gamma(B)$, so $a^{\perp} \subseteq A \cup B \cup \Gamma(A \cup B)$. By symmetry we get the same containment if $a \in B$, so $A \cup B$ is a clump.
Next

(c) C is a maximal clump.

For let $C \subseteq B \subset V$ with B a clump. Then $U \subseteq \Gamma(C) \subseteq B \cup \Gamma(B)$. Suppose $u \in U \cap \Gamma(B)$. Then if $x \in U \cap \Gamma^c(u) - \Gamma(B)$, we have $x \in B$, so $u \in \Gamma(B) \subseteq \Gamma(x)$, contradicting the choice of x. Therefore $\Gamma^c(u) \cap U \subseteq \Gamma(B)$, so as U is a connected subgraph of Γ^c, we conclude $U \subseteq \Gamma(B)$. Then $B \subseteq \Gamma(U) = C$, so $B = C$ as desired.

Thus we may assume $U \subseteq B$. Now $B \neq V$ and Γ is connected, so there exists $x \in \Gamma(B) - B$. Then $x \in \Gamma(B) \subseteq \Gamma(U) = C \subseteq B$, contrary to the choice of x. Thus (c) is established.

Notice (a), (b), and (c) imply (1) and (2). For example if $g \in G$ and $c \in C \cap C^g$ then by (b), $C \cup C^g$ is a clump and then by (c), $C = C \cup C^g = C^g$. So $P = C^G$ is a partition of Γ and hence (2) holds. Then P is a contraction by 6.4 and (a), and by (c), P is maximal. Suppose P_1 is a maximal contraction and let $C_1 \in P_1$. Then $C_1 \cap C^g \neq \emptyset$ for some $g \in G$, so by (b) and (c), $C_1 \subseteq C_1 \cup C^g = C^g$. Hence $P_1 = P$ by maximality of P_1, so P is the unique maximal contraction and (1) holds. By construction, $U^c = \Gamma(C)^c$ is a connected subgraph of Γ^c. Contractions of Γ are contractions of Γ^c, so by symmetry $\Gamma^c(C)$ is a connected subgraph of Γ.

7. Locally conjugate subsets of groups

Let G be a finite group and D a collection of subsets of G. Then D is said to be *locally conjugate* in G if $D^G = D$, $G = \langle D \rangle$, and for all A, B in D, either $[A, B] = 1$ or A is conjugate to B in $\langle A, B \rangle$. Throughout this section we assume D is locally conjugate in G. If $H \leq G$, we write $H \cap D$ for $\{A \in D : A \subseteq H\}$.

Example: Let D be a set of 3-transpositions of G. Claim D is locally conjugate in G. By the 3-transposition hypothesis, $D^G = D$ and $G = \langle D \rangle$. Further if $a, b \in D$ then by the 3-transposition hypothesis, $|ab| = 1, 2,$ or 3. In the first two cases $[a, b] = 1$. In the third, $\langle a, b \rangle \cong S_3$, so $\langle a \rangle$ and $\langle b \rangle$ are Sylow 2-subgroups of $\langle a, b \rangle$ and therefore by Sylow's Theorem, a is conjugate to b in $\langle a, b \rangle$.

A subgroup of G is a *D-subgroup* of G if $H = \langle H \cap D \rangle$. Evidently

(7.1)

  (1) If $E \subseteq D$ with $E = E^E$ then E is locally conjugate in $\langle E \rangle$.
  (2) If H is a D-subgroup of G then $H \cap D$ is locally conjugate in H.

(7.2) *If $\alpha : G \to H$ is a surjective group homomorphism then $D\alpha$ is locally conjugate in H.*

(7.3) *If $D = D_1 + D_2$ is a G-invariant partition of D then D_i is locally conjugate in $G_i = \langle D_i \rangle$, $[G_1, G_2] = 1$, and $G = G_1 G_2$.*

Proof: Let $A \in D_1$ and $B \in D_2$. As our partition is G-invariant, A is not conjugate to B in G and hence not in $\langle A, B \rangle$. So as D is locally conjugate in G, $[A, B] = 1$.

(7.4) *If Δ is an orbit of G on D then $\langle \Delta \rangle$ is transitive on Δ.*

Proof: This follows from 7.3 applied to the partition $D = \Delta + (D - \Delta)$.

Represent G by conjugation on D. A *set of imprimitivity* for G on D is a proper nonempty subset Δ of D such that for each $g \in G$, either $\Delta^g = \Delta$ or $\Delta \cap \Delta^g$ is empty.

(7.5) *If Δ is a set of imprimitivity for G on D then*

  (1) $\Delta^\Delta = \Delta$.
  (2) If $A \in \Delta$ and $B \in \Delta^g \neq \Delta$ such that either $[A, B] = 1$ or $B \subseteq N_G(\Delta)$, then $[\Delta, \Delta^g] = 1$.
  (3) Δ is a clump.

Proof: Let $A \in \Delta$ and $a \in A$. Then $A = A^a \in \Delta^a$, so as Δ is a set of imprimitivity, $\Delta^a = \Delta$. Hence (1) holds. Suppose $B \in \Delta^g \neq \Delta$. If $[A, B] = 1$ then $A = A^b \in \Delta \cap \Delta^b$ for each $b \in B$, so $\Delta^B = \Delta$. So assume $B \subseteq N_G(\Delta)$

and let $H = \langle \Delta, B \rangle$. As $\Delta^B = \Delta$ and Δ normalizes itself by (1), Δ is H-invariant. Therefore $H \cap D = \Delta + (H \cap D - \Delta)$ is an H-invariant partition of $H \cap D$ and hence by 7.3, $[\Delta, B] = 1$. Now by symmetry $[\Delta, \Delta^g] = 1$. So (2) holds and of course (2) implies (3).

(7.6) *Let G be transitive on D and Δ a set of imprimitivity for G on D. Then Δ^G is locally conjugate in G.*

Proof: Suppose $\Delta \neq \Delta^g$ with $[\Delta, \Delta^g] \neq 1$. Let $A \in \Delta$, $B \in \Delta^g$. By 7.5, $[A, B] \neq 1$, so there exists $h \in \langle A, B \rangle$ with $A^h = B$. Then $h \in \langle \Delta, \Delta^g \rangle$ and $\Delta^h = \Delta^g$.

Suppose Δ is a set of imprimitivity for G on D and consider the commuting graph $\mathcal{D}(\Delta^G)$ on Δ^G defined in Section 6. Write D_Δ for the set of vertices distinct from Δ and adjacent to Δ in $\mathcal{D}(\Delta^G)$. Recall $\Delta^\perp = \{\Delta\} \cup D_\Delta$. Let $A_\Delta = \Delta^G - \Delta^\perp$ and $X_\Delta^* = \{A : A \in \Delta^g \in X_\Delta, \ g \in G\}$, for $X = A$ and D.

(7.7) *The complementary graph $\mathcal{D}(D)^c$ to $\mathcal{D}(D)$ is connected if and only if G is transitive on D.*

Proof: If G is intransitive on D then by 7.3, there is a partition $D = D_1 + D_2$ of D with $D \neq D_i$ and $[D_1, D_2] = 1$. Thus D_i is the union of connected components of $\mathcal{D}(D)^c$, so $\mathcal{D}(D)^c$ is disconnected. Conversely if $\mathcal{D}(D)^c$ is disconnected, such a partition exists. As $[D_1, D_2] = 1$, D_i acts on D_{3-i} and then also on $D_i = D - D_{3-i}$, so $G = \langle D_1, D_2 \rangle$ acts on D_i and in particular is intransitive on D.

Theorem 7.8. *Assume G is transitive on D and $\mathcal{D}(D)$ is connected. Let $A \in D$. Then*

   (1) There exists a unique maximal set of imprimitivity α of G on D containing A.
   (2) $A_\alpha \neq \emptyset$ and A_α^* is a connected subgraph of $\mathcal{D}(D)$.
   (3) $D_\alpha \neq \emptyset$ and $\langle D_\alpha \rangle$ is transitive on D_α^* by conjugation.
   (4) $G = \langle D_\alpha, D_\beta \rangle$ for each $\beta \in D_\alpha$.
   (5) If H is a proper D-subgroup of G containing A and D_α then $H \cap D \subseteq \alpha \cup D_\alpha^*$.

Proof: As G is transitive on D, the complementary graph $\mathcal{D}^c$ is connected by 7.7. Hence by 6.5, there is a unique maximal contraction P of $\mathcal{D}$ and $P = \alpha^G$ where $A \in \alpha$ and α is a set of imprimitivity for G on D. On the other hand by

6.4 and 7.5.3, Δ^G is a contraction of D for each set Δ of imprimitivity for G on D. Therefore α is the unique maximal set of imprimitivity containing A.

If $A_\alpha = \emptyset$ then as $\mathcal{D}^c$ is connected, $\alpha = D$, contradicting α proper in D. Similarly $D_\alpha \neq \emptyset$. Now (2) follows from 6.5.3. Similarly by 6.5.3, $\mathcal{D}(D_\alpha^*)^c$ is connected, and then 7.7 implies (3).

Let $L = \langle D_\alpha, D_\beta \rangle$ for some $\beta \in D_\alpha$, and let X be the set of $B \in D$ conjugate to a member of α or β under L. Now $D_A \subseteq \alpha \cup D_\alpha^* \subseteq D_\alpha^* \cup D_\beta^* \subseteq X$ by (3). Similarly $D_B \subseteq X$ for $B \in \beta$. So $D_C \subseteq X$ for each $C \in X$, and hence as $\mathcal{D}(D)$ is connected, $X = D$ and $L = G$. Thus (4) is established.

Assume the hypotheses of (5). Let

$$\Gamma = \{B^h : B \in \beta \in D_\alpha, \ h \in H\}.$$

By (4), $\Gamma \cap A^H$ is empty, so by 7.3, $A^H \subseteq \Gamma^\perp$. But by 6.5, $\Gamma^\perp \subseteq \alpha$. Thus as H acts on A^H, H acts on α, so $H \cap D \subseteq N_D(\alpha)$. Finally by 7.5, $N_D(\alpha) = \alpha \cup D_\alpha^*$.

(7.9) *Assume G is transitive on D, $\mathcal{D}$ is connected, and*

(∗)   *if $B \in D$ and $A, C \in A_B$ with $C \in D_A$,*

                then $B^\perp \cap X \subseteq B^{\langle A^\perp \cap X \rangle}$ for $X = \langle A, B, C \rangle$.

Let α be the unique maximal set of imprimitivity containing A. Then $\langle A^\perp \rangle$ is transitive on $A_A - \alpha$ and G is rank 3 on α^G, with $\{\alpha\}$, D_α, and A_α the orbits of G_α on α^G.

Proof: Let $H = \langle A^\perp \rangle$ and $\Gamma = A_A - \alpha$. By 7.8.2, Γ is connected so to prove H is transitive on Γ it suffices to show that if $B, C \in \Gamma$ with $[B, C] = 1$ then $B \in C^H$. But this follows from (∗).

Therefore H is transitive on Γ. This fact together with 7.8.3 show G is rank 3 on α^G.

Exercises

1. Let X be a set of order 5 and Γ the graph whose vertex set is the set V of all subsets of X of order 2 and with two vertices adjacent if they intersect in the empty set. Thus Γ is the *Petersen graph*. Assume Δ is a graph admitting a rank 3 group G of automorphisms with parameters $n = 10$, $k = 3$, $\lambda = 0$, $\mu = 1$. Prove Δ is isomorphic to the Petersen graph and G is isomorphic to S_5 or A_5.

2. A set D of involutions in a finite group G is a set of *odd transpositions* of G if D is the union of conjugacy classes of G, D generates G, and for all

$a, b \in D$, ab has odd order or order 2. Assume D is a conjugacy class of odd transpositions of G. Prove

(1) D is locally conjugate in G.

(2) If H is a normal subgroup of G then the set of orbits of H on D is a contraction of $\mathcal{D}(D)$.

(3) Let F be a field of order 2^e, $L = SL_2(F)$, and T the set of transvections in L. Prove T is a class of odd transpositions of L. Find the unique maximal contraction of $\mathcal{D}(T)$.

3. Write out a proof of Lemma 6.2.

4. Write out a proof of Lemma 6.4.

5. Let Γ be a finite graph with vertex set X and for $x \in X$ define

$$V_x = \{y \in X : x^\perp = y^\perp\} \quad \text{and} \quad W_x = \{z \in X : \Gamma(x) = \Gamma(z)\}.$$

Prove

(1) $\{V_x : x \in X\}$ and $\{W_x : x \in X\}$ are contractions of Γ.

(2) $V_x \subseteq x^\perp$ and $W_x \subseteq \Gamma^c(x) \cup \{x\}$.

(3) Either $V_x = \{x\}$ or $W_x = \{x\}$.

(4) Assume $\text{Aut}(\Gamma)$ is transitive on X, α is a clump of Γ, and $x \in \alpha$. Then for $U = V$ and W, either $\alpha \subseteq U_x$ or $U_x \subseteq \alpha$.

3

The structure of 3-transposition groups

In this chapter we develop some of the basic properties of 3-transposition groups. For example we see that the nature of the normal subgroups of a group G generated by a class of 3-transpositions is restricted and related to properties of the commuting graph $\mathcal{D}(D)$. In particular in Theorem 9.5 we prove Fischer's basic result that if G is primitive on D of width at least 2 then G is essentially a nonabelian simple group and acts as a rank 3 group by conjugation on D. Almost all results in this chapter were first proved by Fischer in [F1].

8. Basic results on 3-transpositions

In this section D is a set of 3-transpositions in the finite group G and $d \in D$. We call the pair (G, D) a *3-transposition group*. A *morphism* $\alpha : (G, D) \to (\bar{G}, \bar{D})$ of 3-transposition groups is a group homomorphism $\alpha : G \to \bar{G}$ such that $D\alpha \subseteq \bar{D}$. A *D-subgroup* of G is a subgroup H such that $H = \langle H \cap D \rangle$.

(8.1)

   (1) *If $E \subseteq D$ with $E = E^E$ then E is a set of 3-transpositions of $\langle E \rangle$.*

   (2) *If H is a D-subgroup of G then $H \cap D$ is a set of 3-transpositions of H.*

   (3) *If $\alpha : G \to H$ is a surjective homomorphism then $D\alpha$ is a set of 3-transpositions of H and $\alpha : (G, D) \to (H, D\alpha)$ is a morphism of 3-transposition groups.*

(8.2)

   (1) *D is locally conjugate in G.*

   (2) *If $D = D_1 \cup D_2$ is a G-invariant partition of D then D_i is a set of 3-transpositions of $G_i = \langle D_i \rangle$, $[G_1, G_2] = 1$, and $G = G_1 G_2$.*

   (3) *The complementary graph $\mathcal{D}(D)^c$ to the commuting graph $\mathcal{D}(D)$ is connected if and only if G is transitive on D.*

Proof: We saw (1) holds in the Example of Section 7. Then (1) and 7.3 and 7.7 imply (2) and (3), respectively.

As in Section 7, we define:

$$d^\perp = C_D(d),$$
$$D_d = d^\perp - \{d\},$$
$$A_d = D - d^\perp.$$

Similarly for $a_1, \ldots, a_m \in D$ we write

$$D_{a_1, \ldots, a_m} = D_{a_1} \cap \cdots \cap D_{a_m},$$
$$A_{a_1, \ldots, a_m} = A_{a_1} \cap \cdots \cap A_{a_m}.$$

Define a *3-string* to be a triple (a, b, c) in D with $c \in D_a$ and $a, c \in A_b$. Equivalently a 3-string is a subgraph

$$\begin{array}{ccc} a & b & c \\ \circ\!\!-\!\!\!-\!\!\!-\!\!\!\circ\!\!-\!\!\!-\!\!\!-\!\!\!\circ \end{array}$$

of $\mathcal{D}(D)^c$ of type A_3, in the language of Section 5.

(8.3)   *Assume (a, b, c) is a 3-string in D and let $X = \langle a, b, c \rangle$. Then*

(1) There is an isomorphism $(X, X \cap D) \cong (Y, T)$ of 3-transposition groups, where Y is the symmetric group on $\{1, 2, 3, 4\}$ and T is the set of transpositions of Y, mapping a, b, c to $(1, 2), (2, 3), (3, 4)$, respectively.
(2) $b^\perp \cap X = \{b, b^{ac}\}$ and $O_2(X) = \langle (ac)^X \rangle$.

Proof: Let Y be the symmetric group on $\{1, 2, 3, 4\}$, $s_i = (i, i+1) \in Y$, and $S = \{s_1, s_2, s_3\} \subseteq Y$. By 5.3, (Y, S) is a Coxeter system of type A_3. So as a, b, c satisfy the defining relations $a^2 = b^2 = c^2 = (ac)^2 = (ab)^3 = (bc)^3$ of the Coxeter diagram A_3, we conclude from 5.1 that there is a surjective homomorphism $\alpha : Y \to X$ with $s_1\alpha = a$, $s_2\alpha = b$, and $s_3\alpha = c$. To prove α is an isomorphism of groups, it remains to observe $\ker(\alpha) = 1$. This follows as $\langle (s_1 s_3)^Y \rangle$ is the unique minimal normal subgroup of Y while $s_1 s_3 \alpha = ac \neq 1$. So (1) is established with $\alpha^{-1} : (X, X \cap D) \to (Y, T)$ our isomorphism. Using (1), it is easy to calculate that (2) holds inside S_4.

(8.4)   *For distinct $a, b \in D$, $C_D(ab) = C_D(a) \cap C_D(b)$.*

Proof: Evidently $D_{a,b} \subseteq C_D(ab)$. Conversely let $d \in C_D(ab)$. We must show $d \subseteq a^\perp$, so assume not. Then by 8.2.1, a is conjugate to d in $\langle a, d \rangle$. But a inverts

ab and d centralizes ab, so $\langle a, d \rangle \leq N_G(\langle ab \rangle)$. Hence as $C_G(ab) \trianglelefteq N_G(\langle ab \rangle)$, a centralizes ab, so $|ab| = 2$. But now $db = da(ab)$ is of order 6, a contradiction.

Define the *width* of G to be the maximal number of pairwise commuting elements of D. Write $w_D(G)$ for the width of G.

(8.5)   *Let $T \in \mathrm{Syl}_2(G)$. Then*

(1) $\langle D \cap T \rangle$ is abelian.
(2) $|T \cap D| = w_D(G)$ is the width of G.

Proof: For distinct $a, b \in T \cap D$, $\langle a, b \rangle$ is a 2-group so $|ab| = 2$. Thus $ab = ba$, so (1) holds. If E is a set of pairwise commuting members of D, then $\langle E \rangle$ is abelian and hence a 2-group. So by Sylow's Theorem we may assume $\langle E \rangle \leq T$. Thus $E \subseteq T \cap D$, so (2) holds.

(8.6)   *G is of width 1 if and only if $G = \langle d \rangle O_3(G)$ for $d \in D$.*

Proof: If $G = \langle d \rangle O_3(G)$ then by Sylow's Theorem and 8.5, G is of width 1. So assume G is of width 1. Then for each $d \in D$, $\{d\} = d^\perp$. We could now appeal to Glauberman's Z^*-Theorem [G] to complete the proof, but instead we use a more elementary argument due to J. Hall (cf. [H1]) using the Baer–Suzuki Theorem (cf. 2.5) and a result of M. Hall reproduced in Section 19.

The argument goes as follows: Let $d \neq a \in D$ and set $x = da$. Then x is of order 3. We will show $\langle x^G \rangle$ is a 3-group. Then $X = \langle bc : b, c \in D \rangle$ is a normal 3-subgroup of G and $G = \langle D \rangle = \langle d \rangle X$. So it remains to show $\langle x^G \rangle$ is a 3-group. By the Baer–Suzuki Theorem 2.5, it suffices to show $\langle x, x^g \rangle$ is a 3-group for all $g \in G$. Now $x^g = d^g a^g$, so $\langle x, x^g \rangle \leq \langle d, a, d^g, a^g \rangle$. Finally by 19.13 and 19.14, any group H of width 1 generated by four 3-transpositions is of the form $\langle d \rangle O_3(H)$, so the proof is complete.

(8.7)   *If $d \in D$ and $a \in A_d$ then $a^d = d^a$.*

Proof: This is because da is of order 3.

For $d \in D$ define

$$V_d = \{v \in D : v^\perp = d^\perp\},$$
$$W_d = \{w \in D : D_d = D_w\}.$$

(8.8)

   (1) $V_d \subseteq d^\perp$.
   (2) $W_d - \{d\} \subseteq A_d$.
   (3) Either $V_d = \{d\}$ *or* $W_d = \{d\}$.

Proof: Easy; see Exercise 2.5 for a more general statement.

(8.9) *If* $p > 3$ *is prime,* P *is a* p-*subgroup of* G, *and* $d \in N_G(P)$, *then* $[d, P] = 1$.

Proof: First $|\langle d \rangle P| = 2|P|$ so $\langle d \rangle \in \mathrm{Syl}_2(\langle d \rangle P)$. So by 8.5, $\langle d^P \rangle$ is of width 1. Now as $p > 3$, 8.6 says $d^P = \{d\}$. Thus $[d, P] = 1$.

(8.10) Let $a \in D$, $b \in A_a$, $x \in A_{a,b,a^b}$. Then $x^{bax} = x^{ab}$.

Proof: By repeated applications of 8.7, $x^{bax} = b^{xax} = b^{axa} = a^{bxa} = x^{a^b a} = x^{ab}$.

(8.11) *Let* a, b, c *be distinct commuting members of* D. *Then*

   (1) If $abc = 1$ *then* $a^\perp = b^\perp = c^\perp = D$.
   (2) $C_D(abc) = a^\perp \cap b^\perp \cap c^\perp$.

Proof: Certainly $a^\perp \cap b^\perp \cap c^\perp \subseteq C_D(abc)$. Conversely suppose $d \in C_D(abc) - a^\perp$. Then $[d, bc] \neq 1$, so we may also assume $d \in A_b$. Thus (a, d, b) is a 3-string, so $X = \langle a, d, b \rangle \cong S_4$ and abd is of order 4 by 8.3. But $abc \in C_G(X)$ and $(abc)^2 = 1$, so $cd = (abc)(abd)$ is of order 4, a contradiction. This establishes (2), and (2) implies (1).

9. The generalized Fitting subgroup of a 3-transposition group

In this section D is a conjugacy class of 3-transpositions in the finite group G, $d \in D$, and D contains at least two elements. In particular G is transitive on D and $D \neq \{d\}$, so G is nonabelian. Hence A_d is nonempty. Further by 2.3, d does not centralize $F^*(G)$. We use these observations several times later.

(9.1) *Assume* α *is a set of imprimitivity for* G *on* D. *Then* $X = \langle ab : a, b \in \alpha \rangle$ *fixes* α^G *pointwise.*

Proof: By 7.6, α^G is locally conjugate in G in the sense of Section 7, so we can use the theory developed there. In particular considering the commuting graph $\mathcal{D}(\alpha^G)$ we have $\alpha^G = \{\alpha\} \cup D_\alpha \cup A_\alpha$.

For $d \in D$ let α_d be the conjugate of α containing d. Let $a, b \in \alpha$, $g \in G$, and $\beta = \alpha^g$. As α is a set of imprimitivity, a and b and hence also ab act on α by 7.5. If $\beta \in D_\alpha$ then $[\alpha, \beta] = 1$ so certainly ab fixes β. So take $\beta \in A_\alpha$. Then by 7.5, $\beta \subseteq A_{a,b}$. Thus if $c \in \beta$, $a^c = c^a$ by 8.7. Therefore $\beta^a = (\alpha_c)^a = \alpha_{c^a} = \alpha_{a^c} = (\alpha_a)^c = \alpha^c$. Similarly $\beta^b = \alpha^c$. Hence ab fixes β.

(9.2)

   (1) V_d and W_d are sets of imprimitivity for G on D.
   (2) $V_d = D \cap d O_2(G) = d^{O_2(G)}$.
   (3) $W_d = D \cap d O_3(G) = d^{O_3(G)}$.

Proof: Part (1) is trivial. By (1) and 7.5, $\alpha^\alpha = \alpha$ for $\alpha = V_d$ or W_d, so α is a set of 3-transpositions of $\langle \alpha \rangle$ by 8.1.1. Let $X(\alpha) = \langle ab : a, b \in \alpha \rangle$ and $Y(\alpha) = \langle X(\alpha)^G \rangle$. By 9.1, $X(\alpha)$ normalizes $X(\alpha)^g$ for each $g \in G$, so if $X(\alpha)$ is a p-group then $Y(\alpha) \leq O_p(G)$ and $\alpha \subseteq D \cap d O_p(G)$.

By 8.8.1, $V_d \subseteq d^\perp$, so $\langle V_d \rangle$ is a 2-group. Thus by the previous paragraph, $V_d \subseteq D \cap d O_2(G)$. Conversely let $a \in D \cap d O_2(G)$ and $b \in d^\perp$. Then $\langle d, b, O_2(G) \rangle$ is a 2-group containing a so $b \in a^\perp$ by 8.5.1. That is $a \in V_d$. Therefore we have established the first equality in (2).

Similarly let $d \neq w \in W_d$. Then by 8.8.2, $w \in A_d$. Hence W_d is a set of 3-transpositions of $\langle W_d \rangle$ of width 1, so by 8.6, $X(W_d)$ is a 3-group. Then by the first paragraph, $W_d \subseteq D \cap d O_3(G)$. On the other hand let $a \in D \cap d O_3(G)$ and $b \in D_d$. Set $G^* = G/O_3(G)$. Then $a^* = d^*$, so $|a^* b^*| = |d^* b^*| = 2$. Then as $|a^* b^*|$ divides $|ab|$, $|ab| = 2$, so $b \in D_a$. Hence $a \in W_d$ as desired. So the first equality in (3) is established. As $W_d = D \cap d O_3(G)$, $O_3(G)$ is transitive on W_d by Sylow's Theorem, completing the proof of (3).

It remains to show $O_2(G)$ is transitive on V_d. Suppose $d \neq v \in V_d$. Recall there is $a \in A_d$. Then as $v \in V_d$, (v, a, d) is a 3-string, so $X = \langle v, a, d \rangle \cong S_4$ by 8.3. Now $vd \in O_2(G)$, so $O_2(X) = \langle (vd)^X \rangle \leq O_2(G)$. But v is conjugate to d in $O_2(X)$ by 8.3.2, so the proof is complete.

(9.3)   *For each prime $p > 3$, $O_p(G) \leq Z(G)$, and exactly one of the following holds:*

   (1) $O_2(G) \not\leq Z(G)$.
   (2) $O_3(G) \not\leq Z(G)$.
   (3) $E(G) \neq 1$.

Proof: Recall from the opening remarks in this section that d does not centralize $F^*(G)$, while by 8.9, d centralizes $O_p(G)$ for each prime $p > 3$. So at least one of the three alternatives of the lemma holds. Further $G = \langle D \rangle \leq C_G(O_p(G))$ for $p > 3$, so $O_p(G) \leq Z(G)$. Hence we may assume d does not centralize two members K, M of $\{O_2(G), O_3(G), E(G)\}$ and take $K = O_2(G)$ or $E(G)$. Now by 8.6, there is $b \in dK \cap D_d$. Then $bd \in K \leq C_G(M)$, and bd centralizes d, so $D \cap dM \subseteq C_D(bd) \subseteq d^\perp$ by 8.4. Hence $\langle D \cap dM \rangle$ is a 2-group, so $M = O_2(G)$ as $[d, M] \neq 1$. But now reversing the roles of K and M, we have a contradiction.

(9.4) *Assume $O_3(G) \leq Z(G) \geq O_2(G)$. Then*

 (1) *$F^*(G) = LZ(G)$ with L quasisimple.*
 (2) *Either $G = L$ or G/L is of width 1.*
 (3) *If $|G : L| \leq 2$ then G is primitive on D.*
 (4) *If $\mathcal{D}(D)$ is connected then $|G : L| \leq 2$, so G is primitive on D.*

Proof: By 9.3, $F^*(G) = Z(G)E(G)$ and G has a component L.

Suppose $L^d \neq L$. By Burnside's $p^a q^b$-Theorem (cf. 2.7), there is a prime $p > 3$ dividing $|L|$. Let $P \in \mathrm{Syl}_p(L)$. As $L^d \neq L$, d does not centralize the p-group PP^d, contradicting 8.9. Thus $G = \langle D \rangle$ normalizes L.

To complete the proof of (1) and (2), we may assume $G \neq L$ and let $G^* = G/L$. Claim G^* has width 1. For if not there is $a^* \in D_d*$. Then $2 = |a^*d^*|$, so $a \in D_d$. We conclude $[d^L, a] = 1$, so as $L \leq \langle d^L \rangle$, a centralizes L. But then as G is transitive on D and $L \trianglelefteq G$, $G = \langle D \rangle$ centralizes L, a contradiction. Thus (2) is established. Also as G^* has width 1, G^* has order $2 \cdot 3^n$ by 8.6. Thus by Burnside's Theorem 2.7, $L = E(G)$, completing the proof of (1).

Assume $|G : L| \leq 2$, but α is a set of imprimitivity for G containing d and $a \neq d$. By 9.1, ad is in the kernel K of the action of G on α^G. By 8.4, $ad \notin Z(G)$, so $K \not\leq Z(G)$. Hence by (1), $L \leq K$. However as $|G : L| \leq 2$, $G = L\langle d \rangle$, so $G = K$, a contradiction. Thus (3) is established.

If $\mathcal{D}(D)$ is connected then as G/L is of width 1, $|G : L| \leq 2$. So (4) is established.

Theorem 9.5. *Assume G is primitive on D and $w_D(G) > 1$. Then*

 (1) *$F^*(G) = LZ(G)$ with L quasisimple.*
 (2) *$|G : L| \leq 2$.*
 (3) *$\mathcal{D}(D)$ is connected.*
 (4) *G is rank 3 on D and $C_G(d)$ has orbits $\{d\}$, D_d, and A_d. Indeed $\langle D_d \rangle$ is transitive on D_d and $\langle d^\perp \rangle$ is transitive on A_d.*

(5) $\langle d^{\perp} \rangle$ *is a maximal D-subgroup of G.*

Proof: As G is primitive, $V_d = W_d = \{d\}$ by 9.2.1. Then by 9.2.2, $[d, O_2(G)] = 1$, so $G = \langle D \rangle$ centralizes $O_2(G)$. Similarly $O_3(G) \leq Z(G)$, so (1) holds by 9.4.

As G is primitive on D, $C_G(d)$ is maximal in G. As d does not centralize $F^*(G)$, $L \not\leq C_G(d)$, so $G = LC_G(d)$ by maximality of $C_G(d)$. Thus setting $G^* = G/L$, d^* is in the center of G^*, so $G^* = \langle d^* \rangle$. Thus (2) holds.

As the connected components of $\mathcal{D}(D)$ are a system of imprimitivity for G on D, (3) holds. Now by 8.2.1, D is locally conjugate in G, so (3) and 8.3.2 supply the hypotheses of 7.8 and 7.9. Also as G is primitive on D, $\{d\}$ is a maximal set of imprimitivity, so (4) holds by 7.8.3 and 7.9, and (5) holds by 7.8.5.

(9.6) If $d^{\perp} = V_d$ then

 (1) $G = \langle d \rangle PQ$ where P is a d-invariant Sylow 3-subgroup of G and $Q = O_2(G)$.
 (2) $|D| = me$, where $m = |V_d|$ is a power of 2 and $e = |d^P|$ is a power of 3.
 (3) If $e = 9$ and $Q \not\leq Z(G)$ then $P \cong 3^{1+2}$ with $[d, Z(P)] = 1$ and P is faithful on $O_2(G)$.
 (4) If $e = 3$ then $P = \langle g \rangle \cong \mathbf{Z}_3$, $Q = XX^g$ where $X = \langle ab : a, b \in V_d \rangle = [Q, d] = C_Q(d)$, and $\Phi(Q) \leq Z(G) = X \cap X^g$ with $|Q/Z(G)| = m^2$ and $G/Z(G)$ is determined up to isomorphism by m.
 (5) If $m = 2$ then $G \cong S_4$.

Proof: Notice that if $d^{\perp} = V_d \neq D$ then G is forced to be transitive on D; we use this observation implicitly in induction arguments later in place of the assumption that G is transitive on D.

Let $G^* = G/Q$. As $d^{\perp} = V_d$, G^* is of width 1 by 9.2. Thus $G^* = \langle d^* \rangle P^*$, where $P \in \mathrm{Syl}_3(G)$ by 8.6. Hence 9.6 is established if $Q \leq Z(G)$, so from now on assume $Q \not\leq Z(G)$. Hence $O_3(G) \leq Z(G)$ by 9.3.

Let $R \leq P$ be a maximal d-invariant 3-subgroup of G with $[R, d] \neq 1$. As $O_3(G) \leq Z(G)$, $G \neq N_G(R)$, so by induction on the order of G, d acts on a Sylow 3-subgroup T of $N_G(R)$. Thus $T = R$ by maximality of R, so $R = P$, completing the proof of (1).

As V_d is a set of imprimitivity for G, $|D| = me$, where $m = |V_d|$ and e is the number of conjugates of V_d. By 9.2, m is a power of 2. We have seen that $e = |D^*| = |d^P|$, so (2) holds.

Assume $e = 9$. Then d inverts distinct subgroups $\langle x \rangle$ and $\langle y \rangle$ in P, so by 5.5.4 and 5.5.1, $M = \langle d, dx, dy \rangle$ is a homomorphic image of the split extension of 3^{1+2} by $\mathbf{Z}_2$. In particular $|d^M| = 9 = e = |d^P|$, so $M = \langle d^M \rangle = \langle d^P \rangle$. As $G^* = \langle d \rangle P^* \cong \langle d \rangle P$ is generated by D^*, $\langle d \rangle P = \langle d^P \rangle = M$, so $P \cong 3^{1+2}$ or E_9. In the former case if $Z(P)$ centralizes Q, we replace G by $G/Z(P)$ and reduce to the latter case. So we may assume $P \cong E_9$ and it remains to produce a contradiction. Then by Exercise 8.1 in [FGT], we may choose x and y in P so that $[y, C_Q(x)] \neq 1$. But now $H = \langle d^{PC_Q(x)} \rangle$ is a group generated by a class of 3-transpositions with neither $O_3(H)$ nor $O_2(H)$ in the center of H, contradicting 9.3.

Assume $m = 2$. Then $D_d = \{b\}$ and $G = \langle d, b, P \rangle$. Let $d \neq a \in d^P$. If $\langle d, P \rangle = \langle a, d \rangle$ then (d, a, b) is a 3-string and $G = \langle d, b, P \rangle = \langle d, a, b \rangle$, so $G \cong S_4$ by 8.3. So take $c \in d^P - \langle a, d \rangle$. We prove (5) by induction on the order of G, so we may take $\langle d, P \rangle = \langle a, c, d \rangle$, and hence by 5.5, $e = 9$. Therefore by (3), $P \cong 3^{1+2}$ with $[d, Z(P)] = 1 \neq [Q, Z(P)]$.

Now $d^\perp = \{d, b\} \subseteq T \in \mathrm{Syl}_2(G)$ and by 8.5, $T \cap D = \{d, b\}$, so as $Q \leq T$, we have $[Q, d] \leq [T, d] = \langle db \rangle$. Then as $[Z(P), d] = 1$ and $[Q, d] = \langle db \rangle$ is of order 2, we conclude $[Z(P), db] = 1$. But then $Z(P)$ centralizes $\langle d, b, P \rangle = G$, contradicting $[Q, Z(P)] \neq 1$. Therefore (5) is established.

So assume $e = 3$. Then of course $P = \langle g \rangle$ is of order 3. Let $Y = \langle V_d \rangle$ so that Y is elementary abelian and $G = \langle Y, c \rangle$, where $c = d^g$. Let

$$X = \langle ab : a, b \in V_d \rangle.$$

By 9.2.2, $X = Q \cap Y$, so $|Y : X| = 2$. As $G = \langle Y, c \rangle$ with Y abelian and $c \in Y^g$, $Z = Y \cap Y^g$ is centralized by $\langle Y, c \rangle = G$, so $Z \leq Z(G)$ and $Z = X \cap X^g$ as $X = Y \cap Q$. Thus passing to G/Z, we may assume $Z = 1$.

As $V_d = d^Q$, Y is Q-invariant and $X = [Q, d] = [Q, Y]$. Similarly $X^g = [Q, c]$, so $[XX^g, Y] \leq [Q, Y] = X$ and $[XX^g, c] \leq [Q, c] = X^g$ are contained in XX^g. Therefore Y and c act on XX^g, so as $G = \langle Y, c \rangle$, we have $XX^g \trianglelefteq G$ and then $Q = XX^g$. As $X \cap X^g = 1$, $Q = X \times X^g$, and then as $\Phi(X) \leq \Phi(Y) = 1$, $\Phi(Q) = 1$. Now by Exercise 1.1, $X^g = C_Q(c) = [Q, c]$, so $m = |c^Q| = |C_Q(c)|$, and $|Q| = m^2$. Further by Exercise 1.1, the representation of $P\langle d \rangle$ on Q is determined up to equivalence by m, so by 2.1, G is determined up to isomorphism by m. Thus (4) is established.

Exercises

In all the following exercises let D be a set of 3-transpositions of G, $d \in D$, and $H = \langle D_d \rangle$.
1. Prove
   (1) If A_d is nonempty then $V_d = d^{O_2(G)}$.
   (2) $W_d = d^{O_3(G)}$.

2. Assume G is transitive on D and let $b \in D_d$. Prove
   (1) $d * b = V_d \cup b^{O_2(H)}$.
   (2) If $O_2(H) \nleq Z(H)$ then $O_3(G) \leq Z(G)$.
   (3) If $O_3(H) \nleq Z(H)$ then $O_2(G) \leq Z(G)$.

3. Assume G is transitive on D, H is transitive on D_d, $|D_d| > 1$, and $O_3(H) \leq Z(H)$. Prove $O_2(G) \leq Z(G) \geq O_3(G)$.

4. Assume G is transitive on D, $O_2(G) \leq Z(G) \geq O_3(G)$, $T \in \mathrm{Syl}_2(G)$, $X \subseteq T \cap D$ with $|X|$ odd, and $\prod_{x \in X} x = 1$. Prove G is quasisimple.

5. Assume G is transitive on D, $\mathcal{D}(D)$ is disconnected, and $b \in D_d$. Let Δ be a connected component of $\mathcal{D}(D)$ with $d \notin \Delta$. Prove
   (1) $C_\Delta(db)$ is empty.
   (2) $|\Delta|$ is even.
   (3) The width of G is even.

6. Assume G is primitive on D and X is a D-subgroup of G with $O_2(X) \leq Z(X) \geq O_3(X)$, such that $\langle d^\perp \cap X \rangle$ is a maximal D-subgroup of H containing d. Assume Y is a proper D-subgroup of G containing X. Prove
   (1) $F^*(Y) = F^*(X)$.
   (2) If X is quasisimple then $X = Y$.
   (3) If $|G|_3 = |X|_3$ then $X = Y$.
   (4) If $H \cap X$ is maximal in H and $X = \langle N_D(E(X)) \rangle$, then X is a maximal subgroup of G.

7. Assume G is transitive on D and $d \in T \in \mathrm{Syl}_2(G)$. Prove
   (1) $N_G(T)$ is transitive on $T \cap D$.
   (2) If $N_H(T \cap D)$ is t-transitive on $T \cap D - \{d\}$ then $N_G(T)$ is $(t+1)$-transitive on $T \cap D$.
   (3) If $\prod_{d \neq x \in T \cap D} x \in \langle d \rangle$ then $\prod_{x \in T \cap D} x = 1$.

8. If G is transitive on D and G'' is the second term in the derived series of G, then
   (1) G/G'' is trivial or of width 1.
   (2) If G is solvable and $\mathcal{D}(D)$ is connected then $G = \langle d \rangle \cong \mathbf{Z}_2$.

4

Classical groups generated by
3-transpositions

The *classical groups* are the groups of isometries of bilinear and sesquilinear forms on vector spaces. Several families of classical groups are generated by 3-transpositions. In this chapter we consider those groups.

In Section 10 we record the notation, terminology, and elementary results we will need about bilinear forms. A reference for this material is Chapter 7 of [FGT].

In Section 11 we prove that certain classical groups over small fields are generated by 3-transpositions and establish various properties of these groups needed to prove Fischer's Theorem.

10. The classical groups

In this section V is an n-dimensional vector space over a field F and θ is an automorphism of F of order 1 or 2. A *sesquilinear form* on V with respect to θ is a map $f : V \times V \to F$ such that for all $x, y, z \in V$ and all $a \in F$:

$$f(x + y, z) = f(x, z) + f(y, z), \qquad f(ax, y) = af(x, y),$$

$$f(x, y + z) = f(x, y) + f(x, z), \qquad f(x, ay) = a^\theta f(x, y).$$

The form f is said to be *bilinear* if $\theta = 1$. Usually we write (x, y) for $f(x, y)$.

Define f to be *symmetric* if f is bilinear and $f(x, y) = f(y, x)$ for all x, y in V. Define f to be *skew symmetric* or *alternating* if f is bilinear and $f(x, y) = -f(y, x)$ for all x, y in V. Finally f is *hermitian symmetric* if θ is an involution and $f(x, y) = f(y, x)^\theta$ for all x, y in V. We always assume that f has one of these three symmetry conditions. One consequence of this assumption is that for all $x, y \in V$, $f(x, y) = 0$ if and only if $f(y, x) = 0$.

If $f(x, y) = 0$ we write $x \perp y$ and say that x and y are *orthogonal*. For $X \subseteq V$ define

$$X^\perp = \{v \in V : x \perp v \text{ for all } x \in X\}$$

39

and observe that $X^{\perp}$ is a subspace of V and that $X^{\perp} = \langle X \rangle^{\perp}$. $V^{\perp}$ is called the *radical* of V. Write $\mathrm{Rad}(V)$ for $V^{\perp}$. We say f is *nondegenerate* if $\mathrm{Rad}(V) = 0$.

The form f will be said to be *orthogonal* if f is nondegenerate and symmetric, and if in addition when $\mathrm{char}(F) = 2$, $f(x, x) = 0$ for all $x \in V$. The form f is said to be *symplectic* if f is nondegenerate and alternating, and in addition when $\mathrm{char}(F) = 2$, $f(x, x) = 0$ for all $x \in V$. Finally f is said to be *unitary* if f is nondegenerate and hermitian symmetric.

(10.1) *Let f be nondegenerate and $U \leq V$. Then $\dim(U^{\perp}) = \mathrm{codim}(U)$.*

Proof: See 19.2 in [FGT].

A vector $x \in V$ is *isotropic* if $f(x, x) = 0$. If f is alternating and $\mathrm{char}(F)$ is not 2, then every vector is isotropic. Recall that this is part of the defining hypothesis of a symplectic or orthogonal form when $\mathrm{char}(F) = 2$.

A subspace U of V is *totally isotropic* if the restriction of f to U is trivial, or equivalently if $U \leq U^{\perp}$. U is *nondegenerate* if the restriction of f to U is nondegenerate, or equivalently $U \cap U^{\perp} = \mathrm{Rad}(U) = 0$.

Assume f is orthogonal. A *quadratic form* on V associated to f is a map $Q : V \to F$ such that for all $x, y \in V$ and $a \in F$,

$$Q(ax) = a^2 Q(x) \quad \text{and} \quad Q(x + y) = Q(x) + Q(y) + f(x, y).$$

Observe that if $\mathrm{char}(F) \neq 2$, $Q(x) = f(x, x)/2$, so the quadratic form is uniquely determined by f, and hence adds no new information. On the other hand if $\mathrm{char}(F) = 2$ there are many quadratic forms associated to f. Observe however that f is uniquely determined by Q, since

$$f(x, y) = Q(x + y) - Q(x) - Q(y).$$

A *symplectic space* (V, f) is a pair consisting of a vector space V and a symplectic form f on V. A *unitary space* is a pair (V, f) with f a unitary form. An *orthogonal space* is a pair (V, Q) where Q is a quadratic form on V with associated bilinear form f.

In the remainder of this section assume (V, f) is a symplectic or unitary space, or Q is a quadratic form on V with associated orthogonal form f and (V, Q) is an orthogonal space.

A vector $v \in V$ is *singular* if v is isotropic and also $Q(v) = 0$ when V is an orthogonal space. A subspace U of V is *totally singular* if U is totally isotropic and also each vector of U is singular. The *Witt index* of V is the maximum dimension of a totally singular subspace of V. It turns out the Witt index of V is at most $n/2$ (cf. 19.3 in [FGT]).

An *isometry* of spaces (V, f) and (U, g) is a nonsingular linear transformation $\alpha : V \to U$ such that $g(x\alpha, y\alpha) = f(x, y)$ for all $x, y \in V$. Forms f and g (or P and Q) on V are said to be *equivalent* if (V, f) and (V, g) (or (V, Q) and (U, P)) are isometric. Denote by $O(V, f)$ the group of isometries of the space.

Let $X = (x_i : 1 \le i \le n)$ be a basis of V. Define $J = J(X, f)$ to be the n by n matrix $J = (J_{ij})$ with $J_{ij} = f(x_i, x_j)$. Observe that J uniquely determines the form f. Indeed:

(10.2) *A form g on V is equivalent to f if and only if $J(X, g) = AJ(X, f)A^{T\theta}$ for some nonsingular matrix A.*

Proof: See 19.5 in [FGT].

An *orthonormal basis* for V is a basis X such that $J(X, f) = I$. An *orthogonal basis* is a basis X with $J(X, f)$ diagonal.

(10.3) *Assume F is finite. Then*

(1) If V is unitary then V possesses an orthonormal basis.
(2) All unitary forms on V are equivalent.

Proof: See 19.11 in [FGT].

Our space V is a *hyperbolic plane* if $n = 2$ and V possesses a basis $X = (x_1, x_2)$ such that x_1 and x_2 are singular and $(x_1, x_2) = 1$. Such a basis will be termed a *hyperbolic pair*. Define V to be *hyperbolic* if V is the orthogonal direct sum of hyperbolic planes. A hyperbolic basis for a hyperbolic space V is a basis $X = (x_i : 1 \le i \le m)$ such that V is the orthogonal sum of the hyperbolic planes $\langle x_{2i-1}, x_{2i} \rangle$ with hyperbolic pair x_{2i-1}, x_i. V is *definite* if V possesses no nontrivial singular vectors.

(10.4) *All symplectic spaces are hyperbolic. In particular all symplectic spaces are of even dimension and, up to equivalence, each space of even dimension admits a unique symplectic form.*

Proof: See 19.16 in [FGT].

If $\mathrm{char}(F) = 2$ and (V, Q) is orthogonal then (V, f) is symplectic, where f is the bilinear form determined by Q. Hence by 10.4:

(10.5) *If V is orthogonal and $\mathrm{char}(F) = 2$, then V is of even dimension.*

In the remainder of this section assume F is a finite field of characteristic p.

(10.6) *Assume $n = 2$. Then up to equivalence there is a unique nondegenerate definite quadratic form Q on V. Further there is a basis $X = \{x, y\}$ of V such that*

   (1) If p is odd then $(x, y) = 0$, $Q(x) = 1$, and $-Q(y)$ is a generator of $F^\#$.
   (2) If $p = 2$ then $(x, y) = 1$, $Q(x) = 1$, $Q(y) = b$, and $P(t) = t^2 + t + b$ is an irreducible polynomial over F.

Proof: See 21.1 in [FGT].

Denote by V_+ and V_- the (isometry type of the) hyperbolic plane and the 2-dimensional definite orthogonal space over F, respectively. Write $V_+^m V_-^k$ for the orthogonal direct sum of m copies of V_+ with k copies of V_-.

(10.7)

   (1) V_+^m is a hyperbolic space of Witt index m.
   (2) $V_+^{m-1} V_-$ is of Witt index $m - 1$.
   (3) V_+^{2m} is isometric to V_-^{2m}.
   (4) Every 2m-dimensional orthogonal space over F is isometric to exactly one of V_+^m or $V_+^{m-1} V_-$.

Proof: See 21.2 in [FGT].

If $n = 2m$ is even, then 10.7 says that, up to equivalence, there are exactly two quadratic forms on V, and that the corresponding orthogonal spaces have Witt index m and $m - 1$, respectively. Define the *sign* of these spaces to be $+1$ and -1, respectively, and write $\text{sgn}(Q)$ or $\text{sgn}(V)$ for the sign of the space. Thus the isometry type of an even-dimensional orthogonal space over a finite field is determined by its dimension and sign. If V is an orthogonal space of odd dimension over F, then by 10.5, $\text{char}(F)$ is odd. The next lemma handles that situation.

(10.8) *Let F be a field of odd order, n an odd integer, and c a generator of the multiplicative group $F^\#$ of F. Then up to equivalence there are exactly two nondegenerate quadratic forms Q and cQ on an n-dimensional vector space V over F.*

Proof: See 21.4 in [FGT].

The next result is of fundamental importance; it is established in Section 20 of [FGT].

(10.9) Witt's Lemma. *Let V be an orthogonal, symplectic, or unitary space. Let U and W be subspaces of V and suppose* $\alpha : U \to W$ *is an isometry. Then* α *extends to an isometry of V.*

We now consider the isometry groups $O(V, f)$ and $O(V, Q)$, certain normal subgroups of these groups, and the images of such groups under the projective map P of Section 4. Notice that one can also regard the general linear group as the isometry group $O(V, f)$, where f is the trivial form $f(u, v) = 0$ for all $u, v \in V$. The groups G and PG, as G ranges over certain normal subgroups of $O(V, f)$, are called the *classical groups* (where f is trivial, orthogonal, symplectic, or unitary). We'll be concerned with classical groups over fields of order 2, 3, and 4.

Observe that if two spaces are isometric then their isometry groups are isomorphic. Thus in discussing the classical groups we need only concern ourselves with forms up to isometry, and indeed the forms Q and aQ of Lemma 10.8 have isomorphic isometry groups.

We saw in 10.4 that if n is even there is, up to equivalence, a unique symplectic form f on V. Write $\mathrm{Sp}(V)$ for the isometry group $O(V, f)$. $\mathrm{Sp}(V)$ is the symplectic group on V. As V is determined by n and $q = |F|$, we also write $\mathrm{Sp}_n(q)$ for $\mathrm{Sp}(V)$. We call $\mathrm{Sp}_n(q)$ the n-dimensional *symplectic group* over F.

If f is unitary then $O(V, f)$ is called a *unitary group*. Write $GU(V)$ for $O(V, f)$ and $SU(V)$ for $SL(V) \cap O(V, f)$. We call $GU(V)$ the *general unitary group* and $SU(V)$ the *special unitary group*. As f is unitary, F admits an automorphism θ of order 2. Then as F is finite of order q, $q = r^2$ and $\theta : a \mapsto a^r$. We write $GU_n(r)$ and $SU_n(r)$ for $GU(V)$ and $SU(V)$, respectively.

If Q is a nondegenerate quadratic form then $O(V, Q)$ is an *orthogonal group*. Write $\Omega(V, Q)$ for the commutator group of $O(V, Q)$. If n is even then up to equivalence there are just two nondegenerate quadratic forms Q_ϵ on V, distinguished by the sign $\mathrm{sgn}(Q_\epsilon) = \epsilon = +1$ or -1 of the form. Write $O_n^\epsilon(q)$, $\Omega_n^\epsilon(q)$ for the corresponding groups. When n is odd we have seen that up to isomorphism there is a unique n-dimensional orthogonal group $O_n(q)$ over F and we write $\Omega_n(q)$ for its commutator group.

For each group G we can restrict the representation $P : GL(V) \to PGL(V)$ of $GL(V)$ on the projective space $PG(V)$ to G and obtain the image PG of G which is a group of automorphisms of the projective space $PG(V)$. Thus for example we obtain the groups $P\,\mathrm{Sp}_n(q)$, $PSU_n(r)$, $PO_n^\epsilon(q)$, $P\Omega_n^\epsilon(q)$. We write $U_n(r)$ for $PSU_n(r)$.

11. Classical groups generated by 3-transpositions

In this section we continue the hypothesis and notation of the previous section. In addition assume that one of the following holds:

(a) $q = 2$, (V, f) is symplectic, and $v \in V^\#$ is singular;
(b) $q = 4$, (V, f) is unitary, and $v \in V^\#$ is singular;
(c) $q = 2$, (V, Q) is orthogonal, and v is nonsingular;
(d) $q = 3$, (V, Q) is orthogonal, and v is nonsingular.

Recall the definition of a transvection in Section 4. If q is odd and (V, Q) is orthogonal, then a *reflection* on V is an involution r in $O(V, Q)$ such that $[V, r]$ is a point of V. $[V, r]$ is called the *center* of r.

Define $t(v) : V \to V$ by $xt(v) = x - (x, v)v/A$, where $A = Q(v)$ if $q = 3$ and $A = 1$ otherwise.

(11.1) $t(v) \in O(V, f)$ or $O(V, Q)$ and $t(v)$ is the unique transvection or reflection on V with center $\langle v \rangle$.

Proof: See 22.3 and 22.6 in [FGT].

(11.2) Let (V, Q) be an orthogonal space. Then either

(1) $O(V, Q)$ is generated by its transvections or reflections, or
(2) $q = 2$, $n = 4$, and $\operatorname{sgn}(Q) = +1$.

Proof: See 22.7 in [FGT].

(11.3) $\operatorname{Sp}_n(q)$ and $SU_n(q)$ are generated by their transvections unless (V, f) is unitary and $n = 3$.

Proof: See 22.4 in [FGT].

(11.4)

(1) When $p = 2$, $O(V, f)$ and $O(V, Q)$ are transitive on their transvections.
(2) When $q = 3$, $O(V, Q)$ is transitive on the set of reflections $t(v)$ with $Q(v) = \pi$, for $\pi = 1$ and $\pi = -1$.

Proof: This is a consequence of 11.2 and Witt's Lemma, which gives the transitivity of the isometry group on vectors v of the same isometry type.

Let D be the set of transvections when $p = 2$. When $q = 3$ let $D = D_\pi$ be the set of reflections $t(v)$ with $Q(v) = \pi$, for $\pi = \pm1$. Let $G = \langle D \rangle$. If V is symplectic then by 11.3 and 11.4, $G = \operatorname{Sp}(V) = \operatorname{Sp}_n(2)$, and D is a conjugacy class of G. Similarly if V is unitary then either $G = SU(V) = SU_n(2)$, or $n = 3$, where $G = SU_3(2)'$ is of order 54. In either case D is a conjugacy class of G. If $q = 2$ and V is orthogonal then by 11.2 and 11.4, either $G = O(V, Q) = O_n^\epsilon(2)$ and D is a conjugacy class of G, or $n = 4$ and $\epsilon = +1$, where $G \cong S_3 \times S_3$.

Finally assume $q = 3$. Here we need the notion of the *discriminant* $\mu(V)$ of the space (V, Q). Namely define

$$\mu(V) = \det(J(X, f)),$$

where f is the bilinear form of Q, $X = \{x_1, \ldots, x_n\}$ is any basis for V, and $J(X, f) = (f(x_i, x_j))$ is the matrix of f with respect to X defined in the previous section. By Exercise 4.2, this definition is independent of the choice of the basis X and $\mu(V) = \pm 1$. Further by the same exercise, $\mu(V) = (-1)^{n/2} \operatorname{sgn}(V)$ if n is even.

Having defined the discriminant, we can write $O_n^{\mu, \pi}(3)$ for the group $G = \langle D_\pi \rangle$, where $\mu(V) = \mu$. By Exercise 4.10, $O_n^{\mu_1, \pi_1}(3) \cong O_n^{\mu_2, \pi_2}(3)$ if $\mu_1 \pi_1^n = \mu_2 \pi_2^n$. Thus n and $\mu \pi^n$ are the fundamental invariants determining G.

By Exercise 4.9, $G = O_n^{\mu, \pi}(3)$ is of index 2 in $O_n(3)$. Moreover unless $n = 2$ and $\operatorname{sgn}(V) = -1$, G is transitive on D. This follows from 11.2, 11.4, the fact that G is of index 2 in $O(V, Q)$, and the observation that unless $n = 2$ and $\operatorname{sgn}(V) = -1$, an element of D_+ centralizes an element of D_-. This last fact follows in turn from the observation that, aside from the exceptional case, v_+ is orthogonal to some v_-.

(11.5)

    (1) D is a set of 3-transpositions of G.

    (2) For $\langle u \rangle \neq \langle v \rangle$, $[t(u), t(v)] = 1$ if and only if $(u, v) = 0$.

    (3) Either D is a conjugacy class of G or (V, Q) is orthogonal, and

$$(q, n, \operatorname{sgn}(Q)) = (2, 4, +1) \ or \ (3, 2, -1).$$

Proof: When $q = 3$, replacing Q by $-Q$ if necessary, we may assume $Q(v) = 1$. Let $x \in V$. Then

$$(*) \qquad xt(v)t(u) = x - (x, v)v - (x, u)u + (x, v)(v, u)u.$$

In particular if $(v, u) = 0$ then by $(*)$:

$$xt(v)t(u) = x - (x, v)v - (x, u)u = xt(u)t(v),$$

so $t(v)t(u) = t(u)t(v)$ and hence $|t(v)t(u)| = 2$. So assume $(v, u) \neq 0$. Here we show $|t(v)t(u)| = 3$. By 11.1, $t(u) \neq t(v)$, so it suffices to show

$xt(v)t(u)t(v) = xt(u)t(v)t(u)$ for all $x \in V$. But from (*) we conclude

$$xt(v)t(u)t(v) = h(x, v, u) + (x, v)g(x, v, u),$$

where

$$h(x, v, u) = x + (x, v)(v, u)u - (x, u)u + (x, u)(u, v)v - (x, v)v,$$
$$g(x, v, u) = ((v, v) - (v, u)(u, v) - 1)v.$$

Thus as $h(x, v, u) = h(x, u, v)$, it remains to show $g(x, v, u) = 0$. But if $p = 2$ then $(v, v) = 0$ and $(u, v) = (v, u)^\theta$, so

$$(v, u)(u, v) = (v, u)^{1+\theta} = 1,$$

so $g(x, v, u) = 2 = 0$. On the other hand if $q = 3$ then as $Q(v) = 1$, $(v, v) = 2Q(v) = 2$, and $(v, u) = (u, v) = 1$ or -1, so again $g(x, v, u) = 0$.

We have established (1) and (2). Part (3) follows from earlier remarks.

There are various isomorphisms among the classical 3-transposition groups and symmetric groups, which we now record:

$$O_2^+(2) \cong O_2^{-,\pi}(3) \cong \mathbf{Z}_2 \quad \text{and} \quad O_2^{+,\pi}(3) \cong E_4,$$
$$Sp_2(2) \cong U_2(2) \cong O_2^-(2) \cong S_3,$$
$$O_3^{\pi,\pi}(3) \cong PO_3^{\pi,\pi}(3) \cong S_4,$$
$$O_3^{-\pi,\pi}(3) \cong E_8 \quad \text{with } |D| = 3,$$
$$S_5 \cong O_4^-(2) \quad \text{and} \quad O_4^+(2) \cong S_3 \times S_3,$$
$$S_6 \cong Sp_4(2) \cong O_4^{-,\pi}(3) \cong PO_4^{-,\pi}(3),$$
$$O_4^{+,\pi}(3) \cong \mathbf{Z}_2/SL_2(3) * SL_2(3),$$
$$S_8 \cong O_6^+(2),$$
$$O_6^-(2) \cong O_5^{\pi,\pi}(3) \cong PO_5^{\pi,\pi}(3),$$
$$U_4(2) \cong PO_5^{-\pi,\pi}(3).$$

A sketch of the proof of some of these isomorphisms appears in Exercise 7.7 in [FGT]; the other isomorphisms are left as exercises in this chapter. Each isomorphism is an isomorphism of 3-transposition groups, not just an isomorphism of abstract groups.

We now adopt the 3-transposition notation of Section 8 in discussing (G, D). Further we assume that $n \geq 4$ if V is symplectic or unitary and $n \geq 5$ if V is orthogonal. In particular by 11.5, $D_{t(v)}$ consists of those $t(u)$ with $\langle v \rangle \neq \langle u \rangle \leq v^\perp$, and the graph $\mathcal{D}(D)$ is isomorphic to the graph on the points of V isometric to $\langle v \rangle$ obtained by joining $\langle v \rangle$ to its conjugates in $v^\perp$. It is easy to check (cf. Exercise 4.6) that the latter graph is connected and that $O_3(G) \leq Z(G) \geq O_2(G)$. Hence we conclude from 9.4 that

(11.6) $F^*(G) = Z(G)G'$ *with* G' *quasisimple,* $|G : G'| \leq 2$, *and* G *primitive on* D.

It then follows from 11.6 and 9.5 that G is rank 3 on D and $C_G(d)$ has orbits $\{d\}$, D_d, and A_d, where $d = t(v)$. Thus $G_{\langle v \rangle}$ is transitive on the points of V distinct from $\langle v \rangle$ and isometric to $\langle v \rangle$ in $v^\perp$ and $V - v^\perp$. Indeed:

(11.7) $t(w) \in A_{t(v)}$ *if and only if* $\langle v \rangle \neq \langle w \rangle \not\leq v^\perp$, *in which case*

    (1) $\langle v, w \rangle$ *is a nondegenerate line if* $p = 2$.
    (2) $\langle v, w \rangle$ *has sign* -1 *if* (V, Q) *is orthogonal and* $q = 2$.
    (3) $\mathrm{Rad}(\langle v, w \rangle)$ *is a singular point if* $q = 3$.

Proof: The first remark is just 11.5.2. In particular if $t(w) \in A_{t(v)}$ then when v is singular, so is w and hence as $(v, w) \neq 0$, $\langle v, w \rangle$ is a hyperbolic line. Therefore (1) holds in this case, so we may take (V, Q) orthogonal. Similarly when $q = 2$, v is isotropic, so $\langle v, w \rangle$ is hyperbolic line in the symplectic space (V, f), but as v and w are nonsingular, $\langle v, w \rangle$ has sign -1 with respect to Q. Finally when $q = 3$, take $\langle s \rangle$ to be a singular point in $v^\perp$. Then $Q(v + s) = Q(v)$ and $(v + s, v) = (v, v) \neq 0$, so $t(v + s) \in A_{t(v)}$ and then (3) holds by transitivity of $G_{\langle v \rangle}$ on $A_{t(v)}$.

CAUTION: The remaining lemmas in this section are technical and their proofs are often more difficult than most of the results in Part I of this text. Thus those readers who don't feel compelled to read every detail of the proof of Fischer's Theorem are advised to skip the proofs of these lemmas and go on to the next chapter.

(11.8) *Let* $k = |D_{t(v)}|$ *and* $l = |A_{t(v)}|$. *Then*

    (1) *If* (V, f) *is symplectic then* $|D| = 2^n - 1$, $k = 2(2^{n-2} - 1)$, *and* $l = 2^{n-1}$.
    (2) *If* (V, f) *is unitary then* $|D| = u(n)$, $k = 4u(n - 2)$, *and* $l = 2 \cdot 4^{n-2}$, *where* $u(n) = (2^n - (-1)^n)(2^{n-1} - (-1)^{n-1})/3$.

Proof: Suppose (V, f) is symplectic. Then the map $u \mapsto t(u)$ is a bijection of $V^\#$ with D, so $|D| = |V^\#| = 2^n - 1$. Similarly $k = |v^\perp - \langle v \rangle| = 2(2^{n-2} - 1)$ and $l = |V - v^\perp| = 2^{n-1}$.

Assume next that (V, f) is unitary. Here we could calculate the number $s(n)$ of singular points in V directly using the fact that V has an orthonormal basis, but it is probably easier to take a different route. Namely we calculate $s(2) = 3$ and $s(3) = 9$. That is $s(n) = u(n)$ for $n = 2, 3$. Then we proceed by induction on n to show $s(n) = u(n)$ in general. For there are $s(n-2) = u(n-2)$ singular points $\langle u, v \rangle / \langle v \rangle$ in $v^\perp / \langle v \rangle$ and four singular points in $\langle u, v \rangle$ distinct from $\langle v \rangle$, so $k = 4u(n-2)$. Similarly there are 4^{n-2} points $\langle v, w \rangle / \langle v \rangle$ in $V / \langle v \rangle - v^\perp / \langle v \rangle$ and each such line is nondegenerate and hence has $s(2) - 1 = 2$ singular points distinct from $\langle v \rangle$, so $l = 2 \cdot 4^{n-2}$. Then as $u(n) = 1 + 4u(n-2) + 2 \cdot 4^{n-2}$, our proof is complete.

(11.9) *Let (V, f) be symplectic or unitary, and $t(u) \in D_{t(v)}$. Then*

$$t(v) * t(u) = \{t(x) : x \in \langle v, u \rangle\} \text{ is of order } q + 1.$$

Proof: As $t(u) \in D_{t(v)}$, $\langle u, v \rangle = U$ is a singular line. So U contains $q + 1$ singular points. Further the points $\langle x \rangle \neq \langle v \rangle$ of U are the points of V such that $\langle v, u \rangle^\perp = \langle v, x \rangle^\perp$, so the involutions $t(x)$, $x \in U$, are the members of $t(v) * t(u)$.

(11.10) *Let $q = 3$, $n \geq 5$, $t(w) \in A_{t(v)}$, $S = \mathrm{Rad}(\langle v, w \rangle)$, $r \in V - S^\perp$ singular, $W = \langle S, r \rangle$, $K = \langle C_D(W) \rangle$, $\delta : S^\perp \to W^\perp$ the projection with respect to the decomposition $S^\perp = S \oplus W^\perp$, and $M = \langle C_D(S) \rangle$. Then*

(1) *$N_G(S) = P C_G(W) \langle t \rangle$ where $P = C_G(S^\perp / S) \cap C_G(S) \cong E_{3^{n-2}}$ and $t \in \Omega(V)$ is an involution with $W \leq [V, t]$ of dimension 4 and sign +1.*

(2) *The map $h \mapsto [r, h] \delta$ is an equivalence of P with $W^\perp$ as $F C_{O(V)}(W)$-modules.*

(3) *$K \cong O_{n-2}^{-\mu, \pi}(3)$.*

(4) *Either $K = C_G(W)$ and $M = C_G(S)$ or $n = 5$, $\mu\pi = 1$, and $|C_G(W) : K| = 3$.*

(5) *$C_G(W) \langle t \rangle \cong O(W^\perp)$.*

(6) *G is transitive on its D-subgroups J such that $|O_3(J)| \geq |P|$ and $J / O_3(J) \cong K$.*

Proof: Let $g = t(v) t(w)$, so that g is of order 3. Then $t(v)$ and $t(w)$ centralize S and act on the orthogonal space $(S^\perp / S, Q)$ as the reflection with center $\langle w, v \rangle / S$, so g is in the centralizer P of the sections S and $S^\perp / S$. Notice also that P centralizes $V / S^\perp$. Namely if $0 \neq s \in S$ and $h \in P$, then $(s, r) = (sh, rh) = (s, rh)$, so $(s, r - rh) = 0$ and hence $[r, h] = r - rh \in S^\perp$.

Next W is a hyperbolic line, so $K = \langle C_D(W) \rangle$ is the subgroup of $O(W^\perp, Q)$ generated by reflections $t(u)$, $u \in W^\perp$, with $Q(u) = \pi$. By Exercise 4.2.4, $\mu(W) = -\mathrm{sgn}(W) = -1$. Therefore $K \cong O_{n-2}^{-\mu, \pi}(3)$ since $\mu(W^\perp) = \mu(W)\mu(V) = -\mu$ by Exercise 4.2.2. So (3) is established.

Further $C_{O(V)}(W) \cong O(W^\perp)$ acts on P and the map $\gamma : h \mapsto [r, h] + S$ is a $C_{O(V)}(W)$-equivariant map from P into $S^\perp / S$. Now $C_{O(V)}(S^\perp) = 1 = C_{O(V)}(V/S)$ (cf. 22.2 in [FGT]) so P is faithful on $S^\perp$ and γ is an injection. Therefore P is contained in the group of transvections on $S^\perp$ with center S, which is isomorphic to $E_{3^{n-2}}$. On the other hand $C_{O(V)}(W) \cong O(W^\perp)$ is irreducible on

$$W^\perp \cong S^\perp / S \cong E_{3^{n-2}},$$

so as γ is an injection and $1 \neq g \in P$, we conclude γ is an isomorphism. Now an argument like that of 11.8 shows there are 3^{n-2} singular points in $V - S^\perp$, so as $C_P(r) = 1$ and $|P| = 3^{n-2}$, P is regular on these points. Thus $M = PN_M(Fr)$ and $C_G(S) = P(C_G(S) \cap N_G(Fr))$. But

$$N_{O(V)}(S) \cap N_{O(V)}(Fr) = C_{O(V)}(W)\langle -1_W \rangle$$

and $-1_W = t(v)t(u)$ where $W = Fv \perp Fu$ with $Q(u) = -\pi$, so $t(u) \notin G$ (cf. Exercise 4.12). Thus $-1_W \notin G$. On the other hand by Exercise 4.9, $|O(V) : G| = 2$ and the involution t described in 11.10.1 is in $\Omega(V) \leq G$ by Exercise 4.12, so $-1_W \cdot t \in C_{O(V)}(W) - G$ and hence $|C_{O(V)}(W) : C_G(W)| = 2$ and $N_G(Fr) = C_G(W)\langle t \rangle$, completing the proof of (1). Also by Exercise 4.9, unless $n = 5$ and $(-\mu)\pi = -1$, where we have $|C_{O(V)}(W) : K| = 2$, so that $K = C_G(W)$, and in the exceptional case $C_G(W) \cong \mathbb{Z}_2 \times A_4$ and $K = O_2(C_G(W))$ by Exercise 4.11. Therefore (4) is established.

As δ induces an $FC_{O(V)}(W)$-isomorphism of $S^\perp / S$ with $W^\perp$, $\gamma\delta : P \to W^\perp$ is an $FC_{O(V)}(W)$-isomorphism, so (2) holds. Also

$$O(W^\perp) \cong C_{O(V)}(W) = C_G(W)\langle -1_W \cdot t) \rangle \cong C_G(W)\langle t \rangle$$

as -1_W centralizes $C_{O(V)}(W)$, so (5) holds.

Let J be a D-subgroup of G with $R = O_3(J)$ satisfying $|R| \geq |P|$ and $J/R \cong K$. Let $U = C_{[V,R]}(R)$. Then $0 \neq U$ is totally singular, the actions of J on U and $V/U^\perp$ are dual, and $C_{O(V,Q)}(U)$ induces $O(U^\perp/U, Q)$ on $U^\perp/U$ (cf. Exercise 4.4). Let $d \in J \cap D$. As d induces a reflection on V, it is trivial on either U or $V/U^\perp$, so, as the actions of J on U and $V/U^\perp$ are dual, $d \in C_G(U)$. Hence $J = \langle D \cap J \rangle$ centralizes U and $V/U^\perp$. Therefore the kernel J_0 of the action of J on $U^\perp/U$ is contained in R. Now $J/J_0 \leq O(U^\perp/U, Q)$, so as $J/R \cong O_{n-2}^{-\mu,\pi}(3)$, $\dim(U^\perp/U) \geq n - 2$. That is U is a singular point. Thus we may take $S = U$ and $J \leq M$. As $J/R \cong K$, $M = JP$ and $R \leq P$. Therefore as $|R| \geq |P|$, $M = J$, proving (6).

(11.11)   Let $H = \langle D_{t(v)} \rangle$ and (V, Q) orthogonal. Then

(1) If $q = 2$ then $H/Z(H) \cong Sp_{n-2}(2)$.
(2) If $q = 3$ then $H \cong O_{n-1}^{-\pi\mu,\pi}(3)$.

Proof: Part (1) follows from 22.5 in [FGT]. So take $q = 3$. Then $D_{t(v)}$ consists of the reflections $t(u)$ with $u \in v^{\perp}$ and $Q(u) = Q(v)$, so $H \cong O_{n-1}^{\delta,\pi}(3)$, where $\delta = \mu(v^{\perp})$. By Exercise 4.2.2, $\delta = \mu(Fv)\mu = f(v,v)\mu = -Q(u)\mu = -\pi\mu$.

(11.12)   Let $t(w) \in A_{t(v)}$ and $L = \langle D_{t(v),t(w)} \rangle$. Then

(1) $L \cong Sp_{n-2}(2)$ if (V, f) is symplectic.
(2) $L \cong SU_{n-2}(2)$ (or $SU_3(2)'$ when $n = 5$) if (V, f) is unitary.
(3) If (V, Q) is orthogonal and $q = 2$, then $L \cong O_{n-2}^{-\epsilon}(2)$ unless $(n, \epsilon) = (6, -1)$, where $L \cong S_3 \times S_3$.
(4) If (V, Q) is orthogonal and $q = 3$ then $L \cong O_{n-3}^{\pi\mu,\pi}(3)/E_{3^{n-3}}$, except L is S_3 when $(n, \mu\pi^n) = (5, -1)$. In particular when $(n, \mu\pi^n) = (5, 1)$, $L \cong S_3 \times S_3$, whereas when $(n, \mu\pi^n) = (6, -1)$ or $(6, 1)$, then L is $S_3 \times S_3 \times S_3$ or S_4/E_{27}, respectively.

Proof: Let $W = \langle v, w \rangle$. Then $D_{t(v),t(w)}$ is the set of maps $t(u)$ with $u \in W^{\perp}$ and Fu isometric to Fv. Moreover W is described in 11.7. In particular if $p = 2$ then W is a nondegenerate line, with W of sign -1 if (V, Q) is orthogonal, so the lemma holds.

So take $q = 3$. Then $S = \mathrm{Rad}(W)$ is singular, so L is contained in the subgroup KP described in 11.10. From that lemma, $K = \langle C_D(U) \rangle$ for some hyperbolic line U containing S; we pick $U \le v^{\perp}$ and let $Z = U + Fv$, so that Z is a nondegenerate 3-subspace of V with $\mu(Z) = \mu(U)\mu(Fv) = (-1)(-\pi) = \pi$ by Exercise 4.2. Now $g = t(v)t(w) \in P$ and $L \le C_K(t(v))C_P(t(v))$. Also $C_P(t(v)) \cong E_{3^{n-3}}$ and $C_K(t(v)) = K_v$. Further $I = \langle D \cap K_v \rangle \le L$ and $K \cong O_{n-2}^{-\mu,\pi}(3)$ by 11.10.3, so by 11.11.2, $I \cong O_{n-3}^{\pi\mu,\pi}(3)$. Next $C_P(t(v)) \le C_G(\langle t(v), t(w)\rangle) \le N_G(L)$ as $g \in P$ and P is abelian. Also by 11.10.2, P is the natural module for K, so $C_P(t(v))$ is generated by points $\langle x \rangle$ with $Q(x) = \pi$, and hence

$$C_P(t(v)) = \langle [C_P(t(v)), t(u)] : t(u) \in L \cap D_{t(v)} \rangle \le L$$

unless $C_P(t(v))$ is a line of sign $+1$, which is the exceptional case $n = 5$, $\mu\pi = -1$, where $L \cong S_3$ by Exercise 4.11. So this case aside, $P = [P, t(v)]C_P(t(v)) \le L$ and $L = C_P(t(v))I$ is as described in the lemma.

(11.13) *Let U be an n-dimensional vector space over $GF(2)$ and $H \leq GL(U)$ with $H/Z(H) \cong \mathrm{Sp}_n(2)$, and H contains transvections projecting on the transvections of $H/Z(H)$. Then $H = \mathrm{Sp}(U)$ acts naturally on U.*

Proof: By 11.8, $2^n - 1$ divides $|H|$, so as $2^n - 1$ does not divide $|GL_{n-1}(2)|$, H fixes no element of $U^{\#}$. Let T be the set of transvections of $GL(U)$ contained in H and projecting on transvections of $H/Z(H)$. If s,t are distinct in T then $H = \langle C_H(t), C_H(s) \rangle$ ($C_H(t)$ is maximal in H as H is primitive on T), so, as H fixes no point of U, $[U, t] \neq [U, s]$. Then as $|T| = 2^n - 1 = |U^{\#}|$, each point $u \in U^{\#}$ is the center of a unique transvection $t(u)$. Define $g : U \times U \to GF(2)$ by $g(u, v) = 0$ if $[u, t(v)] = 0$ and $g(u, v) = 1$ otherwise. Observe that $0 \neq g$ is symmetric and bilinear and $H \leq O(U, g)$. Then as H is transitive on $U^{\#}$, g is a symplectic form on U and $H = O(U, g) = \mathrm{Sp}(U)$ acts naturally on U.

(11.14) *Let (V, f) be symplectic with $n \geq 6$ and L a D-subgroup of G with $L/Z(L) \cong O_n^\epsilon(2)$. Then $L \cong O_n^\epsilon(2)$ is determined up to conjugation in G by ϵ.*

Proof: Let $E = L \cap D$, $t(v) \in E$, and $L^* = L/Z(L)$. As $L^* \cong O_n^\epsilon(2)$, $L^* = O(U, Q)$ for some quadratic form Q of sign ϵ on an n-dimensional $GF(2)$-space U. For $t(x) \in E$, write $u(x)$ for the generator of $[U, t(x)]$.

Now $t(x) \in E$ is in $D_{t(v)}$ if and only if $f(v, x) = 0$. Let $s = x + v$, $r = u(x) + u(v)$, and $M^* = C_{L^*}(r)$. Then $\{t(v), t(x)\} = t(v)^{O_2(M)}$, so $\{v, x\} = v^{O_2(M)}$ and $O_2(M) \leq C_L(s)$. As

$$C_{L^*}(t(v)) \cong \mathbb{Z}_2 \times \mathrm{Sp}_{n-2}(2),$$

$C_L(t(v))/\langle t(v) \rangle$ acts naturally on $v^\perp/\langle v \rangle$ by 11.13. In particular $\langle s, v \rangle = C_{v^\perp}(O_2(M) \cap C_L(t(v)))$, so $C_V(O_2(M)) = \langle s \rangle$. Thus $M \leq C_G(s)$.

Define $P : V \to GF(2)$ by $P(x) = 1$ if $t(x) \in E$ and $P(x) = 0$ otherwise. Let S, N be the number of Q-singular, Q-nonsingular points, respectively. Then

$$|V^{\#}| = |U^{\#}| = N + S = |L : C_L(d)| + |L : M| = |v^L| + |s^L|,$$

so $V^{\#} = v^L \cup s^L$. Thus v^L, s^L are the sets of P-singular, P-nonsingular points of V, respectively.

A similar counting argument shows $v^\perp = \{v\} \cup x^{C_L(v)} \cup s^{C_L(v)}$. Hence if W is a 2-dimensional subspace of $v^\perp$ containing v then W contains exactly two nonsingular points. Also if $t(y) \in E \cap A_{t(v)}$ then $f(v, y) = 1$ so as $t(v)$ and $t(y)$ are transvections with centers Fv and Fy, $\langle t(v), t(y) \rangle$ induces $L_2(2)$ on $\langle v, y \rangle$ and in particular $v + y = v^{t(v)t(y)} \in v^L$. Therefore if $W \not\leq v^\perp$ is a 2-dimensional subspace containing v then W contains an odd number of nonsingular points.

It remains to check that P is a quadratic form on V with bilinear form f. Then $L = O(V, P) \cong O_n^\epsilon(2)$ and L is determined up to conjugacy in G by ϵ, as G is transitive on quadratic forms of sign ϵ associated to f by 10.7. So let z, w be distinct nonzero vectors in V; we must show $P(z) + P(w) + P(z + w) = f(z, w)$, or equivalently $\langle z, w \rangle$ contains an odd number of singular points if and only if $f(z, w) = 1$. But this follows from the previous paragraph.

(11.15) *Let (V, f) be unitary with $n \geq 3$, U a nondegenerate hyperplane of V, $H = C_G(U^\perp) \cong SU(U)$, and $g \in G - N_G(U)$. Then $G = \langle H, g \rangle$.*

Proof: By Witt's Lemma, H is transitive on the singular points in $V - U$ and on the singular points in U. As $n \geq 3$, g fuses a singular point in $V - U$ to a point $\langle u \rangle$ of U. So, as $t(u) \in H$, $D \subseteq \langle g, H \rangle$. Thus $\langle g, H \rangle = \langle D \rangle = G$.

REMARK: Notice that if (V, f) is unitary then f defines a quadratic form Q on V regarded as a $GF(2)$-space via $Q(v) = f(v, v)$. Hence $GU(V) = O(V, f) \leq O(V, Q)$. In particular $G = SU(V)$ centralizes the subgroup $Z(GU(V))$ of order 3 in $O(V, Q)$ and each $GF(4)$-transvection d in G is of *type a_2* on the orthogonal space (V, Q) over $GF(2)$: That is $[d, V]$ is a totally singular line of (V, Q). The next lemma establishes the converse of this observation.

(11.16) *Assume (V, Q) is orthogonal with $q = 2$ and $H \leq G$ with $H/Z(H) \cong U_{n/2}(2)$, with $Z(H) \leq H'$, and such that each t in the set T of involutions projecting on transvections of $H/Z(H)$ is of type a_2 on V. Then $H \cong SU_{n/2}(2)$ acts naturally on V.*

Proof: Let $m = n/2$. The proof is by induction on m. Check the lemma directly for $m = 2$ and 3. Let U be an m-dimensional unitary space over $GF(4)$ and $\hat{H} = SU(U)$. Let U_i, $i = 1, 2$, be distinct nondegenerate hyperplanes of U with $U_0 = U_1 \cap U_2$ nondegenerate. Identify $\hat{H}/Z(\hat{H})$ with $H/Z(H)$, let $\hat{L}_i = SU(U_i)$, and $Z(H) \leq L_i \leq H$ with $L_i/Z(H) = \hat{L}_i/Z(\hat{H})$. By 11.15, $L_i = \langle L_0, t_i \rangle$, for $t_i \in T \cap L_i - L_0$. Further L_0 centralizes $S = \langle t, s \rangle \cong S_3$ with $t, s \in T$ and $[S, U] = U_0^\perp$. Now $V_0 = [S, V]$ is a nondegenerate 4-dimensional space of sign $+1$ as t, s are of type a_2 on V. Next $r \in L_0 \cap T$ centralizes V_0 as $[V, t, r] = [V, s, r] = 0$. So $[V_0, K_0] = 0$, where $K_0 = \langle T \cap L_0 \rangle$. Therefore as $F^*(L_0) \leq K_0$, L_0 acts faithfully on the nondegenerate $(n - 4)$-dimensional subspace $V_0^\perp$, so by minimality of m, $L_0 \cong SU_{m-2}(2)$ acts naturally on $V_0^\perp$.

Let $M = O^{3'}(C_{O(V,Q)}(L_0))$. Then $M \cong E_{27}$. Indeed $M = N_1 \times N_2$, where $N_2 = Z(GU(V_0^\perp)) \cong Z_3$ and $N_1 = O^{3'}(O(V_0, Q)) \cong E_9$.

Next $|U_m(2)|$ does not divide $|G_v|$ for $v \in V^\#$, so $V = [V, L]$. Hence as $L = \langle t_1, t_2, L_0 \rangle$ by 11.15, $V = [V, t_1] + [V, t_2] + [V, L_0]$, and then as

$\dim([V, t_i]) = 2$ and $[V, L_0] = V_0^{\perp}$ is of dimension $n - 4$, we conclude that for $i = 1, 2$, $[V, L_i] = [V, t_i] \oplus [V, L_0]$ is of dimension $n - 2$ and $C_V(L_i) = C_{V_0}(t_i)$ is a line, so $V = C_V(L_i) \oplus [V, L_i]$ and hence $[V, L_i]$ is nondegenerate. So by minimality of m, $L_i \cong SU_{m-1}(2)$ and $[V, L_i]$ is the natural module. Let α_i be of order 3 in $O(V, Q)$ with $[V, \alpha_i] = [V, L_i]$ and $[\alpha_i, L_i] = 1$. Then $\alpha_i = \gamma \beta_i$ where $N_2 = \langle \gamma \rangle$ and $\beta_i \in N_1$ with $C_V(L_i) = C_{V_0}(\beta_i)$. As $\beta_i \in N_1$ with $C_{V_0}(\beta_i)$ a line, $C_{N_1}([V_0, \beta_i])$ is of order 3 and hence $M_i = O^{3'}(C_{O(V,Q)}(L_i)) = C_{N_1}([V_0, \beta_i])\langle \alpha_i \rangle \cong E_9$. In particular M_1 and M_2 are hyperplanes of M, so there is α of order 3 in $M_1 \cap M_2$. By Lemma 11.15, $H = \langle L_1, L_2 \rangle$, so $[\alpha, H] = 1$. Then H is irreducible on V and $K = GF(4) \leq \mathrm{End}_{GF(2)H}(V)$.

So $V^K = K \otimes_F V = W \oplus W^{\sigma}$ where W is an irreducible KH-module and $\langle \sigma \rangle = \mathrm{Gal}(K/F)$. As $H \leq O(V, f)$, V is isomorphic to its dual V^*, so $V^K \cong (V^K)^*$. Hence $W^* \oplus W^{*\sigma} \cong W \oplus W^{\sigma}$, so $W \cong W^*$ or $W^{*\sigma}$. In the former case $G \leq \mathrm{Sp}(W)$. This is impossible as t is a transvection on W and $O_2(C_H(t))$ is the centralizer in H of the series $0 < [W, t] < C_W(t) < W$. But $O_2(C_H(t))$ is extraspecial (cf. 14.10) whereas the centralizer of this series in $\mathrm{Sp}(W)$ is abelian.

Thus $W \cong W^{*\sigma}$, so H preserves a unitary form on W. This completes the proof.

Exercises

1. Let $G = O_n^{\mu, \pi}(3) \leq O(V, Q)$ with $q = 3$, $t(v) \in D$, and $t(u) \in D_{t(v)}$. Prove $\langle u, v \rangle$ is a nondegenerate line of sign -1.

2. Let (V, Q) be an n-dimensional orthogonal space over $GF(3)$ with bilinear form f. Prove
   (1) $J(X, f) = J(Y, f)$ for all bases X, Y of V.
   (2) Let $V = V_1 \perp \cdots \perp V_m$ be an orthogonal direct sum of subspaces. Then $\mu(V) = \prod_i \mu(V_i)$ and if $\dim(V_i)$ is even for each i then $\mathrm{sgn}(V) = \prod_i \mathrm{sgn}(V_i)$.
   (3) If $X = \{x_1, \ldots, x_n\}$ is an orthogonal basis for V (i.e., $(x_i, x_j) = 0$ for $i \neq j$) then $\mu(V) = (-1)^n \prod_i Q(x_i)$, so $\mu(V) = \pm 1$.
   (4) If n is even then $\mathrm{sgn}(V) = (-1)^{n/2}\mu(V)$.

3. Let $G = U_n(2)$, D the set of transvections in G, and $T \in \mathrm{Syl}_2(G)$. Prove
   (1) $w_D(G) = (4^{n/2} - 1)/3$ if n is even, and $w_D(G) = (4^{(n-1)/2} - 1)/3$ if n is odd.
   (2) $\prod_{x \in T \cap D} x = 1$.
   (3) If $n = 6$ then $A = \langle T \cap D \rangle \cong E_{2^9}$ and $N_G(A)/A \cong L_3(4)$ is 2-transitive on $T \cap D$.

4. Let $q = 3$, $G = O(V, Q)$, $1 \neq R$ a 3-subgroup of G, $U = C_{[V,R]}(R)$, and $M = N_G(U)$. Prove
   (1) U is totally singular.
   (2) The representations of M on U and $V/U^\perp$ are dual; that is $V/U^\perp$ is isomorphic to the dual U^* of U as a G-module.
   (3) $C_M(U)$ induces $O(U^\perp/U, Q)$ on $U^\perp/U$.
5. Prove $\mathrm{Aut}(U_6(2))$ is $PGU_6(2)$ extended by a field automorphism σ. That is σ is an automorphism of the additive group of V fixing each member of an orthonormal basis X of (V, f) such that $(\sum_{x \in X} a_x x)\sigma = \sum_{x \in X} a_x^2 x$ for all $a_x \in GF(4)$.
6. Assume (V, f) or (V, Q) satisfies one of Hypotheses (a)–(d) of Section 11, D is one of the classes of 3-transpositions of $O(V, f)$ or $O(V, Q)$ described in Section 11, and $G = \langle D \rangle$. Assume $n = \dim(V) \geq 4$ if (V, f) is symplectic or unitary and $n \geq 5$ if (V, Q) is orthogonal. Let X be the set of centers of members of D and Γ the graph on X where $x, y \in X$ are adjacent if $x \perp y$. Prove
   (1) $\Gamma \cong \mathcal{D}(D)$ is connected.
   (2) For $x \in X$,

   $$\{x\} = \{y \in X : \{x\} \cup \Gamma(x) = \{y\} \cup \Gamma(y)\} = \{y \in X : \Gamma(x) = \Gamma(y)\}.$$

   (3) $O_2(G) \leq Z(G) \geq O_3(G)$.
7. Each odd-dimensional orthogonal space over a field of odd order has a hyperbolic hyperplane.
8. Let $G = PO_n^{\mu,\pi}(3)$, D the class of reflections defining G, and $T \in \mathrm{Syl}_2(G)$. Prove
   (1) $w_D(G) = n$ when $\mu = (-\pi)^n$ and $w_D(G) = n - 1$ otherwise.
   (2) If $w_D(G) = n$ then $\prod_{d \in T \cap D} d = 1$.
   [*Hint:* Use Exercise 4.2.]
9. $O_n^{\mu,\pi}(3)$ is of index 2 in $O_n^\mu(3)$ if $(n, \mu\pi) \neq (3, -1)$ or $(1, 1)$.
10. Let (V_i, Q_i) be n-dimensional orthogonal spaces over $GF(3)$ of discriminant μ_i and $x_i \in V_i$ with $Q(x_i) = \pi_i$ for $i = 1, 2$. Then
    (1) If $\mu_1 \pi_1^n = \mu_2 \pi_2^n$ then there is a similarity $\alpha : (V_1, Q_1) \to (V_2, Q_2)$ with $x_1\alpha = x_2$.
    (2) If $\mu_1 \pi_1^n = \mu_2 \pi_2^n$ then $O_n^{\mu_1,\pi_1}(3) \cong O_n^{\mu_2,\pi_2}(3)$.
11.
    (1) $O_2^{-,\pi}(3) \cong \mathbf{Z}_2$ and $O_2^{+,\pi}(3) \cong E_4$.
    (2) $O_3^{\pi,\pi}(3) \cong S_4$ and $O_3^{-\pi,\pi}(3) \cong E_8$ with $|D| = 3$.
12. Let V be an n-dimensional orthogonal space over $GF(3)$. Prove
    (1) An involution $t \in O(V)$ is in $\Omega(V)$ if and only if $\dim([V, t])$ is even and $\mu([V, t]) = 1$.

(2) If n is odd and $v \in V$ is nonsingular with $Q(v) = \pi$, then the reflection $t(v)$ induces an inner automorphism on $\Omega(V)$ if and only if $\mu(v^\perp) = 1$ if and only if $\pi\mu(V) = -1$.

[*Hint:* In (1) use Exercise 7.6.1 in [FGT] to conclude $t \in \Omega(V)$ if and only if $t \in SO(V)$ and the spinor norm $\theta(t)$ of t is 1. Then observe that if $\{x_1, \ldots, x_m\}$ is an orthogonal basis of $[V, t]$ then $t = t(x_1) \cdots t(x_m)$, so $t \in SO(V)$ if and only if m is even and from Section 22 of [FGT], $\theta(t) = \prod_i Q(x_i)$.]

13. Let (V, Q) be an n-dimensional orthogonal space over $GF(3)$, $\mu = \mu(V)$, and $\epsilon = \text{sgn}(V)$ if n is even. Let $S(n, \mu)$ be the number of singular points of V and $N(n, \mu, \pi)$ the number of points $\langle v \rangle$ of V with $Q(v) = \pi$. Prove
    (1) $S(n, \mu) = (3^{2m} - 1)/2$ if $n = 2m + 1$ is odd and $S(n, \mu) = (3^m - \epsilon)(3^{m-1} + \epsilon)/2$ if $n = 2m$ is even.
    (2) $N(n, \mu, \pi) = 3^{m-1}(3^m - \epsilon)/2$ if $n = 2m$ is even and $N(n, \mu, \pi) = 3^m(3^m - (-1)^m \mu\pi)/2$ if $n = 2m + 1$ is odd.
    [*Hint:* Prove $S(n, \mu) = 1 + 3S(n - 2, -\mu) + 3^{n-2}$ and $N(n, \mu, \pi) = 1 + N(n - 1, -\pi\mu, \pi) + 2S(n - 1, -\pi\mu)$. Then prove (1) and (2) by induction on n using these formulas.]

14. $O_6^-(2) \cong PO_5^{\pi,\pi}(3)$ and $U_4(2) \cong PO_5^{-\pi,\pi}(3)$ as 3-transposition groups. [*Hint:* $U_4(2) \cong \Omega_6(2) \cong P\Omega_5(3)$ as abstract groups. Then use Exercise 4.12.2 to decide whether a reflection in $PO_5^{\mu,\pi}(3)$ induces inner or outer automorphism on $P\Omega_5(3)$. Or use the uniqueness results in Chapter 5.]

15. Let f be a sesquilinear form on an n-dimensional space V over a field F. Write $\Gamma(V)$ for the group of all *semilinear maps* α of V; that is α is a group automorphism of V and there exists $\sigma(\alpha) \in \text{Aut}(F)$ with $(av)\alpha = a^{\sigma(\alpha)}v\alpha$ for all $a \in F$ and $v \in V$. Write $\Gamma(V, f)$ for the group of all semilinear maps α on V with $f(u\alpha, v\alpha) = f(u, v)^{\sigma(\alpha)}$ for all $u, v \in V$. Prove
    (1) The map $\sigma : \alpha \mapsto \sigma(\alpha)$ is a group homomorphism from $\Gamma(V, f)$ into $\text{Aut}(F)$ with kernel $O(V, f)$.
    (2) For $\sigma \in \text{Aut}(F)$ let $\mathcal{X}(\sigma)$ be the set of bases X of V such that all entries of $J(X, f)$ are in the fixed field $\text{Fix}(\sigma)$ of σ on F. For $X \in \mathcal{X}(\sigma)$ the *field automorphism* of V with respect to X and σ is the map

$$\sigma_X : \sum_{x \in X} a_x x \mapsto \sum_{x \in X} a_x^\sigma x.$$

Show $\sigma_X \in \Gamma(V, f)$ with $\sigma(\sigma_X) = \sigma$.
    (3) For $g \in O(V, f)$, $M_X(g^{\sigma_X}) = M_X(g)^\sigma$, where $(a_{ij})^\sigma = (a_{ij}^\sigma)$. In particular

$$C_{O(V,f)}(\sigma_X) = O(\text{Fix}(\sigma)X, f_X),$$

where $\mathrm{Fix}(\sigma)X$ is the $\mathrm{Fix}(\sigma)$-subspace of V spanned by X and f_X is the restriction of f to $\mathrm{Fix}(\sigma)X$.

(4) If f is a symmetric or alternating bilinear form, so is f_X. If f is unitary and defined by $\theta \in \mathrm{Aut}(F)$ then f_X is symmetric if $fix(\sigma) \leq \mathrm{Fix}(\theta)$ and unitary otherwise. If f is nondegenerate, so is f_X.

(5) For $g \in O(V, f)$, $\sigma_X^g = \sigma_{Xg}$.

(6) For $\sigma, \bar{\sigma} \in \mathrm{Aut}(F)$, $X \in \mathcal{X}(\sigma)$, and $\bar{X} \in \mathcal{X}(\bar{\sigma})$, $\sigma_X = \bar{\sigma}_{\bar{X}}$ if and only if $\sigma = \bar{\sigma}$ and $\mathrm{Fix}(\sigma)X = \mathrm{Fix}(\sigma)\bar{X}$. Moreover the following are equivalent:

   (a) σ_X is conjugate to $\bar{\sigma}_{\bar{X}}$ under $O(V, f)$.

   (b) $\sigma = \bar{\sigma}$ and $\mathrm{Fix}(\sigma)X$ is conjugate to $\mathrm{Fix}(\sigma)\bar{X}$.

   (c) $\sigma = \bar{\sigma}$ and $(\mathrm{Fix}(\sigma)X, f_X)$ is isometric to $(\mathrm{Fix}(\sigma)\bar{X}, f_{\bar{X}})$.

(7) If $\alpha \in \Gamma(V, f)$, $\sigma = \sigma(\alpha)$, and X is a basis for V with $X \subseteq C_V(\alpha)$ then $X \in \mathcal{X}(\sigma)$ and $\alpha = \sigma_X$. Thus α is a field automorphism if and only if $C_V(\alpha)$ contains a basis for V.

16. Assume the hypotheses and notation of Exercise 4.15 and let F be finite of order q^2, $\sigma : a \mapsto a^q$ the automorphism of F of order 2, and $K = \mathrm{Fix}(\sigma)$ the subfield of F of order q. Let T be the set of involutions $t \in \Gamma(V, f)$ with $\sigma(t) = \sigma$. Prove

(1) Each member of T is a field automorphism.

(2) For $t, s \in T$, $s \in t^{O(V, f)}$ if and only if $(C_V(t), f_t)$ is isometric to $(C_V(s), f_s)$ as a K-space with form, where f_t is the restriction of f to $C_V(t)$.

(3) If f is trivial, so that $O(V, f) = GL(V)$, then $GL(V)$ is transitive on T and $C_{GL(V)}(t) \cong GL_n(q)$.

(4) If f is symplectic then $\mathrm{Sp}(V)$ is transitive on T and $C_{\mathrm{Sp}(V)}(t) \cong \mathrm{Sp}_n(q)$.

(5) If f is unitary and q is odd then $GU(V)$ has two orbits T_+ and T_- on T, where for $t_\epsilon \in T_\epsilon$, $C_{GU(V)}(t_\epsilon) \cong O_n^\epsilon(q)$.

(6) If f is unitary and q is even then $GU(V)$ has one or two orbits on T for n odd or even, respectively. If n is odd then $C_{GU(V)}(t) \cong \mathrm{Sp}_{n-1}(q)$, whereas if n is even then the orbits of $GU(V)$ on T have representatives t and s where $C_{GU(V)}(t) \cong \mathrm{Sp}_n(q)$ and $s = td$ for some transvection $d \in C_{GU(V)}(t)$ and $C_G(s) = C_G(t) \cap C_G(d)$.

(7) If f is orthogonal and q is odd then T is empty when $\mu(V) \notin F^2$ whereas $O(V, f)$ has two orbits T_+ and T_- if $\mu(V) \in F^2$, with $C_{O(V, f)}(t_\epsilon) \cong O_n^\epsilon(q)$. Here $\mu(V) = \det(J(X, f))F^2 \in F^\#/F^2$ is the *discriminant* of V.

[*Hint:* Use Exercise 4.15 and in (6) use Exercise 4.17.]

17. Let F be a finite field of even order q, V an n-dimensional vector space over F, and f a nondegenerate symmetric bilinear form on V. Let $U = \{v \in V : f(v, v) = 0\}$. Prove

 (1) (V, f) is symplectic if and only if n is even and $U = V$.

 (2) If n is odd then V is determined up to isometry, U is a hyperplane of V, (U, f) is symplectic, $O(V, f)$ is faithful on U, and $O(V, f) \cong O(U, f) \cong \mathrm{Sp}_{n-1}(q)$.

 (3) If n is even and $U \neq V$ then V is determined up to isometry, $\mathrm{Rad}(U)$ is a point with $U = \mathrm{Rad}(U)^{\perp}$, and $O(V, f)$ is isomorphic to the centralizer in $\mathrm{Sp}_n(q)$ of a transvection.

18. Let $G = O_7^{\mu,\pi}(3)$. Then no 3-local subgroup of G contains a D-section isomorphic to S_8.

 [*Hint:* Use Exercises 7.8 and 14.5 from [FGT] to conclude the maximal parabolics of G are the stabilizers of totally singular subspaces of V and then use 47.8.2 in [FGT] to conclude each 3-local subgroup stabilizes a totally singular subspace.]

5
Fischer's Theorem

In this chapter we prove Fischer's Theorem classifying almost simple groups generated by a class of 3-transpositions. The statement of Fischer's Theorem appears in the Introduction. The isomorphisms in Fischer's Theorem are isomorphisms of 3-transposition groups. For example the statement "$G \cong S_n$ and D is the set of transpositions of G" means there is an isomorphism $(G, D) \cong (S_n, T)$ of 3-transposition groups, where T is the set of transpositions of S_n. Similarly when we say "G is isomorphic to S_n as a 3-transposition group" we mean (G, D) is isomorphic to (S_n, T) in the category of 3-transposition groups.

Assume (G, D) satisfies the hypothesis of the theorem and let $d \in D$, $H = \langle D_d \rangle$, $c \in A_d$, and $L = \langle D_{d,c} \rangle$. The approach is to show $D_{d,c}$ is determined up to conjugacy under $N_{\mathrm{Aut}(H)}(D_d)$, and then that $\mathcal{D}(D)$ is determined up to isomorphism by $(H/Z(H), D_d Z(H)/Z(H))$ and $(L/Z(L), D_{d,c} Z(L)/Z(L))$. As D is a distinguished class of automorphisms of $\mathcal{D}(D)$, this establishes the uniqueness of G.

Theorem D in Section 12 shows that under suitable conditions, $\mathcal{D}(D)$ is determined by $(H/Z(H), D_d Z(H)/Z(H))$ and $(L/Z(L), D_{d,c} Z(L)/Z(L))$. The material in Section 12 comes from R. Weiss in [W1].

Section 13 consists of some preliminary reductions. Then the proof breaks into two cases. In Section 14 we consider the case $O_2(H) \not\leq Z(H)$. The symplectic and unitary groups arise here. This part of the argument is more geometric, focusing on the singular lines of $\mathcal{D}(D)$. Indeed we show easily that D together with the set of singular lines of $\mathcal{D}(D)$ form a *Shult space* in the sense of Exercise 6.1. At this point one could appeal to the Buekenhout–Shult Theorem [BS] to complete the proof, but one object of this treatment is to keep things self-contained.

Section 15 considers the case $O_2(H) \leq Z(H) \geq O_3(H)$. Here when $H/Z(H)$ is of type $M(22)$ or $M(23)$ we need to know D-subgroups of $H/Z(H)$ of type $PO_7^{-\pi,\pi}(3)$ or $PO_8^{+,\pi}(3)$ are determined up to conjugacy in

Aut($H/Z(H)$), respectively. This verification is accomplished in Section 16 without the assumption of the existence of larger Fischer groups. However other information like the group order is derived in Section 16 using either the embedding of the smaller Fischer groups in a group of type $M(24)$ or the hypothesis that groups of type M are of type $\tilde{M}$, as defined in the Introduction. Hence this derivation uses either the existence of groups of type $M(24)$ or the existence of groups of type $\tilde{M}$. The existence of the Fischer groups is eventually verified in Chapter 11. The final details of the proof are tied together in Section 17.

Much of this outline also applies to Fischer's original proof, although the details are often quite different. In a series of papers ([W1], [W2], [SW], [DGW], some written with collaborators) Weiss gives another proof of Fischer's Theorem, again following Fischer's general outline but differing in detail both from Fischer and from the proof given here.

12. A uniqueness lemma

In this section D is a primitive conjugacy class of 3-transpositions of the finite group G. Let $\mathcal{D} = \mathcal{D}(D)$. We say that $\mathcal{D}$ is a *triple graph* if for $d \in D$ and $a \in A_d$, the hyperbolic line $d * a$ equals $\{d, a, d^a\}$. Throughout this section we assume $\mathcal{D}$ is a triple graph. We prove

Theorem D. *Let $D, \bar{D}$ be primitive conjugacy classes of 3-transpositions of the finite groups $G, \bar{G}$, respectively. Assume $d \in D$, $\bar{d} \in \bar{D}$, $x \in A_d$, $\bar{x} \in A_{\bar{d}}$, and $\phi : \mathcal{D}(D_d) \to \mathcal{D}(D_{\bar{d}})$ is a graph isomorphism such that*

(a) $\mathcal{D}(D)$ and $\mathcal{D}(\bar{D})$ are triple graphs.
(b) $\phi(u)^{\phi(v)} = \phi(u^v)$ for all $u, v \in D_d$.
(c) $\phi(D_{d,x}) = D_{\bar{d},\bar{x}}$.

Then ϕ extends to an isomorphism $\phi : \mathcal{D}(D) \to \mathcal{D}(\bar{D})$ and induces an isomorphism of $(G/Z(G), DZ(G)/Z(G))$ with $(\bar{G}/Z(\bar{G}), \bar{D}Z(\bar{G})/Z(\bar{G}))$.

The proof involves a series of lemmas. Let $d \in D$.

(12.1) *d is the unique element of Aut$(\mathcal{D})^{\#}$ fixing $d^{\perp}$ pointwise.*

Proof: Assume $\alpha \in$ Aut$(\mathcal{D})^{\#}$ fixes $d^{\perp}$ pointwise. We first observe that for all $x \in A_d$, α fixes $D_{d,x}$ so α acts on $(d * x) \cap A_d = \{x, x^d\}$. Now if $\alpha \neq d$ then, replacing α by αd if necessary, we may assume α fixes x. By 7.8, A_d is a connected subgraph of $\mathcal{D}$, so it suffices to prove α fixes each $y \in D_x \cap A_d$. If not, $y\alpha = y^d$, so $y^d \in D_x$. But then $d \in y * y^d \subseteq D_x$, a contradiction.

(12.2) *If $\bar{D}$ is a primitive conjugacy class of 3-transpositions of the finite group $\bar{G}$ and $\phi : \mathcal{D}(D) \to \mathcal{D}(\bar{D})$ is a graph isomorphism then*

   (1) $\phi(u)^{\phi(v)} = \phi(u^v)$ for all $u, v \in D$.

   (2) ϕ induces an isomorphism $\alpha : G/Z(G) \to \bar{G}/Z(\bar{G})$ such that

$$\alpha(uZ(G)) = \phi(u)Z(\bar{G}) \quad \text{for all } u \in D.$$

Proof: Let $u, v \in D$. If $v \in u^\perp$ then as ϕ is an isomorphism, $\phi(v) \in \phi(u)^\perp$. Thus $\phi(u^v) = \phi(u) = \phi(u)^{\phi(v)}$.

So assume $v \in A_u$. Then as ϕ is an isomorphism, $\phi(v) \in A_{\phi(u)}$. Also $\mathcal{D}$ and its isomorphic image $\mathcal{D}(\bar{D})$ are triple graphs, so $\{u, v, v^u\} = u * v$ and

$$\phi(u) * \phi(v) = \{\phi(u), \phi(v), \phi(u)^{\phi(v)}\}.$$

Hence $\phi(v^u) = \phi(u)^{\phi(v)}$, establishing (1).

Observe conjugation defines an embedding of $G/Z(G)$ into $\text{Aut}(\mathcal{D})$ and $\mathcal{D} \cong \mathcal{D}(DZ(G)/Z(G))$. So replacing G by $G/Z(G)$, we may assume $Z(G) = 1$ and $G \le \text{Aut}(\mathcal{D})$. Similarly we may assume $Z(\bar{G}) = 1$ and $\bar{G} \le \text{Aut}(\mathcal{D}(\bar{D}))$.

The isomorphism $\phi : \mathcal{D}(D) \to \mathcal{D}(\bar{D})$ induces an isomorphism

$$\alpha : \text{Aut}(\mathcal{D}(D)) \to \text{Aut}(\mathcal{D}(\bar{D}))$$

via $\alpha(\mu) = \phi \circ \mu \circ \phi^{-1}$ for $\mu \in \text{Aut}(\mathcal{D}(D))$. Further by 12.1, $\phi(u)$ is the unique element of $\text{Aut}(\mathcal{D}(\bar{D}))$ fixing $\phi(u)^\perp$ pointwise, so $\alpha(u) = \phi(u)$ and hence $\alpha(G) = \langle \alpha(D) \rangle = \bar{G}$; so (2) is established.

In the remainder of the section let $x \in A_d$.

(12.3) *The map $u \mapsto x^u$ is a bijection of $D_d \cap A_x$ with $D_{x^d} \cap A_d$.*

Proof: Let $u \in D_d \cap A_x$. Then (d, x, u) is a 3-string so by 8.3, $\langle d, x, u \rangle \cong S_4$ and $x^u \in D_{x^d} \cap A_d$. If $v \in D_d$ with $x^u = x^v$ then $x \in C_D(uv)$. But by 8.4, $C_D(uv) = C_D(u) \cap C_D(v)$, contradicting $x \in A_u$. Thus the map of 12.3 is an injection. Finally by 8.7, $d^{dx} = x^d$ and $x^{dx} = d$, so $(D_d \cap A_x)^{dx} = D_{x^d} \cap A_d$ and hence $|D_d \cap A_x| = |D_{x^d} \cap A_d|$, completing the proof of the lemma.

(12.4) *Let $a, b \in D_d \cap A_x$. Then*

   (1) $x^a \in D_{x^b}$ if and only if $a \in D_b$.

   (2) $x^{da} \in D_{x^b}$ if and only if $b^a \in D_x$.

Proof: By 8.7, $d^x = x^d$, $x^a = a^x$, and $x^b = b^x$. Thus $x^a \in D_{x^b}$ if and only if $a^x \in D_{b^x} = (D_b)^x$, so (1) is established.

Further $b^a \in D_x$ if and only if $b = (b^a)^{ad} \in D_{x^{ad}}$. But as (a, x, d) is a 3-string, $x^{ad} \in D_x$ by 8.3, so $b \in D_{x^{ad}}$ if and only if $x * b \subseteq D_{x^{ad}}$ if and only if $x^b \in D_{x^{ad}}$ if and only if $x^{da} \in D_{x^b}$, so that (2) holds.

(12.5) *Let $P(d, x)$ be the set of pairs a, b with $b \in A_a$ and $a * b \subseteq D_d \cap A_x$. Then*

(1) $A_{d,x,x^d} = \{x^{ab} : (a, b) \in P(d, x)\}$.

(2) *For* $(a, b), (u, v) \in P(d, x)$, $x^{ab} = x^{uv}$ *if and only if* $(D_{d,x})^{ab} = (D_{d,x})^{uv}$ *and* $(D_{d,x})^{ba} = (D_{d,x})^{vu}$.

(3) *If* $(a, b) \in P(d, x)$ *then* $D_{x^d,x^{ab}} = (D_{d,x})^{bax}$ *and*

$$A_d \cap D_{x^d,x^{ab}} = \{x^g : g \in (D_{d,x})^{ba} - D_{d,x}\}.$$

(4) *For* $(a, b), (u, v) \in P(d, x)$, $x^{ab} \in D_{x^{uv}}$ *if and only if* $x^{ab} \neq x^{uv}$ *and there exist distinct $g, h \in D_d$ with* $(D_{d,x})^{abg} = (D_{d,x})^{uv}$ *and* $(D_{d,x})^{bah} = (D_{d,x})^{vu}$.

Proof: (3) Let $(a, b) \in P(d, x)$. Then $d^{bax} = d^x = x^d$ and by 8.10, $x^{bax} = x^{ab}$. Thus the first observation in (3) is established. Hence the map $g \mapsto g^x$ is a bijection of $(D_{d,x})^{ba} - D_{d,x}$ with $D_{x^d,x^{ab}} - D_{d,x}$. Further as $g \in A_x, x^g = g^x$, while

$$D_{x^d,x^{ab}} - D_{d,x} = D_{x^d,x^{ab}} \cap A_d$$

since $D_{x^d,x} = D_{x^d,d}$, completing the proof of (3).

(1) First assume $(a, b) \in P(d, x)$. Then $a, b \in D_d$ and $x \in A_d$, so $x^{ab} \in A_d$. Also $a^b \notin D_x$ so by 12.4.2, $x^{db} \in A_{x^a}$ and thus $x^d \in A_{x^{ab}}$. Similarly by 12.4.1, $x^b \in A_{x^a}$, so $x \in A_{x^{ab}}$. Therefore $x^{ab} \in A_{d,x,x^d}$.

Conversely let $z \in A_{d,x,x^d}$; we must produce $(a, b) \in P(d, x)$ with $x^{ab} = z$. Now $x * z = \{x, z, x^z\}$, and $d \neq x, z$ as $x, z \in A_d$. Further if $d = x^z$ then $x = d^z = z^d$, so $z = x^d$, contradicting $z \in A_{x^d}$. Therefore

$$d \notin x * z = \bigcap_{w \in D_{x,z}} w^{\perp}$$

so there is $w \in A_d \cap D_{z,x}$. By 12.3, the map $a \mapsto x^{da}$ is a bijection of $D_d \cap A_{x^d}$ with $D_x \cap A_d$, so there is $a \in D_d \cap A_{x^d} = D_d \cap A_x$ with $x^{da} = w$. Similarly there is $b \in D_d \cap A_w$ with $w^{db} = z$. Then $z = w^{db} = x^{dadb} = x^{ab}$, so to complete the proof of (1) it remains to show $(a, b) \in P(d, x)$.

If $b \in D_a$ then $ad \in C_G(b)$ so, as $w \in A_b, x = w^{ad} \in A_b$. Then (a, x, b) is a 3-string so by 8.3, $x^{ab} \in D_x$. But $x^{ab} = z \in A_x$, a contradiction. Therefore $b \in A_a$.

By choice of a, b, $a \in D_d \cap A_x$ and $b \in D_d \cap A_w$. So as $w = x^{da}$,

$$b^a \in D_d \cap A_{x^d} = D_d \cap A_x.$$

Finally suppose $b \in D_x$. Then $(b^a)^a = b \in D_x$ so by 12.4.2, $x^{da} \in D_{x^{aba}}$. But then $x^d \in D_{x^{ab}} = D_z$, a contradiction. So (1) is established.

(2) We next prove (2). Let $(a, b), (u, v) \in P(d, x)$ and $z = x^{ab}$. If $z = x^{uv}$ then certainly $(D_{d,x})^{uv} = D_{d,z} = (D_{d,x})^{ab}$. Also $(D_{d,x})^{vu} = D_{x^{dx},z^x} = (D_{d,x})^{ba}$ by (3).

Conversely assume $(D_{d,x})^{ab} = (D_{d,x})^{uv}$ and $(D_{d,x})^{ba} = (D_{d,x})^{vu}$; we must show $z = x^{uv}$. As

$$d * x = \bigcap_{y \in D_{d,x}} y^{\perp} \quad \text{and} \quad D(D_{d,x})^{uv} = (D_{d,x})^{ab}$$

we have

$$(d * x)^{uv} = (d^*x)^{ab} = \{d, x, x^d\}^{ab} = \{d, x^{ab}, x^{dab}\}$$

as $\mathcal{D}$ is a triple graph, so we conclude $z = x^{ab} = x^{uv}$ or x^{duv}, and we may assume the latter, so that $z^d x^{uv}$. Also by (3), $D_{x^d,x^{ab}} = D_{x^d,x^{uv}}$, so $z = x^{ab} = x^{uv}$ or x^{uvy}, where $y = x^d$, and we may assume $z = x^{uvy}$, so $z^d = x^{uv} = z^y$. But then by 8.4, $z \in C_D(dy) = D_{d,y}$, a contradiction. Thus (2) is established.

(4) It remains to prove (4). Assume $(a, b), (u, v) \in P(d, x)$ and let $x^{ab} = z$ and $w = x^{uv}$.

Suppose first that $z \in D_w$. Let $g = d^{wz}$ and $k = x^{dwz}$. Then (w, d, z) and (w, x^d, z) are 3-strings by (1), so by 8.3, $g \in D_d$, $k \in D_{x^d}$, $z^{dg} = w$, and $z^{dxdk} = w$. Then

$$(D_{d,x})^{abg} = (D_{d,x^d})^{abg} = D_{d,z^{dg}} = D_{d,w} = (D_{d,x})^{uv},$$

and using (3),

$$(*) \qquad (D_{d,x})^{baxk} = (D_{x^d,z})^{k} = (D_{x^d,z^{dxd}})^{k}$$
$$= D_{x^d,z^{dxdk}} = D_{x^d,w} = (D_{d,x})^{vux}.$$

Now set $h = k^x$. Then by (*),

$$(D_{d,x})^{bah} = (D_{d,x})^{baxkx} = (D_{d,x})^{vu}$$

and $h = k^x \in D_{x^{dx}} = D_d$.

So we must show $g \ne h$. First $z^{dg} = w = z^{dxdk}$. Thus if $g = k$ then $z \in C_D(xd) = D_{x,d}$, a contradiction. So $g \ne k$. Next

$$g^k = d^{wz(zwd)x(dwz)} = d^{xdwz} = x^{wz} \ne x$$

as $x \notin D_{w,z}$. So $g \ne x^k = k^x = h$.

Conversely assume $z \neq w$ and $g, h \in D_d$ are distinct such that $(D_{d,x})^{abg} = (D_{d,x})^{uv}$ and $(D_{d,x})^{bah} = (D_{d,x})^{vu}$. As $(D_{d,x})^{abg} = (D_{d,x})^{uv} = D_{d,w}$,

$$d * (z^g) = (d^{abg}) * (x^{abg}) = (d * x)^{abg} = d * w,$$

so the triple graph property implies $z^g = w$ or w^d. Next $k = h^x \in D_{x^d}$ and by (3), $D_{x^d,z} = (D_{d,x})^{bax}$ and $D_{x^d,w} = (D_{d,x})^{vux}$, so

$$D_{x^d,z^k} = (D_{d,x})^{baxk} = (D_{d,x})^{bahx} = (D_{d,x})^{vux} = D_{x^d,w},$$

and hence $(x^d) * (z^k) = (x^d) * w$, so $z^k = w$ or w^{dxd}.

Claim $g \in A_w$. Assume otherwise. Then $g \in C_G(w)$ so, as $z^g = w$ or w^d, and as $z \neq w$, we conclude $z = w^d$. In particular $w \in A_z$. As $w \notin D_d$, $w \notin C_D(kd)$, so $w^k \neq w^d = z$ and hence $z^{kdxd} = w$. Next observe $k \in A_z$. For if not, $z = z^k$ and as $z^k = w$ or w^{dxd} while $z \neq w$, we have $z = w^{dxd}$. Then $w^d = z = w^{dxd}$, contradicting $w \in A_d$ by (1). Therefore $k \in A_z \cap D_{x^d}$ and $z \in A_{x^d}$ by (1), so (k, z, x^d) is a 3-string. But then $w = z^{kdxd} \in D_z$ by 8.3, contrary to an earlier remark.

Thus $g \in A_w$. Hence (g, w, d) is a 3-string, so $w^{dg} \in D_w$ by 8.3. Thus if $z^g = w^d$ we are done, so we may assume $z^g = w$. Similarly we may assume $z^k = w$. But then $w \in C_D(kg) - D_g$, so $k = g$ by 8.4. Thus $k \in D_{d,x^d} \subseteq D_x$, so $h = k^x = k = g$, contrary to assumption.

We are now in a position to prove Theorem D. Assume the hypothesis of Theorem D and extend ϕ to a map $\phi : D \to \bar{D}$ by

$$\phi(d) = \bar{d}, \qquad \phi(x) = \bar{x}, \qquad \phi(x^d) = \bar{x}^{\bar{d}},$$

$$\phi(u) = \phi(u), \quad \text{for } u \in D_d,$$

$$\phi(x^a) = \bar{x}^{\phi(a)}, \qquad \phi(x^{da}) = \bar{x}^{\bar{d}\phi(a)}, \quad \text{for } a \in D_d \cap A_x,$$

$$\phi(x^{ab}) = \bar{x}^{\phi(a)\phi(b)}, \quad \text{for } a, b \in P(d, x).$$

We first check that

(12.6) ϕ *is a well defined bijection.*

Proof: Certainly $\phi : d^\perp \mapsto \bar{d}^\perp$ is a well defined bijection as $\phi : \mathcal{D}(D_d) \to \mathcal{D}(D_{\bar{d}})$ is a graph isomorphism. Next A_d is partitioned by $A_d \cap D_x$, $A_d \cap D_{x^d}$, and A_{d,x,x^d}, and of course we have the corresponding partition of $A_{\bar{d}}$. By 12.3, $\phi : A_d \cap D_x \to A_{\bar{d}} \cap D_{\bar{x}}$ and $\phi : A_d \cap D_{x^d} \to A_{\bar{d}} \cap D_{\bar{x}^{\bar{d}}}$ are well defined bijections.

Similarly by 12.5.1 and 12.5.2,

$$\phi : A_{d,x,x^d} \to A_{\bar{d},\bar{x},\bar{x}^{\bar{d}}}$$

is a well defined bijection. Namely for $(a, b), (u, v) \in P(d, x)$, $(\phi(a), \phi(b))$, $(\phi(u), \phi(v)) \in P(\bar{d}, \bar{x})$ as $\phi : \mathcal{D}(D_d) \to \mathcal{D}(D_{\bar{d}})$ is a graph isomorphism with $\phi(D_{d,x}) = D_{\bar{d},\bar{x}}$. By 12.5.2, $x^{ab} = x^{uv}$ if and only if $(D_{d,x})^{ab} = (D_{d,x})^{uv}$ and $(D_{d,x})^{ba} = (D_{d,x})^{vu}$. Now by hypotheses (b) and (c) of Theorem D,

$$\phi((D_{d,x})^{ab}) = \phi(D_{d,x})^{\phi(a)\phi(b)} = (D_{\bar{d},\bar{x}})^{\phi(a)\phi(b)}.$$

Thus $x^{ab} = x^{uv}$ if and only if $\bar{x}^{\phi(a)\phi(b)} = \bar{x}^{\phi(u)\phi(v)}$.

(12.7) ϕ and ϕ^{-1} are morphisms of graphs.

Proof: By symmetry it suffices to show that $\phi(D_r) \subseteq D_{\phi(r)}$ for all $r \in D$. By construction this holds for $r = d, x$, and x^d. As $\phi : \mathcal{D}(D_d) \to \mathcal{D}(D_{\bar{d}})$ is a graph isomorphism, $\phi(D_{d,r}) = D_{\bar{d},\phi(r)}$ for $r \in D_d$. On the other hand if $r \in A_d$ then $r = x^a$ or x^{da}, for $a \in D_d \cap A_x$, or $r = x^{ab}$ for some $(a, b) \in P(d, x)$. In the first two cases

$$\phi(D_{d,r}) = \phi((D_{d,x})^a) = \phi(D_{d,x})^{\phi(a)} = (D_{\bar{d},\bar{x}})^{\phi(a)} = D_{\bar{d},\phi(r)}.$$

Similarly in the third case $\phi(D_{d,r}) = D_{\bar{d},\phi(r)}$. So we may assume $r, s \in A_d$ with $s \in D_r$ and it remains to show $\phi(s) \in D_{\phi(r)}$.

Suppose $r, s \in D_{x^d}$. Then $r = x^a$, $s = x^b$ for some $a, b \in D_d \cap A_x$, and by 12.4.1, $a \in D_b$. Then $\phi(a) \in D_{\phi(b)}$, so again by 12.4.1,

$$\phi(s) = \bar{x}^{\phi(b)} \in D_{\bar{x}^{\phi(a)}} = D_{\phi(r)}.$$

Similarly if $r, s \in D_x$ then $\phi(s) \in D_{\phi(r)}$.

Suppose $r \in D_{x^d}$ and $s \in D_x$. Then $r = x^{da}$ and $s = x^b$ and by 12.4.2, $b^a \in D_{d,x}$. Then $\phi(b)^{\phi(a)} \in D_{\bar{d},\bar{x}}$, so by 12.4.2, $\phi(s) \in D_{\phi(r)}$.

So without loss $r \in A_{x,x^d}$. Thus $r = x^{ab}$ for some $(a, b) \in P(d, x)$. Then by 12.5.3, $D_{x^d,r} \cap A_d = \{x^g : g \in (D_{d,x})^{ba} - D_{d,x}\}$. Therefore

$$\phi(D_{x^d,r} \cap A_d) = \{\bar{x}^h : h \in (D_{\bar{d},\bar{x}})^{\phi(b)\phi(a)} - D_{\bar{d},\bar{x}}\} \subseteq D_{\phi(r)}.$$

Similarly if $s \in D_x$ then $\phi(s) \in D_{\phi(r)}$.

So we may assume $s = x^{uv}$ for some $(u, v) \in P(d, x)$. Then by 12.5.4, there are distinct $g, h \in D_d$ with $(D_{d,x})^{abg} = (D_{d,x})^{uvh}$. Hence applying ϕ and appealing once again to 12.5.4, we conclude $\phi(s) \in D_{\phi(r)}$, completing the proof of 12.7.

Notice that 12.6, 12.7, and 12.2.2 establish Theorem D.

13. Primitive 3-transposition groups

In this section, D is a set of 3-transpositions in the finite group G with G primitive on D. Let $d \in D$ and $H = \langle D_d \rangle$. By 9.5, $\mathcal{D}(D)$ is connected, G is rank 3 on D, and $C_G(d)$ has orbits $\{d\}$, D_d, and A_d. Indeed by 9.5.4, H is transitive on D_d and $\langle d \rangle H$ is transitive on A_d. So we can choose $b \in D_d$ and $c \in A_d$. For $x \in D$, write H_x for $\langle D_x \rangle$.

(13.1) *Assume $\mathcal{D}(D_d)$ is disconnected. Then*

(1) *$D_{d,c}$ has width 1, and G is transitive on triples (x, y, z) with $x \in D$, $y \in A_x$, and $z \in D_{x,y}$.*

(2) *$\mathcal{D}(D_{d,b})$ is complete.*

Proof: As $\mathcal{D}(D_d)$ is disconnected, so is $\mathcal{D}(D_b)$, so there is $a \in D_b$ in a different connected component of $\mathcal{D}(D_b)$ than d. In particular $D_{a,b,d}$ is empty. Let $\Gamma = D_{a,d}$. Then $\Gamma \cap D_b$ is empty, so by 8.5, $\langle \Gamma \rangle$ is of width 1 and then by 8.6, $\langle \Gamma \rangle$ is transitive on Γ. Thus (1) holds as G is rank 3 on D.

Suppose $\mathcal{D}(D_{d,b})$ is not complete. Then there exists $u, v \in D_{d,b}$ with $v \in A_u$. But by (1), $D_{u,v}$ is of width 1, whereas $d, b \in D_{u,v}$, a contradiction.

(13.2) *Let $Q = O_2(H)$. Then*

(1) *$d * b = \{d\} \cup b^Q$.*

(2) *$d * b = u * v$ and $d^\perp \cap b^\perp = u^\perp \cap v^\perp$ for each pair of distinct $u, v \in d * b$.*

(3) *$N_G(d * b)$ is 2-transitive on $d * b$.*

(4) *If $a \in D$ and $|(d * b) \cap D_a| > 1$, then $d * b \subseteq a^\perp$.*

Proof: By 6.1.4, $d * b = \{d\} \cup V_{d,b}$, where

$$V_{d,b} = \{x \in D_d : x^\perp \cap D_d = b^\perp \cap D_d\}.$$

Next applying 9.2.1 to H, $V_{d,b} = b^Q$, so part (1) holds. Part (2) follows from 6.1.1 and 6.1.5, part (3) follows from 6.1.2, and part (4) follows from part (2).

(13.3) *Let $Q = O_2(H)$ and $|d * b| = m + 1$. Assume Q is not in the center of H, $H = Q\langle D_{d,c} \rangle$, and $m \leq 4$. Then*

(1) *$m = 2$ or 4.*

(2) *$Q\langle d \rangle$ is regular on A_d.*

(3) *$|D_{d,c} \cap (d * b)| = 1$.*

(4) *If $N_{Q\langle d \rangle}(\langle D_{d,c} \rangle) \leq \langle d \rangle$ then $\mathcal{D}(D)$ is a triple graph.*

Proof: As Q is not in the center of H, $[b, Q] \neq 1$, so by 13.2.1, $m > 1$ and m is a power of 2. Thus (1) holds.

Let $\Gamma = D_{d,c}$. We have already observed that $H\langle d\rangle$ is transitive on A_d, so as $H = Q\langle\Gamma\rangle$, it follows that $P = \langle d\rangle Q$ is transitive on A_d. Similarly as H is transitive on D_d, $b^Q \cap \Gamma$ is nonempty. Suppose u, v are distinct members of $(d*b) \cap \Gamma$. Then by 13.2.4, $d*b \subseteq c^\perp$, contradicting $c \in A_d$. Thus (3) is established.

We next claim that $C_P(b)$ fixes each point of $d*b$. If $m = 2$ this is clear, so take $m = 4$. We will show $N_G(d*b)$ induces S_5 or A_5 on $d*b$. Then as $P \trianglelefteq C_G(d)$, P^{d*b} is semiregular, so the claim holds. Now by 13.2, $N_G(d*b)^{d*b}$ is 2-transitive, so if it is not A_5 or S_5 then $N_G(d*b)^{d*b}$ is of order 20 with cyclic Sylow 2-groups. So we can obtain a contradiction by showing there is $g \in C_G(d)$ interchanging a and b for each $a \in d*b - \{d,b\}$. Indeed let $e \in D_d \cap A_b$ (notice such an element exists as H is transitive on D_d). $X = \langle a,b,e\rangle$, $f \in D_e \cap X$, and $g = fe$. Then (a,e,b) is a 3-string, so by 8.3.2, g interchanges a and b, completing the proof of the claim.

We now complete the proof of (2) by showing $C_P(c) = 1$. Indeed if $y \in P$ fixes c then y fixes the unique point of $(d*b) \cap D_c$ for each $b \in D_d$. Then as $C_P(b)$ fixes $d*b$ pointwise, y fixes $d^\perp$ pointwise. But then by 9.5.5, y fixes D pointwise, so $y = 1$.

Finally suppose $N_{Q\langle d\rangle}(\langle\Gamma\rangle) \leq \langle d\rangle$ and recall the definition of a triple graph in Section 12. We assume $x \in d*c$ and $x \neq d, c$, or c^d, and produce a contradiction. As $H = \langle D_{d,c}\rangle Q$, $D_{d,c} \nsubseteq b^Q = D_d \cap b^\perp$, so by 6.2.3, $N_G(d*c)$ is 2-transitive on $d*c$. Hence $|\Gamma| = |D_{d,x}|$, so as $\Gamma \subseteq D_{d,x}$, we conclude $\Gamma = D_{d,x}$. Next by (2) there is $y \in Q\langle d\rangle$ with $c^y = x$, and as $x \neq c$ or c^d, $y \notin \langle d\rangle$. But as $\Gamma = D_{d,x}$, $y \in N_{Q\langle d\rangle}(\langle\Gamma\rangle) \leq \langle d\rangle$, a contradiction.

(13.4) *If $O_3(H)$ is not in the center of H then $G \cong S_5$ as a 3-transposition group.*

Proof: Set $Q = O_3(H)$, $Q_{dg} = Q^g$ for $g \in G$, and assume Q is not in the center of H.

Suppose first that $\mathcal{D}(D_d)$ is connected and let $e \in D_{d,b}$. Then $b^Q \subseteq D_e$ by 9.2, so $b^Q \subseteq C_G(\Delta)$, where $\Delta = d^{Qe}$, again by 9.2. Assume $\Delta \neq d^{Qb}$ and let $H_b^* = H_b/Q_b$. Now $\Delta \subseteq D_b$ and as $\Delta \neq d^{Qb}$, $\{d^*\} \neq \Delta^*$ and $\Delta^* \subseteq d^* O_3(C_{H_b^*}(e^*))$. Hence by Exercise 3.2, $O_2(H_b^*) \leq Z(H_b^*)$. So as $\mathcal{D}(D_b)$ is connected, H_b^* is primitive on D_b^* by 9.4.4. Thus $H_b = \langle d^\perp \cap D_b, d^{Qe}\rangle$ by 7.8.5. Thus $H_b \subseteq C_G(b^Q)$, so $b^Q \subseteq W_b = \{b\}$, a contradiction.

We have shown $d^{Qb} = d^{Qe}$ for each $e \in D_{d,b}$. Therefore as $\mathcal{D}(D_d)$ is connected, $d^{Qb} = d^{Qx}$ for each $x \in D_d$. But then $d^{Qb} \subseteq W_d = \{d\}$, a contradiction.

Therefore $\mathcal{D}(D_d)$ is disconnected. So by 13.1, $C_G(d)$ is transitive on pairs $x, y \in D_d$ with $x \in A_y$. Hence as $x^Q - \{x\} \subseteq A_x$ and $Q \trianglelefteq C_G(d)$, we conclude $A_x \cap D_d \subseteq x^Q$. Therefore as H is transitive on D_d, $H = Q\langle b\rangle$ is of width 1.

It follows that (in the notation of 6.3) $k = |D_d| = |b^Q|$ is a power of 3 and $\lambda = 0$. Similarly by 13.1.1, $\mu = |D_{d,c}|$ is a power of 3 and $l = |A_d| = |H\langle d\rangle : \langle D_{d,c}\rangle| = 2 \cdot 3^a$. But by 6.3, $l = k(k-1)/\mu$, so $k - 1 = 2$. Hence $k = 3$ and as $\mu < k$, $\mu = 1$. Therefore $|D| = 10$. It is now easy to check that $\mathcal{D}(D)$ is the Petersen graph and $G \cong S_5$ (cf. Exercise 2.1).

Theorem 13.5. *If $\mathcal{D}(D_d)$ is disconnected then $G/Z(G) \cong S_5$, S_6, $U_4(2)$, or $U_5(2)$ as a 3-transposition group.*

Proof: Let $\Gamma = D_{d,c}$, and take $b \in \Gamma$. By 13.1, Γ is of width 1 and $\mathcal{D}(D_{d,b})$ is complete. Then by 9.6, $H = QP\langle b\rangle$ where P is a b-invariant Sylow 3-subgroup of H and $Q = O_2(H)$. By 13.4 we may assume $O_3(H) \leq Z(H)$, so by 9.3, $Q \not\leq Z(H)$. Adopt the notation of 6.3 and let $m + 1 = |d * b|$ and $e = |b^P|$. By 9.6, $k = me$ with m a power of 2 and e a power of 3. Similarly as $H\langle d\rangle$ is transitive on A_d, $l = |A_d| = 3^\alpha 2^\beta$. As $\mathcal{D}(D_{d,b})$ is complete, $D_{b,d} = d * b - \{d, b\}$ so $\lambda = m - 1$, whereas $\mu = |\Gamma|$ is a power of 3 as Γ is of width 1. Now by 6.3,

$$l = \frac{k(k - \lambda - 1)}{\mu} = \frac{m^2 e(e - 1)}{\mu},$$

so as $l = 3^\alpha 2^\beta$, $e - 1$ is a power of 2, and hence $e = 3$ or 9.

CASE I, $e = 9$. Assume $e = 9$. Then by 9.6, $P \cong 3^{1+2}$ is faithful on Q with $P = [P, b]$ and $[Z(P), b] = 1$. As $\mathcal{D}(D_d)$ is disconnected, so is $\mathcal{D}(N_{D_d}(P))$, so $N_H(P)$ is transitive on $N_{D_d}(P)$ and hence $C_Q(b)$ is transitive on the set S of b-invariant Sylow 3-subgroups of H. Hence $|S|$ is a power of 2.

Let $Z = Z(P)$. Then $\langle d\rangle Z \leq C_G(b) \leq N_G(H_b)$, so, as $|S|$ is a power of 2, Z acts on some d-invariant Sylow 3-subgroup P_b of H_b. Then by the Thompson $A \times B$-lemma (cf. 2.4) $[C_{P_b}(Z), d] \neq 1$. Hence $C_D(Z)$ is not contained in $d^\perp$. Let $K = \langle d^{\langle C_D(Z)\rangle}\rangle$.

Indeed we claim $C_D(P)$ is not contained in $d^\perp$. Suppose $b \notin d^K$. Then by 8.2.2, $C_D(Z) = D_1 \cup D_2$ where $D_1 = d^K$ and $b \in D_2$, and $K = K_1 K_2$ with $K_i = \langle D_i\rangle$ and $[K_1, K_2] = 1$. Now $P \leq C_G(Z\langle d\rangle) \leq N_G(K_1)$, so

$$P = [P, b] \leq [P, K_2] \leq K_2 \leq C_G(K_1).$$

But, as $D_2 \subseteq d^\perp$ whereas $C_D(Z) \not\subseteq d^\perp$, $D_1 \not\subseteq d^\perp$, so $C_D(P) \not\subseteq d^\perp$, contrary to our assumption.

So assume $b \in d^K$. Thus $b^P \subseteq d^K$. If $O_3(K) \not\leq Z(K)$ the claim follows from the $A \times B$-lemma, 2.4, so take $O_3(K) \leq Z(K)$. Notice $P\langle b \rangle = \langle D_d \cap K \rangle$ by 9.6.3 and as $Z \leq Z(K)$. Thus by Exercise 3.2.3, $O_2(K) \leq Z(K)$. Now 13.4 and 9.4.4 applied to the connected component E of $\mathcal{D}(d^K)$ say $\langle E \rangle \cong S_5$. This is impossible as $P = [P, b] \leq \langle E \rangle$.

So the claim is established. Hence by transitivity of $C_G(d)$ on A_d, we may take $c \in C_D(P)$. Then $b^P \subseteq \Gamma$, so as Γ is of width 1 and $H = \langle b \rangle PQ$, we conclude $\Gamma = b^P$. Therefore $H = \langle \Gamma \rangle Q$ and $\mu = 9$.

We next claim that $m \leq 4$. First $\langle \Gamma \rangle = \langle b, e, f \rangle$ for some $e, f \in \Gamma$. Let $a \in f * d - \{f, d\}$. By 13.2.2, $f^\perp \cap a^\perp = d^\perp \cap b^\perp$ and $f * a = d * b$. But $d^\perp \cap b^\perp = \{d\} \cup b^Q = d * b$, so $D_{f,a} = f * a - \{f, a\} \subseteq d^\perp$. Hence $D_{b,e,f,a} = \{d\}$ since $D_{b,e}$ is of width 1, and $D_{f,a} \subseteq d^\perp$. However $\langle b, e, a \rangle \cong P\langle b \rangle$ by 5.5.2, so $\langle b, e, a \rangle$ is conjugate to $P\langle b \rangle = \langle \Gamma \rangle$ by Sylow's Theorem. Thus $|D_{b,e,a}| \geq 3$, since $C_D(P) \not\leq d^\perp$ and $C_D(P) = C_D(P\langle b \rangle)$. Then

$$\bigcup_{a \in f * d - \{d\}} D_{b,e,a} \subseteq D_{b,e}$$

with

$$D_{b,e,a} \cap D_{b,e,f} = \{d\} \quad \text{and} \quad |D_{b,e,a} - \{d\}| \geq 2,$$

so $9 = |D_{b,e}| \geq 1 + 2m$ since $|f * d - \{d\}| = m$. That is $m \leq 4$.

We have achieved the hypotheses of 13.3, so we can apply that lemma to conclude $Q\langle d \rangle$ is regular on A_d and $m = 2$ or 4. Also $\Gamma = b^P$, so

$$\mu = |\Gamma| = |b^P| = e = 9.$$

Thus

$$|Q\langle d \rangle| = l = \frac{m^2 e(e-1)}{\mu} = 8m^2 = 32 \text{ or } 128.$$

Recall that P is faithful on Q. But the minimal degree of a faithful $GF(2)P$-module is 6 and this representation is determined up to equivalence (cf. 34.9 in [FGT]) so $m = 4$ and P is irreducible on $Q\langle d \rangle / \langle d \rangle$. Hence $\mathcal{D}(D)$ is a triple graph by 13.3.4.

Let $M = H\langle d \rangle$ and $M^* = M/\langle d \rangle$. Suppose $(\bar{G}, \bar{D})$ is a second pair satisfying the hypothesis of (G, D), let $\bar{d} \in \bar{D}$, $\bar{H} = \langle D_{\bar{d}} \rangle$, and so on. As the representation of $P\langle b \rangle^*$ on Q^* is determined up to equivalence and $\langle b^* \rangle$ is a Sylow 2-subgroup of $N_{M^*}(P^*)$, by 2.1 there is an isomorphism $\alpha : M^* \to \bar{M}^*$ with $\alpha(D^*) = \bar{D}^*$. As $D \cap b\langle d \rangle = \{b\}$, α induces an isomorphism $\phi : \mathcal{D}(D_d) \to \mathcal{D}(D_{\bar{d}})$ with $\phi(u)^{\phi(v)} = \phi(u^v)$ for all $u, v \in D_d$. Further $\Gamma = N_D(P)$, so $\phi(\Gamma) = D_{\bar{d}, \bar{c}}$ for some $\bar{c} \in A_{\bar{d}}$.

Thus by Theorem D, ϕ induces an isomorphism of $(G/Z(G), DZ(G)/Z(G))$ with $(\bar{G}/Z(\bar{G}), \bar{D}Z(\bar{G})/Z(\bar{G}))$. That is $(G/Z(G), DZ(G)/Z(G))$ is determined up to isomorphism. But by 11.9 and 11.12, the transvections in $U_5(2)$ satisfy our hypothesis, so $G/Z(G) \cong U_5(2)$ as a 3-transposition group.

CASE II, $e = 3$. As $e = 3$, $l = 6m^2/\mu$. Then as μ is a power of 3 and m is a power of 2, $\mu = 1$ or 3. Assume first that $\mu = 1$. Recall $\lambda = m - 1$ and $k = me = 3m$, so

$$(\mu - \lambda)^2 + 4(k - \mu) = m^2 + 8m.$$

Hence by 6.3, $m^2 + 8m = \delta^2$ is a square.

Let $s = (\delta, m)$. If $s < m$ then $\delta/s = t$ is odd and $r = m/s$ is even, so

$$\frac{8m}{s^2} = \frac{\delta^2}{s^2} - \frac{m^2}{s^2} = t^2 - r^2$$

is odd. Thus as m is a power of 2, $1 = t^2 - r^2 = (t - r)(t + r)$, a contradiction. Therefore $s = m$ divides δ, so $\delta^2 = m(m + 8)$ is divisible by m^2. Hence $m \leq 8$. Then as $1 \neq m$ is a power of 2, we check directly that $m(m + 8)$ is not a square.

Therefore $\mu = 3$. In particular we may take $\Gamma = b^P$ and $H = \langle \Gamma \rangle Q$. Recall by 6.3,

$$(\mu - \lambda)^2 + 4(k - \mu) = \delta^2 \quad \text{is a square,}$$

so as $\lambda = m - 1$, $k = 3m$, and $\mu = 3$, $\delta = m + 2$. Then by 6.3, $m + 2$ divides

$$2k + (\lambda - \mu)(k + l) = 6m + m(2m + 3)(m - 4) \equiv -24 \bmod(m + 2).$$

So $m = 2$ or 4. Therefore 9.6 and 13.3 apply.

Let $V = \langle d * b \rangle$, $M = H\langle d \rangle$, and $M^* = M/Z(M)$. By 9.6.4, M^* is the split extension of $O_2(M^*) = Q^* \cong E_{m^2}$ by $P\langle b \rangle^* \cong S_3$ with $V^* \cong E_{2m}$, $Z(M) = V \cap V^g$ for $g \in D_d \cap A_b$, and M^* is determined up to isomorphism by m. Indeed if $m = 2$ then by 9.6.5, $M^* \cong H \cong S_4$ so $\langle d \rangle = Z(M)$.

So assume $m = 4$. Then $d * b = \{d\} \cup b^Q = \{b_0, \ldots, b_4\}$ is of order $m + 1 = 5$ with $b^{*Q} = V^* - Q^*$. In particular $1 = b_1^* \cdots b_4^*$, so $x = b_0 \cdots b_4 \in Z(M)$. By symmetry, x centralizes H_b, so by 9.5.5, x centralizes $\langle H, H_b \rangle = G$. Passing to $G/Z(G)$, we may assume $Z(G) = 1$, so $1 = x = b_0 \cdots b_4$. Thus $d = b_0 = b_1 \cdots b_4 \in H$ and $V = \langle b_1, \ldots, b_4 \rangle$ is of rank at most 4. So as $V^* \cong E_8$ and $d \in V$, $Z(M) = V \cap V^g = \langle d \rangle$.

Hence in either case $Z(M) = \langle d \rangle$ and M^* is determined up to isomorphism by m. As $\langle d \rangle = Z(M)$, $P\langle b \rangle = \langle \Gamma \rangle$, and $C_{Q^*}(P^*) = 1$, we have $N_{Q\langle d \rangle}(\langle \Gamma \rangle) = \langle d \rangle$, so $\mathcal{D}(D)$ is a triple graph by 13.3.4. Then arguing as in Case I, $G/Z(G)$ is determined up to isomorphism as a 3-transposition group. Now by 11.9 and

11.12, S_6 and $U_4(2)$ satisfy our hypothesis for $m = 2$ and $m = 4$, respectively, so our proof is complete.

(13.6) *If $D_{d,c}$ is complete then $G \cong S_5$ as a 3-transposition group.*

Proof: The proof is by induction on the order of G, so let G be a minimal counterexample. Let M_0 be the largest normal 2,3-subgroup of H, $M = [H, M_0]$, and $H^* = H/M$. By 13.5 we may assume $\mathcal{D}(D_d)$ is connected, whereas as $\mathcal{D}(D)$ is connected, $D_d \neq \{b\}$. So by Exercise 3.8, H is not solvable and then by 9.4.4, H^* is primitive on D_d^*. Next let $e \in D_d \cap A_b$; then $D_{d,b,e} \subseteq D_{b,e}$, so as $\mathcal{D}(D_{b,e}) \cong \mathcal{D}(D_{d,c})$ is complete, $\mathcal{D}(D_{d,b,e})$ is complete and indeed $D_{b,e} = \{d\} \cup D_{d,b,e}$, so $\mu = |D_{d,c}| = 1 + |D_{d,b,e}|$. Hence by minimality of G, $H^* \cong S_5$. Therefore H^* has width 2 and $D_{d,b,e}^* = \{f^*\}$. Now $fM \cap D \subseteq D_{b,e}$ so $\mathcal{D}(fM \cap D)$ is complete and hence $[f, M]$ is a 2-group. Then as $M = [M_0, H]$, $M = O_2(H)$. So by 13.2, $d * f = \{d\} \cup (fM \cap D)$.

If $M \leq Z(H)$ then in the language of 6.3, $k = 10$, $\lambda = 3$, and $\mu = 2$, so by 6.3.3, $\delta^2 = 33$ is not a square, a contradiction to 6.3.3.

So $O_2(H) = M \nleq Z(H)$. Then by 13.2, $|d * f| > 2$. So as H^* has width 2 and $\mu = |D_{d,c}| = 1 + |D_{d,b,e}| \geq |d * f|$, $D_{d,c}$ intersects some singular line $d * b$ in at least two points, and hence $d \in D_c$ by 13.2.4, a contradiction.

(13.7) *If $O_3(H/O_2(H)) \neq Z(H/O_2(H))$ then $G \cong S_5, S_6, U_4(2)$, or $U_5(2)$ as a 3-transposition group.*

Proof: Let $Q = O_2(H)$ and $P/Q = O_3(H/Q)$. By 13.4 we may assume $O_3(H) \leq Z(H)$, and then assume that $Q \nleq Z(H)$ and $P/Q \nleq Z(H/Q)$. In particular by 13.2, $|d * b| > 2$ for each singular line $d * b$.

By 13.5 we may assume $\mathcal{D}(D_d)$ is connected, so H/P has width greater than 1. Hence there exists $b' \in D_{d,b} - b^P$. Let $x \in bP \cap A_b$ and $x' \in b'P \cap A_{b'}$. As the singular line $b * b'$ has at least three elements there is $y \in b * b' - \{b, b'\}$. Then y is not in b^P or $(b')^P$, say the former. Then $b', y \in D_x$. But now by 13.2.4, $b \in D_x$, a contradiction.

14. The case $O_2(\langle D_d \rangle) \nleq Z(\langle D_d \rangle)$

In this section, D is a conjugacy class of 3-transpositions of the finite group G with $Z(G) = 1 = O_3(G) = O_2(G)$. Let $d \in D$ and $H = \langle D_d \rangle$. In addition in this section we assume

Hypothesis Q. Let $Q = O_2(H)$. Assume $Q \nleq Z(H)$.

We prove:

Theorem Q.   *Assume Hypothesis Q. Then there is an isomorphism of 3-transposition groups $G \cong Sp_n(2)$ or $U_n(2)$ with $n \geq 4$, and D mapped to the set of transvections.*

The proof of Theorem Q is by induction on the order of G, so we take G to be a minimal counterexample to the theorem. We begin a series of reductions.

(14.1)   *G is primitive on D.*

Proof: Assume not. By 9.4.4, $\mathcal{D}(D)$ is disconnected. Let E be the connected component of $\mathcal{D}(D)$ containing d, $M = \langle E \rangle$, and $K = F^*(G)$. As $Z(G) = 1$, K is nonabelian simple and $M = \langle d \rangle K$ by 9.4. Notice $F^*(M) = K$, so $Z(M) = 1$. Thus by minimality of G, $M \cong Sp_n(2)$ or $U_n(2)$, $n \geq 4$, and E is the set of transvections in M. Therefore unless $M \cong Sp_4(2) \cong S_6$, $d \in K$, in which case $M = K$. If $M = K$ then certainly $M \trianglelefteq G$, whereas if $M \cong S_6$ this follows from Exercise 16.2 in [FGT]. Thus in any case $M \trianglelefteq G$.

As $M \trianglelefteq G$, $D = d^G \subseteq M$, so $G = M$. Hence $d^G = d^M = E$, contradicting $\mathcal{D}(D)$ disconnected.

In the remainder of this section let $c \in A_d$ and set $\Gamma = D_{c,d}$ and $L = \langle \Gamma \rangle$. Let $b \in \Gamma$ and $X = \langle D_b \cap \Gamma \rangle$. By 13.2, $d * b = \{d\} \cup b^Q$. Let $m = |b^Q|$, so that m is a power of 2, $|d * b| = m + 1$, and as $Q \not\leq Z(H)$, $m \geq 2$. Notice that if $u \in D_{b,c,d}$ then $b * u \subseteq \Gamma$ by definition of $b * u$. Write $b \circ u$ for the singular line through b and u defined by Γ. Recall

$$b * u = \bigcap_{y \in b^\perp \cap u^\perp} y^\perp \quad \text{and} \quad b \circ u = \bigcap_{y \in b^\perp \cap u^\perp \cap \Gamma} y^\perp \cap \Gamma.$$

Now $b^\perp \cap u^\perp \cap \Gamma \subseteq b^\perp \cap u^\perp$, so

$$b * u = \bigcap_{y \in b^\perp \cap u^\perp} y^\perp \subseteq \bigcap_{y \in b^\perp \cap u^\perp \cap \Gamma} y^\perp$$

so, as $b * u \subseteq \Gamma$, $b * u \subseteq b \circ u$. In particular $|b \circ u| > 2$. For $x \in D$, let $H_x = \langle D_x \rangle$ and $Q_x = O_2(H_x)$.

(14.2)

(1) $\mathcal{D}(D_d)$ is connected.
(2) $\mathcal{D}(\Gamma)$ is not complete.
(3) $O_3(H/Q) \leq Z(H/Q)$.

Proof: Suppose $\mathcal{D}(D_d)$ is disconnected or $\mathcal{D}(\Gamma)$ is not complete or $O_3(H/Q) \not\leq Z(H/Q)$. Then by 13.5, 13.6, and 13.7, $G \cong S_5$, S_6, $U_4(2)$, or $U_5(2)$. But if $G \cong S_5$ then $H \cap S_3$, so Hypothesis Q is not satisfied. Therefore as $\mathrm{Sp}_4(2) \cong S_6$ as a 3-transposition group, Theorem Q holds, contrary to the choice of G as a counterexample.

(14.3) L *is transitive on* Γ.

Proof: Assume otherwise and let $H^* = H/Q$. By 14.2, $\mathcal{D}(D_d)$ is connected and $O_3(H^*) \leq Z(H^*)$. Hence by 9.4.4, H^* is primitive on D_d^* and hence b^Q is a maximal set of imprimitivity for H on D_d. Then by 7.9, G is transitive on triples (x, y, z) with $z \in D$, $x, y \in D_z$, and $y \in A_x$. Hence $N_G(\Gamma)$ is transitive on Γ.

Now if L is not transitive on Γ then by 8.2.2, there is a partition $\Gamma = \Gamma_1 \cup \Gamma_2$ with $[\Gamma_1, \Gamma_2] = 1$. Let $b_i \in \Gamma_i$. By 14.2, $\mathcal{D}(\Gamma)$ is not complete, so there is $a_i \in A_{b_i} \cap \Gamma_i$. By 13.2, $\{b_2\} = D_{b_1} \cap (a_1 * b_2)$, so $a_1 * b_2 - \{b_2\} \subseteq \Gamma_1$. But by symmetry, $a_1 * b_2 - \{a_1\} \subseteq \Gamma_2$, contradicting $|a_1 * b_2| > 2$.

(14.4) $O_3(L) \leq Z(L) \geq O_2(L)$.

Proof: By 14.3, G is transitive on triples (x, y, z) from D with $x, y \in D_z$ and $y \in A_x$. Hence as $\mathcal{D}(D_b)$ is connected there is $u \in D_{b,c,d}$. We observed earlier that $b * u \subseteq b \circ u$ and $|b \circ u| > 2$. By Exercise 3.2.1 and 9.2.2, $b \circ u = b^{O_2(L)} \cup u^{O_2(X)}$, so as $|b \circ u| > 2$, either $O_2(L) \not\leq Z(L)$ or $O_2(X) \not\leq Z(X)$. Therefore by 9.3 and Exercise 3.2.2, $O_3(L) \leq Z(L)$.

So we may assume $O_2(L) \not\leq Z(L)$, and it remains to derive a contradiction. By 9.2,

$$b^{O_2(L)} = \{y \in \Gamma : b^\perp \cap \Gamma = y^\perp \cap \Gamma\},$$

so $b^{O_2(L)} = b \circ y$ for $b \neq y \in b^{O_2(L)}$. Thus $b * y \subseteq b \circ y = b^{O_2(L)}$. So $\Delta = b^{O_2(L)} - \{b\}$ is partitioned by the sets $\Delta(y) = b * y - \{b\}$ for $y \in \Delta$. However $|\Delta(y)| = m \geq 2$ is a power of 2, so $|b^{O_2(L)}| \equiv 1 \bmod m$, contradicting $|b^{O_2(L)}|$ a power of 2.

(14.5) *Either*

   (1) $L/Z(L) \cong \mathrm{Sp}_n(2)$ *with* $n \geq 4$ *and* $m = 2$, *or*
   (2) $L/Z(L) \cong U_n(2)$ *with* $n \geq 4$ *and* $m = 4$.

In any case b projects on a transvection of $L/Z(L)$, $b * u = b \circ u$, *and* $O_2(X) \leq Q_b Z(L)$.

Proof: Again pick $u \in D_{b,c,d}$. As $O_2(L) \leq Z(L)$, $b \circ u = \{b\} \cup u^{O_2(X)}$ by Exercise 3.2.1 and 9.2.2. Then as $|b \circ u| > 2$, $O_2(X) \not\leq Z(X)$. So by 14.4 and minimality of G, $L/Z(L) \cong \mathrm{Sp}_n(2)$ or $U_n(2)$ with b projecting on a transvection. By 11.9, $|b \circ u| = 3, 5$ for $L/Z(L)$ symplectic, unitary, respectively. If $O_2(X) \leq Q_b Z(L)$ then $b \circ u = \{b\} \cup u^{O_2(X)} \subseteq \{b\} \cup u^{Q_b} = b * u$, so as $b * u \subseteq b \circ u$, we have $b * u = b \circ u$. Then $m = 2, 4$ for the respective choice of $L/Z(L)$.

So it remains to show $O_2(X) \leq Q_b Z(L)$. But as $L/Z(L) \cong \mathrm{Sp}_n(2)$ or $U_n(2)$, X is irreducible on $O_2(X)Z(L)/Z(L)\langle b \rangle$ unless $L/Z(L) \cong U_4(2)$, where at least $C_L(b)$ is irreducible; for example when $n \geq 6$ this follows from 14.8 by induction on the order of G or in general (and in particular in the remaining cases) from the fact that $O_2(X)Z(L)/Z(L)\langle b \rangle$ is the natural module for $X/O_2(X)Z(L) \cong \mathrm{Sp}_{n-2}(2)$ or $SU_{n-2}(2)$. Also as $b * u \subseteq b \circ u$, $[O_2(X) \cap Q_b, u] \neq 1$. So $O_2(X) \leq Q_b Z(L)$.

(14.6) $H = LQ$.

Proof: Let $Z_b/Q_b = Z(H_b/Q_b)$ and $\bar{H}_b = H_b/Z_b$. As $L/Z(L) \cong \mathrm{Sp}_n(2)$ or $U_n(2)$, we have $X/O_2(X)Z(L) \cong \mathrm{Sp}_{n-2}(2)$ or $SU_{n-2}(2)$. Let $u \in D_{d,b,c}$. If $u \in d * b$ then as $u, b \in D_c$, we get $d \in u * b \subseteq D_c$ by 13.2.4, a contradiction. So $u \not\subseteq b * d$, and hence $(d * u) \cap (d * b) = \{d\}$. Thus the image $\overline{d * u}$ of $d * u$ in $\bar{H}_b$ has order $m + 1 > 2$. So as $\overline{d * u} \subseteq \bar{d} * \bar{u}$, we have $|\bar{d} * \bar{u}| > 2$. By 14.2.3, $O_3(\bar{H}_b) = 1$ and of course $O_2(\bar{H}_b) = 1 = Z(\bar{H}_b)$. As $|\bar{d} * \bar{u}| > 2$, $O_2(\langle \bar{D}_{d,b} \rangle) \not\leq Z(\langle \bar{D}_{d,b} \rangle)$ by Exercise 3.2. Hence by minimality of G, $\bar{H}_b \cong \mathrm{Sp}_m(2)$ or $U_m(2)$. Also $\langle \bar{D}_{b,d,c} \rangle = \bar{X}$ and $O_2(X) = Q_b O_2(L) \cap X$ by 14.5, so $\mathrm{Sp}_{n-2}(2), U_{n-2}(2) \cong X/X_0 \cong \bar{X}/Z(\bar{X})$, where $X_0/O_2(X) = Z(X/O_2(X))$. So by 11.12, $m = n$; that is $\bar{H}_b \cong \mathrm{Sp}_n(2)$ or $U_n(2)$, respectively. Therefore $H/Z_d \cong \bar{H} \cong L/Z(L)$, so as $L \leq H$, we have $H = LQ$.

Let $P = Q\langle d \rangle$ and $\tilde{P} = P/\langle d \rangle$. Similarly for $x \in D$ let $P_x = Q_x \langle x \rangle$.

(14.7) P *is regular on* A_d *and* $|P| = 2m^n$.

Proof: By 14.5 and 14.6, the hypotheses of 13.3 are satisfied, so by 13.3, P is regular on A_d. So in the notation of 6.3, $l = |P|$. Also by 6.3, $l = k(k-\lambda-1)/\mu$, so it remains to calculate k, λ, and μ. Of course $\mu = |\Gamma| = v(n)$, where

$$v(n) = 2^n - 1 \quad \text{or} \quad \frac{(2^n - (-1)^n)(2^{n-1} - (-1)^{n-1})}{3}$$

for $L/Z(L)$ symplectic or unitary, respectively (cf. 11.8). Again by 13.3, Γ intersects each line through d in a unique point, so $k = m\mu$. Similarly $D_{d,b}$

consists of $d*b-\{d,b\}$ of order $m-1$ together with the ν lines $d*u, u \in D_b \cap \Gamma$. So $\lambda = m - 1 + m\nu$. Notice $\nu = m\nu(n-2)$. Now

$$l = \frac{k(k-\lambda-1)}{\mu} = m^2(\mu - \nu - 1) = 2m^n.$$

(14.8) $O_2(L) = 1$ and L acts faithfully and irreducibly on $\tilde{P}$ with $\Phi(\tilde{P}) = 1$ and $|[b, \tilde{P}]| = m$.

Proof: Certainly L acts on P and $\langle d \rangle$, and hence also on $\tilde{P}$. Let $u, w \in D_b \cap \Gamma$ with $w \in A_u$. Then uw is faithful on $O_2(X)$, so as $L/Z(L)$ is simple or S_6, $C_L(Q) \le Z(L)$. As $H = LQ$, we have $O_2(L) \le C_Q(c)$ and $C_L(Q) \le Z(H)$. But $C_P(c) = 1$ as P is regular on A_d. Thus $O_2(L) = 1$. But $O^2(C_L(\tilde{P})) \le C_L(P) \le Z(H)$, and therefore $C_L(\tilde{P}) = O^2(C_L(\tilde{P}))$ fixes $c^P = A_d$ and $d^\perp$ pointwise. That is $C_L(\tilde{P}) = 1$.

Let $a, b, r, s \in \Gamma$ with $r \in A_a$, $s \in A_b$, and $a, r \in D_{b,s}$. Set $g = bs$ and $h = ar$. Then by induction, $N_L(\langle g \rangle) \cap C_G(a)$ is irreducible on $[(P_a \cap L)/\langle a \rangle, g]$ of order m^2 and X is irreducible on $(P_b \cap L)/\langle b \rangle$ of order m^{n-2} unless $L/Z(L) \cong U_4(2)$, where at least $C_L(b)$ is irreducible. So as $|\tilde{P}| = m^n$ by 14.7, $P_b \cap L = C_{P_b}(cd)$ and $C_G(\langle b, cd \rangle)$ is irreducible on $C_{P_b}(cd)/\langle b \rangle$. Also $C_G(\langle cd, a \rangle) \cap N_G(\langle g \rangle)$ is irreducible on $[C_{P_a}(cd)/\langle a \rangle, g]$ of order m^2. Hence conjugating in G, $C_G(\langle d, g \rangle)$ is irreducible on $C_{\tilde{P}}(g)$ and $C_G(\langle d, g \rangle) \cap N_G(\langle h \rangle)$ is irreducible on $[C_{\tilde{P}}(g), h]$ of order $m^2 = |\tilde{P} : C_{\tilde{P}}(g)|$. So Exercise 1.4 says L is irreducible on $\tilde{P}$ and $\Phi(\tilde{P}) = 1$.

(14.9) If $L/Z(L) \cong \mathrm{Sp}_n(2)$ then $G \cong \mathrm{Sp}_{n+2}(2)$ and D is the set of transvections in G.

Proof: By 14.8 and 11.13, $\tilde{P}$ is the natural module for $L \cong \mathrm{Sp}_n(2)$. In particular $N_P(L) = \langle d \rangle$, so by 13.3.4, $\mathcal{D}(D)$ is a triple graph. Also by 2.1, $H\langle d \rangle/\langle d \rangle$ is determined up to isomorphism. So arguing as in the last two paragraphs of Case I of 13.5, (G, D) is determined up to isomorphism as a 3-transposition group. But by 11.9 and 11.12, $\mathrm{Sp}_{n+2}(2)$ satisfies our hypothesis, so $G \cong \mathrm{Sp}_{n+2}(2)$ as a 3-transposition group.

(14.10) If $L/Z(L) \cong U_n(2)$ then $G \cong U_{n+2}(2)$, P is extraspecial, and D is the set of transvections in G.

Proof: We first show P is extraspecial. For $\langle b \rangle = \Phi(O_2(X))$, so $\langle d \rangle \le \Phi(P)$. But by 14.8, $\Phi(\tilde{P}) = 1$, so $\langle d \rangle = \Phi(P)$. Further by 14.8, L is irreducible on $\tilde{P}$. Hence P is extraspecial.

Now by 23.10 in [FGT], $\tilde{P}$ is an orthogonal space over $GF(2)$. So by 14.8 and 11.16, $\tilde{P}$ is the natural module for $L \cong SU_n(2)$. Now complete the proof as in 14.9.

Notice that 14.9 and 14.10 complete the proof of Theorem Q.

15. The case $O_2(\langle D_d \rangle) \le Z(\langle D_d \rangle) \ge O_3(\langle D_d \rangle)$

In this section D is a set of 3-transpositions of the finite group G and as usual $d \in D$ and $H = \langle D_d \rangle$. In addition in this section we assume

Hypothesis Z. *G is primitive on D, $Z(G) = 1$, and $O_3(H) \le Z(H) \ge O_2(H)$.*

We prove:

Theorem Z. *Assume Hypothesis Z. Then one of the following holds:*

(1) $G \cong S_n$, $n \ge 7$, and D is the set of transpositions of G.

(2) $G \cong O_n^\epsilon(2)$, $n \ge 6$, and D is the set of transvections of G.

(3) $G \cong PO_n^{\mu,\pi}(3)$, $n \ge 5$, and D is a conjugacy class of reflections.

(4) G is of type $M(22)$, $M(23)$, or $M(24)$, G is determined up to isomorphism, and D is determined up to conjugation in $\mathrm{Aut}(G)$.

The proof of Theorem Z is by induction on the order of G, so we assume all proper D-sections of G satisfying Hypothesis Z also satisfy one of the conclusions of Theorem Z. We begin a series of reductions.

Let $c \in A_d$, $\Gamma = D_{c,d}$, $L = \langle \Gamma \rangle$, $b \in \Gamma$, and $X = \langle D_b \cap \Gamma \rangle$.

(15.1)

(1) H is primitive on D_d.

(2) $\mathcal{D}(\Gamma)$ is not complete.

Proof: By Hypothesis Z, $O_2(H) \le Z(H) \ge O_3(H)$. However if $\mathcal{D}(D_d)$ is disconnected then by 13.5, G is S_5, S_6, $U_4(2)$, or $U_5(2)$. On the other hand if G is S_5 then $O_3(H) \not\le Z(H)$, and in the remaining cases, $O_2(H) \not\le Z(H)$, a contradiction. Thus $\mathcal{D}(D_d)$ is connected and $O_2(H) \le Z(H) \ge O_3(H)$, so H is primitive on D_d by 9.4.4. Part (2) holds by 13.6.

(15.2) $N_G(\Gamma)$ *is transitive on* Γ.

Proof: By 9.5.4 and 15.1.1, $C_H(b)$ is transitive on $D_d \cap A_b$, so G is transitive on triples (x, y, z) with $z \in D$, $x, y \in D_z$, and $y \in A_x$.

(15.3) *If L is not transitive on* Γ *then* $H/Z(H)$ *is* S_6, $U_4(2)$, $U_5(2)$, $O_6^-(2)$, *or* $PO_6^{-,\pi}(3)$ *and L is isomorphic to* $S_3 \times S_3$, $S_3 \times S_3$, $SU_3(2)' * SU_3(2)'$, $(S_3)^3$, *or* $(S_3)^4$, *respectively.*

Proof: Assume L is not transitive on Γ. Let $\Gamma_1, \ldots, \Gamma_r$ be the orbits of L on Γ and $L_i = \langle \Gamma_i \rangle$. Choose notation so that $b \in \Gamma_1$. By 15.2, $N_G(L)$ is transitive on $\{\Gamma_1, \ldots, \Gamma_r\}$ so $L_i \cong L_1$ for all i. By 15.1.2, $|L_i| > 2$. We claim

(15.3.1) *Either*

  (a) $r = 2$ *and* $X = L_2$ *has width 1, or*
  (b) X *is not transitive on* $D_b \cap \Gamma$.

For, assume X is transitive on $D_b \cap \Gamma$. Then as $\Gamma_i \subseteq D_b \cap \Gamma$ for $i > 1$, $r = 2$ and $\Gamma_2 = D_b \cap \Gamma$. Thus L_1 has width 1, so as $L_2 \cong L_1$, (a) holds.

Let $w \in D_d \cap A_b$, $\Delta = D_{d,b,w}$, and $Y = \langle \Delta \rangle$. Then $Y \cong X$, so by 15.3.1, either Y has width 1 or Y is not transitive on Δ.

If $H/Z(H) \cong S_5$ then $Y \cong Z_2$, so Y is transitive on Δ and hence 15.3.1.a holds with $X \cong Y \cong \mathbf{Z}_2$. But then $|\Gamma| = 2$, contradicting 15.1.2. Therefore $H/Z(H) \neq S_5$, so by 13.4, $O_3(\langle D_{d,b} \rangle) \leq Z(\langle D_{d,b} \rangle)$.

If $O_2(\langle D_{d,b} \rangle) \not\leq Z(\langle D_{d,b} \rangle)$ then by Theorem Q, $H/Z(H) \cong Sp_n(2)$ or $U_n(2)$. So by 14.3, Y is transitive on Δ, and hence Y has width 1. Therefore $H/Z(H) \cong S_6$, $U_4(2)$, or $U_5(2)$, and $Y \cong S_3$, S_3, or $SU_3(2)'$, respectively (cf. 11.12). Hence the lemma holds by 15.3.1.

This leaves the case $O_p(\langle D_{d,b} \rangle) \leq Z(\langle D_{d,b} \rangle)$, for $p = 2, 3$, where by minimality of G, $H/Z(H)$ is on the list of Theorem Z. In particular Y has width greater than 1, so Y is not transitive on Δ and hence (cf. 15.6 applied in an inductive context to H) $H/Z(H) \cong O_6^-(2)$ or $PO_6^{-,\pi}(3)$ and $Y \cong (S_3)^2$ or $(S_3)^3$, respectively. But $L_2 \cdots L_r$ is isomorphic to a normal subgroup of Y, so $L_i \cong S_3$ and $L \cong S_3 \times Y$, completing the proof.

(15.4) $d * c = \{d, c, c^d\}$, *so* $\mathcal{D}(D)$ *is a triple graph.*

Proof: Let $w \in A_b \cap D_d$, $\Delta = \{b, w, w^b\}$, and write $b \circ w$ for the line through b and w determined by D_d. Thus $\Delta \subseteq b * w \subseteq b \circ w$ (cf. an argument following 14.1). Now if $H/Z(H) \cong S_5$, $Sp_n(2)$, or $U_n(2)$ then $b \circ w = \Delta$, so the lemma holds. If $H/Z(H)$ is none of these groups then by 15.1.1, 13.4, and Theorem Q, H, D_d, satisfy Hypothesis Z, so by minimality of G, we again have $b \circ w = \Delta$.

(15.5) $\langle d \rangle = C_G(H\langle d \rangle)$.

Proof: This is a consequence of 15.4 and 12.1.

(15.6) *One of the following holds:*

   (1) L is transitive on Γ.

   (2) $H/Z(H) \cong S_6$, $L \cong S_3 \times S_3$, and $G \cong O_6^-(2)$.

   (3) $H/Z(H) \cong O_6^-(2)$, $L \cong S_3 \times S_3 \times S_3$, and $G \cong PO_6^{-,\pi}(3)$.

Proof: Assume otherwise and let $K = H\langle d \rangle$ and $K^* = K/\langle d \rangle$. Then K^* and L are described in 15.3.

We first claim that Γ^* is determined up to conjugation in K^*. For example if K^* is not $U_5(2)$ then $L = L_1 \times \cdots \times L_r$ with $L_i \cong S_3$ and $L_1 = \langle b, w \rangle$. As H is primitive on D_d, L_1 is determined up to conjugation by 9.5.4; then $\langle D_{d,b,w} \rangle = L_2 \cdots L_r$, so the claim holds. On the other hand if $K^* \cong U_5(2)$, then H has no $SU_3(2)' * SU_3(2)'$ subgroup, since a Sylow 3-subgroup of $U_5(2)$ is isomorphic to $\mathbf{Z}_3 \times (\mathbf{Z}_3 \ wr \ \mathbf{Z}_3)$. So K^* is not $U_5(2)$.

As (K^*, D_d^*) is determined up to isomorphism as a 3-transposition group and Γ^* is determined up to conjugation in K^*, it follows from Theorem D, 12.2, and 15.4 that G is determined up to isomorphism as a 3-transposition group by K^* and Γ^*. If $K^* \cong S_6$ or $O_6^-(2) \cong O_5^{\pi,\pi}(3)$ (cf. Exercise 4.14) then $O_6^-(2)$ or $PO_6^{-,\pi}(3)$ satisfies our hypothesis, respectively (cf. 11.11 and 11.12), so $G \cong O_6^-(2)$, $PO_6^{-,\pi}(3)$, respectively.

Suppose $K^* \cong U_4(2)$. Then, in the notation of 6.3, $k = 45$, $\lambda = 12$, and $\mu = 6$ (cf. 11.8) so $\Delta = (\mu - \lambda)^2 + 4(k - \mu)$ is not a square, contradicting 6.3. Similarly if $K^* \cong PO_6^{-,\pi}(3)$ then by Exercise 4.2.4 the sign ϵ of K^* is $+1$, so by Exercise 4.13.2, $k = 117$, $\lambda = 36$, and $\mu = 12$, so Δ is not a square.

In the remainder of this section let Δ be the connected component of $\mathcal{D}(\Gamma)$ containing b, $w \in D_d \cap A_b$, $K = \langle \Delta \rangle$, and $Y = \langle D_{d,b,w} \rangle$. Thus $X \cong Y$.

(15.7) *If L is transitive on Γ and Y is transitive on $D_{d,b,w}$, $|D_{d,b,w}| > 1$, and $O_3(Y) \leq Z(Y)$, then $O_3(K) \leq Z(K) \geq O_2(K)$, $F^*(K) = F^*(L) = E(L)Z(L)$ with $E(L)$ quasisimple, and K is primitive on Δ.*

Proof: As $X \cong Y$ and $O_3(Y) \leq Z(Y)$, we have $O_3(X) \leq Z(X)$. Similarly Y is transitive on $D_{d,b,w}$ and $|D_{d,b,w}| > 1$, so X is transitive on $D_b \cap \Gamma$ and $|D_b \cap \Gamma| > 1$. Therefore by Exercise 3.3, $O_3(L) \leq Z(L) \geq O_2(L)$, so by 9.4, $F^*(L) = E(L)Z(L)$ with $E(L)$ quasisimple. Then $F^*(L) \leq \langle \Delta \rangle = K$, so $F^*(K) = F^*(L)$ and as $\mathcal{D}(\Delta)$ is connected, K is primitive on Δ by 9.4.4.

(15.8) *Assume $H/Z(H) \cong S_n$. Then either*

   (1) $L/Z(L) \cong S_{n-1}$, $G \cong S_{n+2}$, and D is the set of transpositions, or

   (2) $L \cong S_3 \times S_3$, $G \cong O_6^-(2)$, and D is the set of transvections.

Proof: Recall case (2) arises in 15.6; because of 15.6 we may assume L is transitive on Γ. Also $X \cong Y$ and $Y/Z(Y) \cong S_{n-3}$, so either $O_3(Y) \leq Z(Y)$ or $n \leq 6$. Hence by 15.7, either $O_p(K) \leq Z(K)$ for $p = 2$ and 3, and K is primitive on Δ, or $n \leq 6$. If $n > 6$ then as $X/Z(X) \cong Y/Z(Y) \cong S_{n-3}$, by minimality of G, either $K/Z(K) \cong S_{n-1}$ or $n = 9$ and $K/Z(K) \cong O_6^-(2)$. As $|O_6^-(2)|$ does not divide $|S_9|$, the last case is out. So $K/Z(K) \cong S_{n-1}$. Similarly if $n = 6$ and $O_3(K) \leq Z(K)$, then by Exercise 3.2, $O_p(K) \leq Z(K)$ for $p = 2$ and 3, so K is primitive on Δ by 9.4.4, and then $K/Z(K) \cong S_5$ by 13.4. Finally if $n = 5$ then $b^\perp \cap \Gamma = \{a, b\}$ is of order 2 and for $e \in A_b \cap \Gamma$, (a, e, b) is a 3-string, so $K_0 = \langle a, e, b \rangle \cong S_4$ by 8.3. Then as $K_0 Z(H)/Z(H)$ is maximal in $H/Z(H) \cong S_5$, $K_0 = K$. So in each of these cases, $K/Z(K) \cong S_{n-1}$ and then by Exercise 5.8 in [FGT], K is determined up to conjugation as a D-subgroup of H and K is a maximal D-subgroup of H, so $L = K$. Now Theorem D (Section 12) implies G is determined up to isomorphism, so (1) holds.

So assume $n = 6$ and $O_3(K) \not\leq Z(K)$. Then $O_3(L) \not\leq Z(L)$, so as L is a D-subgroup of H, $L \leq \langle N_D(U) \rangle$ for some $U \in \mathrm{Syl}_3(H)$. But then $L \cong S_3 \times S_3$, so L is not transitive on Γ, contrary to paragraph one of this proof.

(15.9)   *If $H/Z(H) \cong \mathrm{Sp}_n(2)$ with $n \geq 6$ then $L \cong O_n^\epsilon(2)$, $G \cong O_{n+2}^{-\epsilon}(2)$, and D is the set of transvections.*

Proof: By 11.12, $Y/Z(Y) \cong \mathrm{Sp}_{n-2}(2)$, so by 15.7, $O_2(K) \leq Z(K) \geq O_3(K)$ and K is primitive on Δ. Then by minimality of G, $K/Z(K) \cong O_n^\epsilon(2)$. By 11.14, $K\langle d \rangle$ is determined up to conjugation in $H\langle d \rangle$, so G is determined up to isomorphism by Theorem D. Finally by 11.11 and 11.12, $O_{n+2}^{-\epsilon}(2)$ satisfies our hypothesis, so the lemma holds.

(15.10)   *Let T be the set of transvections in $U \cong U_6(2)$. Then up to conjugation in U there are three T-subgroups B of U with $B/Z(B) \cong PO_6^{+,\pi}(3)$, and $\mathrm{Aut}(U)$ permutes the three classes as S_3.*

Proof: This follows from 29.9.3.

(15.11)   *If $H/Z(H) \cong U_n(2)$ for $n \geq 5$, then $n = 6$, $L/Z(L) \cong PO_6^{+,\pi}(3)$, G is of type $M(22)$, G is determined up to isomorphism, and D is determined up to conjugation in $\mathrm{Aut}(D)$.*

Proof: By 11.12, $Y/Z(Y) \cong U_{n-2}(2)$, so by 15.7, either $O_2(K) \leq Z(K) \geq O_3(K)$ and K is primitive on Δ, or $n = 5$. In the latter case $O_3(Y) \not\leq Z(Y)$, so by Exercise 3.2.3, $O_2(L) \leq Z(L)$. Then $O_3(L) \not\leq Z(L)$ by 13.4. But L is a D-subgroup of H with $X \cong SU_3(2)'$ and L is transitive on $L \cap D$, so by Exercise 5.2, $L \cong S_5/E_{81}$ and $|\Gamma| = 30$. But $C_H(\langle b, w \rangle)$ contains a Q_8-subgroup irreducible on $O_3(Y)/Z(Y)$, so $C_H(\langle c, b \rangle)$ contains a Q_8-subgroup R irreducible on $O_3(X)/Z(X)$. Then R acts on L, so, as $O_3(X) \cap O_3(L) \not\leq Z(X)$, $O_3(X) = [O_3(X) \cap O_3(L), R] \leq O_3(L)$, contradicting $O_3(X) \cong 3^{1+2}$ and $O_3(L)$ abelian.

So $n > 5$, $O_2(K) \leq Z(K) \geq O_3(K)$, and K is primitive on Δ. Then by 11.12, $X/Z(X) \cong Y/Z(Y) \cong U_{n-2}(2)$ with $n - 2 \geq 4$, so by minimality of G, $n = 6$ or 8. Suppose first $n = 6$. Then as $U_4(2) \cong PO_5^{-\pi,\pi}(3)$ by Exercise 4.14, we conclude from 15.13 applied in an inductive setting that $K/Z(K) \cong PO_6^{+,\pi}(3)$. As $|U_6(2)|_3 = 3^6 = |PO_6^{+,\pi}(3)|_3$, $L = K$. Hence G is of type $M(22)$. By 15.10, $K\langle d \rangle/\langle d \rangle$ is determined up to conjugacy in $\mathrm{Aut}(H\langle d \rangle/\langle d \rangle)$, so by Theorem D, G is determined up to isomorphism and D is determined up to conjugation in $\mathrm{Aut}(G)$.

This leaves the case $n = 8$, where by minimality of G, K is of type $M(22)$. But then by Remark 16.8, 13 divides $|K|$, impossible as $K \leq H$, and 13 does not divide the order of $U_8(2)$.

(15.12) $H/Z(H)$ *is not isomorphic to* $O_n^\epsilon(2)$ *for* $n \geq 8$.

Proof: Assume otherwise. Then by 11.12, $Y/Z(Y) \cong O_{n-2}^{-\epsilon}(2)$. So by 15.7, $O_2(K) \leq Z(K) \geq O_3(K)$ and K is primitive on Δ. Then by minimality of G, $n = 8$ and either $\epsilon = +1$, $X/Z(X) \cong O_6^-(2) \cong O_5^{\pi,\pi}(3)$, and $K/Z(K) \cong PO_6^{-,\pi}(3)$, or $\epsilon = -1$, $X/Z(X) \cong O_6^+(2) \cong S_8$, and $K/Z(K) \cong S_{10}$. In either case $|K/Z(K)|$ does not divide $|H/Z(H)|$, a contradiction.

(15.13)   *If* $H/Z(H) \cong PO_n^{\nu,\pi}(3)$ *with* $n \geq 5$ *then* $L \cong O_{n-2}^{-\nu,\pi}(3)/E_{3^{n-2}}$, *and* $G \cong PO_{n+1}^{-\nu\pi,\pi}(3)$.

Proof: Notice the case $H/Z(H) \cong PO_5^{\pi,\pi}(3) \cong O_6^-(2)$ occurs in both 15.6.3 and 15.13 as $O_3^{-\pi,\pi}(3)/E_{27} \cong S_3^3$ by 11.12. Thus we assume L is transitive on Γ.

By 11.12, if $(n, \nu\pi) \neq (5, -1)$, then $Y/Z(Y) \cong O_{n-3}^{\pi\nu,\pi}(3)/E_{3^{n-3}}$. Further Y is isomorphic to S_3, S_3^2, S_3^3 for $(n, \nu\pi^n)$ equal to $(5, -1)$, $(5, 1)$, $(6, -1)$, respectively, by 11.12, and in all other cases Y is transitive on $D_{d,b,w}$. So in any case $O_3(X) \not\leq Z(X)$ and hence by Exercise 3.2.3, $O_2(L) \leq Z(L)$.

Then by 13.4, $O_3(L) \not\leq Z(L)$ unless possibly $(n, \nu\pi) = (5, -1)$ and $L \cong S_5$. In the latter case in the notation of 6.3, $k = 45$, $\lambda = 12$ (cf. Exercise 4.13.2) and $\mu = 10$, so by 6.3, $|A_d| = k(k - \lambda - 1)/\mu = 144$. But by Exercise 5.3, $|L^H| = 216$, whereas $N_H(L) = N_H(d * c)$ is of index $|A_d|/2 = 72$ in H by 15.4, a contradiction.

So $O_3(L) \nleq Z(L)$. Let $L^* = L/O_3(L)$, and for $x \in \Gamma$ write

$$\mathcal{W}_x = \{a \in \Gamma : \Gamma \cap D_x = \Gamma \cap D_a\}.$$

Then $\mathcal{W}_x = x^{O_3(L)}$ by 9.2.3. Take $x \in D_b \cap \Gamma$; then $\mathcal{W}_x \subseteq x^{O_3(X)}$ by Exercise 3.1.2. But $X \cong Y$ and $H/Z(H) \cong PO_n^{\nu,\pi}(3)$, so $|y^{O_3(Y)}| = 3$ for $y \in D_{d,b,w}$ by 11.12.4. Thus $|x^{O_3(X)}| = 3$, so $x^{O_3(L)} = \mathcal{W}_x = x^{O_3(X)}$ is of order 3. Further

$$O_3(X) = \langle xy : x \in D_b \cap \Gamma, \ y \in \mathcal{W}_x \rangle,$$

so $O_3(X) \leq O_3(L)$ and $X^* \cong X/O_3(X)$. Hence by a remark in paragraph two of the proof of this lemma, $X^* \cong O_{n-3}^{\nu\pi,\pi}(3)$ unless $(n, \nu\pi) = (5, -1)$, where $X^* \cong \mathbf{Z}_2$. Similarly there are $w \in \mathcal{W}_b$ and $bw \in C_{O_3(L)}(O_3(X))$, so $|O_3(L)| \geq 3|O_3(X)| \geq 3^{n-2}$.

If Y is transitive on $D_{d,b,w}$ then X^* is transitive on $(D_b \cap \Gamma)^*$, so as $O_3(X^*) = 1$, Exercise 3.3 says $O_2(K^*) \leq Z(K^*) \geq O_3(K^*)$, and then by 9.4.4, K^* is primitive on Δ^*. Next if $n > 7$, then $X^* \cong O_{n-3}^{\nu\pi,\pi}(3)$ satisfies $O_3(X^*) = O_2(X^*) = 1$, so by minimality of G, $K^*/Z(K^*) \cong PO_{n-2}^{-\nu,\pi}(3)$. Similarly if $(n, \nu\pi^n) = (7, -1)$, then $\nu = -\pi$ and $X^* \cong O_4^{-1,\pi}(3) \cong S_6$, so by 15.8,

$$K^* \cong O_6^-(2) \cong O_5^{\pi,\pi}(3) \cong O_5^{-\nu,\pi}(3)$$

(with the isomorphism $O_6^-(2) \cong O_5^{\pi,\pi}(3)$ coming from Exercise 4.14) or $K^* \cong S_8$. The case $K^* \cong S_8$ is impossible as $L \leq H$, whereas no 3-local in $H/Z(H) \cong PO_7^{-\pi,\pi}(3)$ contains a D-section isomorphic to S_8 by Exercise 4.18.

If $(n, \nu\pi^n) = (7, 1)$ or $(6, 1)$ then $\nu = \pi$ or 1, respectively, and

$$\{b^*\} = \{a^* \in \Gamma^* : \Gamma^* \cap b^{*\perp} = \Gamma^* \cap a^{*\perp}\},$$

so by 9.2.2, $O_2(K^*) \leq Z(K^*)$. Then by Theorem Q, $K^*/Z(K^*) \cong U_4(2) \cong PO_5^{-\nu,\pi}(3)$ (cf. Exercise 4.14) or $S_6 \cong PO_4^{-\nu,\pi}(3)$.

In the remaining cases where $(n, \nu\pi^n) = (5, -1)$, $(5, 1)$, or $(6, -1)$, we saw in paragraph two of the proof of this lemma (keeping in mind that $O_3(X) = O_3(L)$, so $O_3(X^*) = 1$) that X^* is E_{2j}, where $j = |(D_b \cap \Gamma)^*| = 1$, 2, or 3, respectively. So by 9.6, K is a $\{2, 3\}$-group and $\mu = 3me$, where $m = j + 1$ is a power of 2 and $1 < e$ is a power of 3. As m is a power of 2, $j \neq 2$, so $(n, \nu\pi) \neq (5, 1)$. If $(n, \nu\pi) = (5, -1)$ then K^* contains an S_4-subgroup and $|O_3(L)| \geq 27$, so K contains a subgroup $K_0 \cong S_4/E_{27}$. Then as $H/Z(H) \cong PO_5^{-\pi,\pi}(3)$, 11.10.6 says $K_0\langle d \rangle/\langle d \rangle$ is determined up to conjugacy in $H/\langle d \rangle$ and is a maximal proper D-subgroup. The latter fact says $L = K = K_0$. A similar argument shows that if $(n, \nu\pi^n) = (6, -1)$ then $K/Z(K) \cong O_4^{-\nu,\pi}(3)/E_{3^4}$ and $L = K$.

So in all cases, 11.10.6 says $\langle d \rangle K/\langle d \rangle$ is determined up to conjugation under the group $\mathrm{Aut}(H\langle d \rangle/\langle d \rangle)$, and is a maximal D-subgroup of $H\langle d \rangle/\langle d \rangle$. Thus $L = K$ and by Theorem D, G is determined up to isomorphism. Hence the lemma is established.

(15.14)

   (1) *If $H/Z(H)$ is of type $M(22)$ then $L/Z(L) \cong PO_7^{-\pi,\pi}(3)$, G is of type $M(23)$, G is determined up to isomorphism, and D is determined up to conjugacy in* $\mathrm{Aut}(G)$.

   (2) *If $H/Z(H)$ is of type $M(23)$ then $L/Z(L) \cong S_3/P\Omega_8^+(3)$, $\mathcal{D}(\Gamma)$ is disconnected, G is of type $M(24)$, G is determined up to isomorphism, and D is determined up to conjugacy in* $\mathrm{Aut}(G)$.

   (3) *$H/Z(H)$ is not of type $M(24)$.*

Proof: First by definition of the Fischer groups, $Y/Z(Y) \cong PO_6^{+,\pi}(3)$, $PO_7^{-\pi,\pi}(3)$, or $S_3/P\Omega_8^+(3)$, respectively, so by 15.7, $O_\infty(K) = Z(K)$ and K is primitive on Δ. Then by minimality of G and 15.13, $K/Z(K) \cong PO_7^{-\pi,\pi}(3)$ or $PO_8^{+,\pi}(3)$ in the first two cases. In the third as $\mathcal{D}(\Gamma \cap D_b)$ is disconnected, 13.5 applied to K supplies a contradiction, establishing (3).

Let $H^* = H/Z(H)$. Consider the case H^* of type $M(22)$. Here $K^* \cong PO_7^{-\pi,\pi}(3)$ so by 16.3 and 16.5, K^* is a maximal subgroup of H^* determined up to conjugacy in $\mathrm{Aut}(H^*)$. So $L = K$ and Theorem D completes the proof as usual.

This leaves the case H^* of type $M(23)$. There, by 16.6 and 16.11, K^* is determined up to conjugacy in H^* with $N_{H^*}(F^*(K^*))$ a D-subgroup equal to K^* or isomorphic to $S_3/P\Omega_8^+(3)$. So it remains to assume $K = L$ and derive a contradiction. In that event by 16.7, 16.9, and 16.10,

$$k = 31,671 = 3^4 \cdot 17 \cdot 23 \quad \text{and} \quad \lambda = 3,510 = 2 \cdot 3^3 \cdot 13 \cdot 5,$$

and by Exercise 4.13.2, $\mu = 1,080 = 2^3 \cdot 3^3 \cdot 5$. But then

$$\delta^2 = (\lambda - \mu)^2 + 4(k - \mu)$$

is divisible by 27 but not 81, whereas by 6.3, δ is an integer.

We can now complete the proof of Theorem Z. Let $H^* = H/Z(H)$. By 15.1, H is primitive on D_d^* and by Hypothesis Z, $O_3(H^*) = O_2(H^*) = 1$. If $O_2(X^*) \not\le Z(X^*)$ then by Theorem Q, $H^* \cong \mathrm{Sp}_n(2)$ or $U_n(2)$ as a 3-transposition group, so that Theorem Z holds by 15.8 (recall $\mathrm{Sp}_4(2) \cong S_6$), 15.9, 15.11, and 15.13. (Recall $U_4(2) \cong PO_5^{-\pi,\pi}(3)$.)

So assume $O_2(X^*) \le Z(X^*)$. By 13.4 and 15.8, we may assume $O_3(X^*) \le Z(X^*)$. Hence H^* satisfies Hypothesis Z, so by induction on the order of G, H^* is on the list of Theorem Z. Now 15.8, 15.12, 15.13, and 15.14 complete the proof. (Recall $O_6^-(2) \cong PO_5^{\pi,\pi}(3)$ and $O_6^+(2) \cong S_8$.)

16. The Fischer groups

In this section D is a set of 3-transpositions of G with $Z(G) = 1$, $d \in D$, $H = \langle D_d \rangle$, $c \in A_d$, $\Gamma = D_{d,c}$, and $L = \langle \Gamma \rangle$.

We establish various properties of the Fischer groups. For example we calculate the number of classes of D-subgroups L of G with $L/Z(L) \cong PO_7^{-\pi,\pi}(3)$, $PO_8^{+,\pi}(3)$, when G is of type $M(22)$, $M(23)$, respectively (under the assumption that such subgroups exist). Recall that to prove Theorem Z we needed to know that $N_{\text{Aut}(G)}(D)$ was transitive on such D-subgroups. We also calculate the order of the Fischer groups and the order of D, although the calculation of the group order assumes either that groups of type $M(24)$ exist or that groups of type M are also of type $\tilde{M}$.

(16.1) If $G \cong U_6(2)$ and B is a D-subgroup of G with $B/Z(B) \cong PO_6^{+,\pi}(3)$ then $C_G(B) = 1$ and B is a maximal subgroup of G.

Proof: Regard G as the image of $\hat{G} = SU(V)$, where V is a 6-dimensional unitary space over the field F of order 4, and let $\hat{B}$ be the preimage of B in $\hat{G}$. By 29.9.1, $\hat{B}$ is irreducible on V with $End_{F\hat{B}}(V)^{\#} = Z(\hat{G})$, so $Z(\hat{G}) = C_{\hat{G}}(\hat{B})$ and hence $C_G(B) = 1$.

As $C_G(d) = C_B(d)O_2(C_G(d))$ and $C_B(d)$ is irreducible on $O_2(C_G(d))/\langle d \rangle$, $C_B(d)$ is maximal in $C_G(d)$. So to prove B maximal in G, it suffices by Exercise 3.6 to observe $|U_6(2)|_3 = |O_6^{+,\pi}(3)|_3$.

(16.2) Let G be of type $M(22)$ and $A = N_{\text{Aut}(G)}(D)$. Then

  (1) $A = \text{Aut}(\mathcal{D}(D))$.
  (2) $C_A(d)/\langle d \rangle$ is the split extension of $U_6(2)$ by a field automorphism.
  (3) $C_A(d)$ is transitive on the two H-classes of D-subgroups B of H such that $B \notin L^H$ and $B/Z(B) \cong PO_6^{+,\pi}(3)$.

Proof: Part (1) follows from 12.1. Let $C_A(d)^* = C_A(d)/\langle d \rangle$. By 12.1, $\langle d \rangle = C_A(H)$. Then by Exercise 4.5 and 15.10, $N_{\text{Aut}(H^*)}(L^*)$ is the split extension of L^* by a field automorphism. Further $C_A(d)^* \le H^* N_{\text{Aut}(H^*)}(L^*)$. Conversely assume $\phi \in N_{\text{Aut}(H^*)}(L^*)$. Now D_d^* is a primitive conjugacy class of 3-transpositions of H^* and ϕ induces an automorphism of $\mathcal{D}(D^*) = \mathcal{D}(D)$ via conjugation with $\phi(u^v) = \phi(u)^{\phi(v)}$ for all u, $v \in D_d$. By 15.4, $\mathcal{D}(D)$ is a triple graph and as ϕ acts on L^*, $\phi(D_{d,c}) = D_{d,c}$. Hence by Theorem D, $\phi = a^*$ for some $a^* \in C_A(d)^*$. Hence $C_A(d)^* = H^* N_{\mathrm{Aut}(H^*)}(L^*)$, so that (2) holds. Finally (2) and 15.10 imply (3).

(16.3) *Let G be of type $M(22)$, $A = N_{\mathrm{Aut}(G)}(D)$, and assume K is a D-subgroup of G with $K/Z(K) \cong PO_7^{-\pi,\pi}(3)$. Then*

(1) $C_A(K) = 1$.
(2) K is simple.
(3) K is a maximal subgroup of G.

Proof: Take $d \in K \cap D$ and let $B = \langle K \cap d^\perp \rangle$. By 11.11.2, $B/\langle d \rangle \cong PO_6^{+,\pi}(3)$ and of course $B \leq H\langle d \rangle$ with $H\langle d \rangle/\langle d \rangle \cong U_6(2)$. By 16.2, $C_A(d) = H\langle d, a \rangle$, where a induces a field automorphism on H and a is faithful on B. Hence by 16.1, $C_A(d) \cap C_A(B) = \langle d \rangle$. Then $C_A(K) \leq C_A(d) \cap C_A(B) = \langle d \rangle$, so (1) holds.

Let $d \in S \in \mathrm{Syl}_2(K)$. By (1) and Exercise 4.8, $|S \cap D| = 7$ and $\prod_{x \in S} x = 1$. Then by (1) and Exercise 3.4, (2) holds. Finally by 16.1, B is a maximal subgroup of H. By (2), K is simple, so $O_2(K) = O_3(K) = 1$ and by Exercise 3.6.2, K is a maximal D-subgroup of G, so in particular $K = \langle N_D(E(K)) \rangle$. Therefore Exercise 3.6.4 says (3) holds.

We can milk more from this argument, namely:

(16.4) *Let G be of type $M(22)$ and assume a group of type $M(23)$ exists. Then*

(1) G is simple, and
(2) H is quasisimple with $\langle d \rangle = Z(H)$.

Indeed if a group of type $M(23)$ exists then the D-subgroup K of G of 16.3 and its Sylow 2-subgroup S exists by the definition of groups of type $M(22)$ and the uniqueness of such groups established in Theorem Z. Then applying Exercise 3.4 to $S \cap D$ and G we conclude G is simple, whereas applying that exercise to $S \cap D$ and H we conclude H is quasisimple. Further $d = \prod_{d \neq x \in S \cap D} x \in H$, so by 16.2.2, $\langle d \rangle = Z(H)$.

(16.5) *Let G be of type $M(22)$. Then G has at most two classes of D-subgroups K with $K/Z(K) \cong PO_7^{-\pi,\pi}(3)$, and all classes are fused in $A = N_{\mathrm{Aut}(G)}(D)$.*

Proof: Let $\mathcal{X}$ be the set of such D-subgroups and pick $K \in \mathcal{X}$ with $d \in K$. Let $a \in A_d \cap K$, $b \in D_{d,a} \cap K$ and set $Y = \langle D_d \cap K \rangle$ and $Y_b = \langle D_b \cap K \rangle$.

Then $Y/Z(Y) \cong Y_b/Z(Y_b) \cong PO_6^{+,\pi}(3)$ by 11.11. Hence applying 11.12.4
to $Y_b/Z(Y_b)$, $W = \langle D_{a,d} \cap Y_b \rangle \cong S_4/E_{27}$. Hence as S_4 has 6-transposition,
$|D_{a,b,d} \cap K| = 6N$, where $N = |O_3(W) : O_3(W) \cap C(u)| = 3$ for $u \in W \cap D$.
Finally we saw during the proof of 11.10 that $N = 3$, so $|K \cap D_{a,b,d}| = 18$.

Suppose $Y = L$. Then $D_{a,b,d} \cap K = D_{a,b} \cap Y = D_{c,d,a,b}$ is of order 18.
Now $K = \langle Y, a \rangle \not\leq \langle D_c \rangle$ as $|PO_7^{-\pi,\pi}(3)|$ does not divide $|U_6(2)|$. Therefore
$a \in A_c$. But $\{c, d, a\} \subseteq D_b$, so by Exercise 5.4, $|D_{c,d,a,b}| = 9, 13$, or 45.
Hence $Y \neq L$. Thus by 16.2, Y is determined up to conjugacy in $C_A(d)$.

Now let $d \in X \in \mathcal{X}$ and $x \in A_d \cap X$. We wish to show X is conjugate to
K in $C_A(d)$. Conjugating in $C_A(d)$ we may take $Y = \langle D_d \cap X \rangle$. By 11.10.6,
$D_a \cap Y$ is conjugate to $D_x \cap Y$ in Y, so without loss, $D_a \cap Y = D_x \cap Y$.
Then $Y \cap D_{a,b} = Y \cap D_{x,a,b}$. But we just saw $|D_{x,d,a,b}| = 9, 13$, or 45, so
as $|Y \cap D_{x,a,b}| = 18$, we conclude $|D_{x,d,a,b}| = 45$. Then by Exercise 5.4,
$x \in a * d$, and hence $X = \langle Y, x \rangle = \langle Y, a \rangle = K$.

We have shown that A is transitive on $\mathcal{X}$. But by 16.2.2, $|G : A| \leq 2$, so G
has at most two orbits on $\mathcal{X}$, completing the proof of the lemma.

(16.6) *Let G be of type $M(23)$. Then G is transitive on D-subgroups K with*
$K/Z(K) \cong PO_8^{+,\pi}(3)$.

Proof: The proof is essentially a repeat of 16.5. Let $\mathcal{X}$ be the set of D-subgroups
K with $K/Z(K) \cong PO_8^{+,\pi}(3)$, $d \in K \in \mathcal{X}$, $a \in A_d \cap K$, and $Y = \langle D_d \cap K \rangle$.
Then $Y/Z(Y) \cong PO_7^{-\pi,\pi}(3)$ by 11.11. Further if (b, u) is an edge in $\mathcal{D}(D_a \cap Y)$
then an observation in the proof of the previous lemma applied to $\langle D_u \cap K \rangle$
shows $|Y \cap D_{a,b,u}| = 18$, whereas $|D_{c,d,a,b,u}| = 9, 13$, or 45 for $c \in D_{b,u} \cap A_a$.
Thus $L \neq Y$, so by 16.5, Y is determined up to conjugation in H.

Thus in proving $X \in \mathcal{X}$ is conjugate to K in G, we may choose $d \in X$ and
$Y = \langle X \cap D_d \rangle$. Also as in the previous lemma, we may take $x \in A_d \cap X$ with
$D_x \cap Y = D_a \cap Y$. Then we conclude $|D_{x,a,d,b,u}| = 45$, so by Exercise 5.4,
$x \in a * d$, and hence $X = K$.

(16.7) *Let G be of type $M(22)$ and $A = N_{\mathrm{Aut}(G)}(D)$. Assume a group of type
$M(24)$ exists. Then*

(1) $|A : G| = 2$.
(2) $H = C_G(d)$.
(3) $|D| = 3,510 = 2 \cdot 3^3 \cdot 5 \cdot 13$.
(4) $|G| = 2^{17} \cdot 3^9 \cdot 5^2 \cdot 7 \cdot 11 \cdot 13$.
(5) G has width 22 and $\prod_{x \in T \cap D} x = 1$ for $T \in \mathrm{Syl}_2(G)$.
(6) G has 2 classes of D-subgroups isomorphic to $PO_7^{-\pi,\pi}(3)$.

Proof: If (2) fails then by 16.2.2, there is $\alpha \in C_G(d)$ inducing a field automorphism on $H/\langle d \rangle$. Let E be a set of 3-transpositions of a group M of type $M(23)$. By definition of such groups and the uniqueness of groups of type $M(22)$ established in Theorem Z, we may regard G as $\langle E_e \rangle / Z(\langle E_e \rangle)$ for some $e \in E$. Then for $f \in A_e$, the image K of $\langle E_{e,f} \rangle$ in G is isomorphic to $PO_7^{-\pi,\pi}(3)$. Next the existence of a group of type $M(24)$ ensures there exists an E-subgroup X of M with $X/Z(X) \cong PO_8^{+,\pi}(3)$ and $e \in X \cap E$, and by the proof of 16.6, the image J of $\langle X \cap E_e \rangle$ in G is also isomorphic to $PO_7^{-\pi,\pi}(3)$, but is not conjugate to K in G. In particular (6) holds by 16.5 and 16.5 also contradicts $\alpha \in G$, so (2) holds. Indeed let $B = N_{\text{Aut}(M)}(E)$; we have shown no element of B induces an outer automorphism on $\langle E_e \rangle / Z(\langle E_e \rangle)$, a fact we use in the proof of the next lemma.

Anyway (2) and 16.2.2 imply (1). By (2), $|C_G(d)| = |H| = 2^{16} \cdot 3^6 \cdot 5 \cdot 7 \cdot 11$. By 11.8, $|D_d| = (2^6 - 1)(2^5 + 1)/3 = 693$. Further by 16.1, $\langle d \rangle L = N_H(L)$, so $L = C_H(c)$, and hence $|A_d| = |H : L| = 2,816$. Therefore (3) holds. Then as $|G| = |H||D|$, (4) holds. Finally by Exercise 4.3, $H/\langle d \rangle$ is of width 21 and $\prod_{x \in T \cap D_d} x \in \langle d \rangle$ for $d \in T \in \text{Syl}_2(G)$. Therefore G has width 22 and 3.7.3 implies (5).

REMARK 16.8: If we drop the assumption in 16.7 that groups of type $M(24)$ exist, then the proof of 16.7 shows that (3) and (5) hold for G of type $M(22)$, that $|A| = 2^{18} \cdot 3^9 \cdot 5^2 \cdot 7 \cdot 11 \cdot 13$, and that $|A : G| \leq 2$. For by 12.1 and 16.2.2, $|C_A(d) : H| = 2$. Further if G is also of type $\tilde{M}(22)$ (cf. the Introduction) then by hypothesis, $C_G(d)$ is quasisimple with $\langle d \rangle = Z(H)$, so (1), (2), and (4) also hold. In the opposite direction, 16.7.2 and 16.4.2 say that if a group of type $M(24)$ exists then groups of type $M(22)$ are of type $\tilde{M}(22)$.

(16.9) *Let G be of type $M(23)$ and assume groups of type $M(24)$ exist. Then*

(1) *G is simple.*
(2) *$G = N_{\text{Aut}(G)}(D)$.*
(3) *H is quasisimple with $\langle d \rangle = Z(H)$.*
(4) *$H = C_G(d)$.*
(5) *$|D| = 31,671 = 3^4 \cdot 17 \cdot 23$.*
(6) *$|G| = 2^{18} \cdot 3^{13} \cdot 5^2 \cdot 7 \cdot 11 \cdot 13 \cdot 17 \cdot 23$.*
(7) *G has width 23 and $\prod_{x \in T \cap D} x = 1$ for $T \in \text{Syl}_2(G)$.*

Proof: Let $A = N_{\text{Aut}(G)}(D)$. We saw in the proof of the previous lemma that $C_A(d) = HC_A(H)$. But by 12.1, $C_A(d^\perp) = \langle d \rangle$, so $H\langle d \rangle = C_A(d)$. So by a Frattini argument, $A = GC_A(d) = G$.

Next 16.7.5 and Exercise 3.7.3 imply (7). Then (1) and (3) follow from (7) and Exercise 3.4.

By 16.7.3, $|D_d| = 3,510$. By 16.3.3, $L = C_H(c)$, so as $|L| = 2^9 \cdot 3^9 \cdot 5 \cdot 7 \cdot 13$, we conclude with 16.7.4 that $|A_d| = |H : L| = 2^9 \cdot 5 \cdot 11 = 28,160$. Hence (4) holds and then as $|G| = |H||D|$, also (6) holds.

REMARK 16.10: If we drop the assumption in 16.9 that groups of type $M(24)$ exist, then the proof of 16.9 shows that (1), (3), (5), and (7) hold for G of type $M(23)$ and that $|G| = 2^{18+\epsilon} \cdot 3^{13} \cdot 5^2 \cdot 7 \cdot 11 \cdot 13 \cdot 17 \cdot 23$, for $\epsilon = 0$ or 1. Namely Remark 16.8 determines $|N_{\text{Aut}(H)}(D_d)|$ and by 12.1. $C_G(d)/\langle d \rangle \leq N_{\text{Aut}(H)}(D_d)$. Indeed if G is also of type $\tilde{M}(23)$ (cf. the Introduction) then (2) holds by hypothesis, and then (6) holds too. In the opposite direction, if groups of type $M(24)$ exist then by Remark 16.8, $H/\langle d \rangle$ is of type $\tilde{M}(22)$, whereas by (3) and (4), $H = C_G(d)$ is quasisimple, so G is of type $\tilde{M}(23)$. That is if groups of type $M(24)$ exist then groups of type M are of type $\tilde{M}$, for $M = M(22)$ and $M(23)$.

(16.11) *Let G be of type $M(23)$ and suppose K is a D-subgroup of G with $K/Z(K) \cong PO_8^+(3)$. Then*

 (1) $C_G(K) = 1$.
 (2) $M = N_G(E(K))$ *is a D-subgroup maximal in G and $M/E(K) \cong S_3$ or* $\mathbf{Z}_2$.
 (3) *If $M(24)$ exists then $M/E(K) \cong S_3$.*

Proof: Part (1) follows from 16.3.1 via the usual argument. Let

$$X = \langle N_D(E(K)) \rangle;$$

if a group of type $M(24)$ exists then $X \neq K$ by definition of groups of type $M(24)$. In any case $|K|_3 = 3^{12}$, so by Remark 16.10, $|G|_3 = 3|K|_3$, and then as $|X/E(K)| = 2 \cdot 3^a$ by 8.6 and 9.4, we conclude that $X/E(K) \cong S_3$ or $\mathbf{Z}_2$. Now 16.3.3 and the usual argument show $M = X$ is maximal in G.

(16.12) *Let G be of type $M(24)$. Then*

 (1) $|G : E(G)| = 2$.
 (2) $G = N_{\text{Aut}(G)}(D)$.
 (3) $H = C_G(d)$.
 (4) $H \cong \mathbf{Z}_2 \times M(23)$.
 (5) $|D| = 306,936 = 2^3 \cdot 3^3 \cdot 7^2 \cdot 29$.
 (6) $|G| = 2^{22} \cdot 3^{16} \cdot 5^2 \cdot 7^3 \cdot 11 \cdot 13 \cdot 17 \cdot 23 \cdot 29$.

(7) Let $T \in \text{Syl}_2(G)$ and $V = \langle T \cap D \rangle$. Then G is of width 24, $\prod_{x \in T \cap D} x = 1$, $V \cong E_{2^{12}}$, and $N_G(V)/V \cong M_{24}$ acts 5-transitively on $T \cap D$.

Proof: By 16.9.2 and the usual argument, (2) holds and $C_G(d) = H\langle d \rangle$. Let $d \in \Delta = T \cap D$. By 16.9.7 and Exercise 3.7, G is of width 24 and $\prod_{x \in \Delta} x = 1$. In particular $d \in H$, so (3) holds. In addition for $b, u \in \Delta$ distinct from d, $\langle d, b, u \rangle \cong E_8$ by 8.11.1, and by Exercise 4.3, $V/\langle d, b, u \rangle \cong E_{2^9}$ and $N_G(V)_{d,b,u}/V \cong L_3(4)$ acts 2-transitively on $\Delta - \{d, b, u\}$. Then $V \cong E_{2^{12}}$ and by repeated applications of Exercise 3.7.2, $N_G(V)$ is 5-transitive on Δ. Next as $N_G(V)_{d,b,u}/V \cong L_3(4)$, we conclude from Exercise 5.5 that $N_G(V)/V \cong M_{24}$. Also V is the image of the permutation module for M_{24} on Δ, so from the structure of that permutation module described in Section 19 of [SG], V has a $N_G(V)$-invariant subgroup U of index 2. Then as the multiplier of M_{24} is of odd order (cf. Exercise 7.10), $N_G(V)/U$ splits over V/U, so $N_G(V)$ has a subgroup M of index 2 with $M \cap D$ empty. However $T \leq N_G(V)$, so by Thompson transfer (cf. 37.4 in [FGT]) $d \notin [G, G]$. Hence (1) follows from 9.5. Then (1) and (3) imply (4).

Next by 16.9.5, $|D_d| = 31,671$. By 16.11, $L = C_H(c)$, so as $|L| = 3^{13} \cdot 2^{13} \cdot 5^2 \cdot 7 \cdot 13$, $|A_d| = |H : L| = 2^6 \cdot 11 \cdot 17 \cdot 23 = 275,264$, using 16.9.6. Then (5) and (6) follow as usual.

17. Epilogue: The proof of Fischer's Theorem

In this section D is a conjugacy class of 3-transpositions of the finite group G and $d \in D$. We gather together the various strands from previous sections that establish Fischer's Theorem. We begin with the following observation:

(17.1)   *Assume $Z(G) = 1$ and D_d is nonempty. Then the following are equivalent:*

(1) $G' = [G, G]$ is simple.

(2) G is primitive on D.

(3) $O_3(G) = O_2(G) = 1$ and $\mathcal{D}(D)$ is connected.

Proof: As $G = \langle D \rangle$ and D is a conjugacy class of G, $|G : G'| \leq 2$. So if (1) holds then either $O_2(G) = O_3(G) = 1$, or G' is of order 2 or 3. As $Z(G) = 1$, G' is not of order 2, whereas if G' is of order 3 then $G \cong S_3$, contradicting D_d nonempty. So $O_2(G) = O_3(G) = 1$ and then by 9.4.3, (2) holds.

If (2) holds then as $Z(G) = 1$, (1) and (3) hold by 9.5. On the other hand (3) implies (2) by 9.4.

Now to the proof of Fischer's Theorem. By the hypothesis of Fischer's Theorem, $Z(G) = 1$ and G' is simple. If D_d is empty then G is of width 1, so by 8.6, G' is a 3-group. Thus as G' is simple, $G \cong S_3$, one of the conclusions of Fischer's Theorem.

So we may assume D_d is nonempty. Then by 17.1, G is primitive on D, so the hypotheses of Section 13 are satisfied. Let $H = \langle D_d \rangle$. By 13.4 we may assume $O_3(H) \leq Z(H)$. Further if $O_2(H) \not\leq Z(H)$ then Hypothesis Q of Section 14 is satisfied, and then Theorem Q in Section 14 says one of the conclusions of Fischer's Theorem holds. So we may assume $O_2(H) \leq Z(H)$.

Now Hypothesis Z of Section 15 is satisfied, so Theorem Z completes the proof.

Observe that Fischer's Theorem shows that D is determined up to conjugation under $\mathrm{Aut}(G)$. For suppose E is a second class of 3-transpositions of G. By Fischer's Theorem, there is an isomorphism of 3-transposition groups $\alpha : (G, E) \to (\bar{G}, \bar{D})$ with $(\bar{G}, \bar{D})$ on the list of Fischer's Theorem. Thus G is isomorphic to $\bar{G}$ as an abstract group. But then there exists an isomorphism of 3-transposition groups $\beta : (\bar{G}, \bar{D}) \to (G, D)$, since the only isomorphisms of abstract groups among the groups on Fischer's list are those recorded in Section 11, and as noted there, in each case there exists an isomorphism of 3-transposition groups. Finally $\alpha\beta : (G, E) \to (G, D)$ is an isomorphism, so $\gamma = \alpha\beta \in \mathrm{Aut}(G)$ with $E\gamma = D$.

So we have demonstrated that D is unique up to conjugation under $\mathrm{Aut}(G)$. Indeed usually D is the unique class of 3-transpositions in G. For example we will see in Exercise 11.1 that each Fischer group has a unique class of 3-transpositions. This is also true in the symmetric groups and the classical groups over fields of order 2 or 4, except that $S_6 \cong \mathrm{Sp}_4(2)$ has two classes. This is a special case of the fact that the orthogonal group $PO_n^{\mu,\pi}(3)$ has one class unless n is even and $\mu = -1$, where there are two classes.

We close this chapter with a description of the almost simple groups G generated by a class D of 3-transpositions such that $\mathcal{D}(D)$ is disconnected.

(17.2) *Let D be a conjugacy class of 3-transpositions of the finite group G with $Z(G) = O_3(G) = O_2(G) = 1$ and $\mathcal{D}(D)$ disconnected. Then either*

(1) $F^*(G) \cong \Omega_8^+(2)$ and $G/F^*(G) \cong S_3$, or
(2) $F^*(G) \cong P\Omega_8^+(3)$ and $G/F^*(G) \cong S_3$.

Proof: By 9.4, $F^*(G) = X$ is simple and G/X is of width 1 with $|G : X| > 2$. In particular $d \notin X$, so d induces an outer automorphism on X. Further $\mathrm{Out}(X)$ involves S_3.

Let $G_0 = X\langle d \rangle$. Then $X = F^*(G_0) = [G_0, G_0]$, so G_0 is described in Fischer's Theorem. Then as d induces an outer automorphism on X, we conclude G_0 is a symmetric or orthogonal group, or G_0 is of type $M(24)$. If $G_0 \cong S_n$ then $X \cong A_n$ and hence $\text{Out}(X)$ is a 2-group (cf. Exercise 16.2 in [FGT]), contradicting the previous paragraph. If G_0 is of type $M(24)$ then by 37.1, $\text{Out}(X) \cong \mathbf{Z}_2$ for the same contradiction. Finally if X is orthogonal, appeal to Exercises 5.6 and 5.7.

Exercises

1. $U_4(2)$ is generated by five transvections.
2. Let $G = U_5(2)$, D the set of transvections in G, and L a D-subgroup of G such that $O_3(L) \not\le Z(L)$, L is transitive on $D \cap L$, and $\langle D_d \cap L \rangle \cong SU_3(2)'$ for $d \in L \cap D$. Prove $L \cong S_5/E_{81}$ and $|L \cap D| = 30$.
   [*Hint:* $|U_5(2)|_3 = 3^5$. Use this to show $|\{a \in L \cap D$ with $D_d \cap L = D_a \cap L\}| = 3$ and then apply 13.4 to $L/O_3(L)$.]
3. Let $G = O_5^{-\pi,\pi}(3)$, D the set of reflections in G, and $L \cong S_5$ a D-subgroup of G. Prove $|L^G| = 216$.
   [*Hint:* Prove L stabilizes a nondegenerate hyperplane of sign -1 of the space of G.]
4. Let $G = U_6(2)$, D the set of transvections of G, $d \in D$, and $c, a \in A_d$. Prove
   (1) $|D_{d,c,a}| = 9$, 13, or 45.
   (2) If $|D_{d,c,a}| = 45$ then $a = c^d$.
5. Let G be a t-transitive permutation group such that the stabilizer in G of $t - 2$ points is $L_3(4)$ on 21 letters. Prove G is a Mathieu group M_{22}, M_{23}, or M_{24}, for $t = 3, 4, 5$, respectively.
   [*Hint:* Use Exercise 6.5 in [SG] and the results on Steiner systems in Section 18 of [SG].]
6. Let $G = O_n^\epsilon(2)$, $n \ge 6$, $L = E(G)$, $A = \text{Aut}(L)$, $U \in \text{Syl}_2(L)$, V the orthogonal space defining G, $0 \ne v \in C_V(U)$, and $P = C_G(v)$. Prove
   (1) $P = QC_P(w)$, where $w \in V - v^\perp$ is singular, $Q = C_P(v^\perp) \cong E_{2^{n-2}}$, $C_P(w) \cong O_{n-2}^\epsilon(2)$, and the map $\alpha : h \mapsto [h, w]$ is an equivalence of Q with $\langle v, w \rangle^\perp$ as $\mathbf{F}_2 C_P(w)$-modules.
   (2) If $G \ne O_8^+(2)$ then $A = LN_A(P)$, whereas if $G = O_8^+(2)$ then $|A : LN_A(P)| \le 3$.
   (3) $LN_A(P) = G$.
   [*Hint:* To prove (2), use 47.7 in [SG] to prove $N_A(U)$ permutes the maximal parabolics over U, and use the fact that the maximal parabolics over U are the stabilizers of the totally singular subspaces of V fixed by U. Then show that if $G \ne O_8^+(2)$, $P \cap L$ is the unique maximal

parabolic P_0 with $P_0/O_2(P_0) \cong \Omega^\epsilon_{n-2}(2)$, whereas if $G = O_8^+(2)$ there are three such parabolics. Then to prove (3), show $N_A(P)$ preserves the quadratic form on Q induced by α.]

7. Let $G = PO^\epsilon_n(3)$, $n \geq 6$, $L = E(G)$, $A = \mathrm{Aut}(L)$, $U \in \mathrm{Syl}_3(L)$, V the orthogonal space defining G, $0 \neq v \in C_V(U)$, and $P = C_G(v)$. Prove

   (1) If $G \neq PO_8^+(3)$ then $A = LN_A(P)$, whereas if $G = PO_8^+(3)$ then $|A : LN_A(P)| \leq 3$.

   (2) $LN_A(P)$ is the group of similarities of V preserving the form up to ± 1. [*Hint:* Argue as in the hint for the previous problem. Notice the structure of P is more or less worked out in 11.10.]

6

The geometry of 3-transposition groups

We have already made extensive use of finite geometry in our study of groups generated by 3-transpositions. In this chapter we consider another type of geometry associated with a 3-transposition group G. This geometry is the *Fischer space* of G. The notion is due to F. Buekenhout [Bu].

Fischer spaces are a special class of *partial linear spaces*. *Steiner triple systems* are also partial linear spaces, and the Fischer space of a 3-transposition group of width 1 is a triple system. In Section 19 we prove a result of M. Hall [H4] that can be interpreted as classifying all 4-generator 3-transposition groups of width 1, or as classifying certain Steiner triple systems. Recall we used Hall's result to prove Lemma 8.6.

Fischer's Theorem gives a complete description of finite almost simple 3-transposition groups. Section 20 consists of a survey of results in the literature on 3-transposition groups that are not finite or not almost simple. Much of this work uses Fischer spaces, so Section 20 contains further discussion of those objects. Section 20 also contains some discussion of generalizations of the 3-transposition condition. In particular we find that the long root elements of groups of Lie type are naturally described by one such generalization.

18. Fischer spaces

A *partial linear space* is a pair $(X, \mathcal{L})$ where X is a set and $\mathcal{L}$ is a collection of subsets of X such that each pair of distinct elements x, y of X is in at most one member of $\mathcal{L}$. We refer to the members of X as *points* and the members of $\mathcal{L}$ as *lines*. We say points x and y are *collinear* if there is a line $l \in \mathcal{L}$ with $x, y \in l$. In that event our axiom says l is uniquely determined, and we denote l by $x * y$. We usually denote $(X, \mathcal{L})$ by X. A *morphism* $\alpha : (X, \mathcal{L}) \to (X', \mathcal{L}')$ is a function $\alpha : X \to X'$ such that $\mathcal{L}\alpha \subseteq \mathcal{L}'$.

Examples:

(1) The points and lines in the projective plane $PG(V)$ of a 3-dimensional vector space V over a field F form a partial linear space in which each pair of points is on a line. The plane $PG(V)$ is said to be of *order q* if F is of order q. Equivalently each line has $q + 1$ points, there are $q + 1$ lines through each point, and there are $q^2 + q + 1$ points and lines.

(2) Let p be a prime. The *affine plane of order p* is the partial linear space $(V, \mathcal{L})$, where V is a 2-dimensional vector space over the field of order p and $\mathcal{L}$ is the set of all cosets $U + v$ as U ranges over all 1-dimensional subspaces of V and v ranges over V. Thus the affine plane of order p has p^2 points and $p(p + 1)$ lines, each line contains p points, and each point is on $p + 1$ lines.

(3) The *dual* of a partial linear space $(X, \mathcal{L})$ is the partial linear space whose point set is $\mathcal{L}$ and with lines L_x, $x \in X$, where L_x is the set of all lines of X through x. The *dual affine plane of order p* is the dual of the affine plane of order p. Thus the dual affine plane of order p has $p(p + 1)$ points and p^2 lines, each line contains $p + 1$ points, and each point is on p lines. Notice the dual affine plane of order 2 is isomorphic to the partial linear space obtained by deleting a point x and all lines through x from the projective plane of order 2.

(4) Let D be a set of 3-transpositions of a group G and

$$\mathcal{L}(D) = \{\{x, y, x^y\} : x \in D, y \in A_d\}.$$

Then $D = (D, \mathcal{L}(D))$ is a partial linear space and G is represented as a group of automorphisms of D via conjugation. The kernel of this representation is $Z(G)$, so passing to $\bar{G} = G/Z(G)$, we have $\bar{G}$ faithful on $\bar{D}$ and $(D, \mathcal{L}(D)) \cong (\bar{D}, \mathcal{L}(\bar{D}))$.

A *subspace* of a partial linear space $(X, \mathcal{L})$ is a subset Y of X such that whenever x and y are distinct points of Y, then $x * y \subseteq Y$. In that event $(Y, \mathcal{Y})$ is a partial linear space, where $\mathcal{Y}$ consists of all lines $x * y$ as (x, y) varies over all pairs of distinct collinear points in Y. We denote this space by Y. Observe the intersection of any collection of subspaces is a subspace. Thus if $S \subseteq X$ we can define the *subspace generated by* S to be the intersection $\langle S \rangle$ of all subspaces of X containing S.

A *Fischer space* is a partial linear space $X = (X, \mathcal{L})$ such that

(F1) Each line contains exactly three points.

(F2) If l, m are distinct lines with a common point then the subspace $\langle l, m \rangle$ generated by l and m is isomorphic to the dual affine plane of order 2 or the affine plane of order 3.

Let X be a Fischer space. We say X is *nondegenerate* if each point of X is on at least one line of X. For $x \in X$, define $t(x) : X \to X$ by setting $yt(x) = z$ if $y \neq x$ is collinear with x and $x * y = \{x, y, z\}$, and setting $yt(x) = y$ otherwise. Write $T(X) = \{t(x) : x \in X\}$.

Examples:

(5) Let $G = S_4$ be the symmetric group of degree 4 and D the set of transpositions in G. Then $(D, \mathcal{L}(D))$ is the dual affine plane of order 2 and $d = t(d)$ for each $d \in D$.

(6) Let G be the split extension of $V \cong E_9$ by an involution inverting V. Then D is a set of 3-transpositions of G and $(D, \mathcal{L}(D))$ is isomorphic to the affine plane of order 3 with $t(d) = d$ for each $d \in D$.

(18.1) *Let X be a nondegenerate Fischer space and $x \in X$. Then*

(1) $t(x)$ is an automorphism of X of order 2.

(2) $t(x)$ is the unique automorphism of X fixing all lines through x whose fixed point set consists of x together with all points not collinear with x.

(3) $T(X)$ is a set of 3-transpositions of $\langle T(X) \rangle$.

(4) $X \cong (T(X), \mathcal{L}(T(X)))$.

Proof: By construction of $t(x)$, $t(x)$ is a permutation of X with $t(x)^2 = 1$. As X is nondegenerate, $t(x)$ is an involution. Further if $t(x)$ is an automorphism of X then visibly (2) holds. Thus to establish (1) and (2), we must show that if $l \in \mathcal{L}$ then $lt(x) \in \mathcal{L}$.

If $x \in l$ or x is collinear with no point of l then $l = lt(x)$. So assume $x \notin l$ but x is collinear with $y \in l$. Let $m = x * y$ and $U = \langle l, m \rangle$. As X is a Fischer space, U is a dual affine plane of order 2 or an affine plane of order 3. But then by the discussion in Examples (5) and (6), the restriction of $t(x)$ to U is an automorphism of U, so indeed $lt(x) \in \mathcal{L}$.

Thus (1) and (2) are established. We now prove (3). Let y be a point distinct from x. Then by (2), $t(y)^{t(x)} = t(yt(x))$. Therefore if x and y are not collinear then $yt(x) = y$ so $[t(x), t(y)] = 1$ and hence $t(x)t(y)$ is of order 2. So assume x and y are collinear. Then

$$t(y)t(x)t(y) = t(x)^{t(y)} = t(xt(y))$$
$$= t(z) = t(yt(x)) = t(y)^{t(x)} = t(x)t(y)t(x),$$

where $x * y = \{x, y, z\}$. Therefore $(t(x)t(y))^3 = 1$, completing the proof of (3).

Finally to prove (4) we show the map $\alpha : x \mapsto t(x)$ is an isomorphism of the Fischer space X with $(T(X), \mathcal{L}(T(X)))$. By (2), α is a bijection. Further $(x * y)\alpha = \{t(x), t(y)y, t(x)^{t(y)}\}$, so α and α^{-1} are morphisms.

(18.2)   *Let D be a set of 3-transpositions of G. Then $(D, \mathcal{L}(D))$ is a Fischer space.*

Proof: We have already observed that $D = (D, \mathcal{L}(D))$ is a partial linear space. By construction, all lines have order 3. Let $a * b$ and $b * c$ be lines in D, $S = \langle a, b, c \rangle$ the subgroup of G generated by a, b, and c, and $U = \langle a * b, b * c \rangle$ the subspace of D generated by the lines $a*b$ and $b*c$. Thus $U \cong (D \cap S, \mathcal{L}(D \cap S))$. If $c \in D_a$ then (a, b, c) is a 3-string, so $S \cong S_4$ by 8.3. Then by Example (5), U is the dual affine plane of order 2. Thus we may assume $c, c^b \in A_a$. Then by 5.5.4 and 5.5.1, $S/Z(S)$ is the split extension of $V \cong E_9$ by an involution inverting V, so by Example (6), U is an affine plane of order 3.

Example:

(7) A *Steiner triple system* is a partial linear space in which all lines are of order 3 and each pair of points is collinear. Denote by $\mathcal{S}$ the class of all Steiner triple systems in which each triple of points not on a line generates an affine plane of order 3. Observe that a Steiner system S is a Fischer system if and only if $S \in \mathcal{S}$. Conversely a Fischer space X is a Steiner triple system if and only if the set $T(X)$ of 3-transpositions of the subgroup $G = \langle T(X) \rangle$ of $\mathrm{Aut}(X)$ is of width 1. In that event by 8.6, $G = \langle t(x) \rangle O_3(G)$.

So the study of $\mathcal{S}$ is equivalent to the study of 3-transposition groups of width 1. Notice that Theorem 19.14 in the next section describes all 4-generator 3-transposition groups of width 1. Recall we used Theorem 19.14 to prove Lemma 8.6. Marshall Hall proved Theorem 19.14 to describe the Steiner systems in $\mathcal{S}$ generated by 4 points.

19. A result of M. Hall, Jr.

In this section we prove a result of Marshall Hall, describing the largest 3-transposition group G of width 1 generated by four 3-transpositions. We follow the proof of Lemma 4.2 in Hall [H4]. As we saw in Section 18, one can also view this result as describing all 4-generator Steiner triple systems in which each triple of points not on a line generates an affine plane of order 3.

Throughout the section we assume G is a group, $S = \{s_1, s_2, s_3, s_4\}$ is a set of involutions in G, and $G = \langle S \rangle$. Let $s = s_4$, $x_i = s s_i$, $x_{ij} = [x_i, x_j]$, $x_{ijk} = [x_{ij}, x_k]$, and $y_{ijk} = [x_{ij}, x_k^{-1}]$ for $1 \leq i, j, k \leq 3$ and i, j, k distinct. In addition we assume that S satisfies the Hall relations:

(H1) $s_i^2 = (s_i s_j)^3 = 1$ *for distinct $i, j, 1 \leq i, j \leq 4$.*
(H2) $(x_i x_j)^3 = 1$ *for all distinct $i, j \in I = \{1, 2, 3\}$.*

(H3) $x_{ijk}^3 = (x_{ji}x_k x_{ij} x_k^\epsilon)^3 = 1$ *for* $\epsilon = 1, -1$, *and for all distinct* $i, j, k \in I$.

(H4) $(x_k^\epsilon x_i^\alpha x_j^\beta x_i^\alpha)^3 = 1$ *for* $\epsilon, \alpha, \beta \in \{\pm 1\}$ *and all distinct* $i, j, k \in I$.

A *Hall system of rank* 4 is a pair $(\bar{G}, \bar{S})$ such that $\bar{G} = \mathrm{Grp}(\bar{S} : (H))$ is the group generated by $\bar{S}$ subject to the Hall relations (H). Thus if $(\bar{G}, \bar{S})$ is a Hall system of rank 4, then our group G satisfying the Hall relations is a homomorphic image of $\bar{G}$.

Let $X = \{x_i, x_{ij}, x_{ijk}, y_{ijk} : i, j, k\}$. Denote by $\mathcal{F}$ the set of subgroups $F = \langle s, x, y \rangle$ with

$$x^3 = y^3 = (x^\epsilon y)^3 = 1 \quad \text{for } \epsilon = \pm 1,$$

and with x, y inverted by s. By 5.5:

(19.1)   *Each* $F \in \mathcal{F}$ *is an image of the group* $W(3, 3)$ *of 5.5.*

Notice that the Hall relations $(H1)$ and $(H2)$ imply $\langle s, x_i, x_j \rangle \in \mathcal{F}$ for all $i, j \in I$. Hence by 19.1 there is a surjective group homomorphism α : $W(3, 3) \to \langle s, x_i, x_j \rangle$, and by 5.5, $W(3, 3) = \langle t \rangle \langle r_i, r_j \rangle$ is the split extension of $\langle t \rangle \cong \mathbb{Z}_2$ by $\langle r_i, r_j \rangle \cong 3^{1+2}$ with t inverting r_l and centralizing $Z(\langle r_i, r_j \rangle = \langle [r_i, r_j] \rangle$, and with $t\alpha = s$ and $r_l\alpha = x_l$. In particular as t centralizes $[r_i, r_j]$, s centralizes $[x_i, x_j] = x_{i,j}$. Also there exist an automorphism ξ of $W(3, 3)$ fixing t and r_j and inverting r_i and $[r_i, r_j]$, so as ξ acts on each normal subgroup of $W(3, 3)$, ξ induces an automorphism ξ_{ij} of $\langle s, x_i, x_j \rangle$ fixing s and x_j, mapping x_i to x_i^{-1}, and inverting x_{ij}.

Similarly as s inverts x_k and centralizes x_{ij}, s inverts $x_k^{x_{ij}}$. This observation together with the Hall relation $(H3)$ implies $\langle s, x_k, x_k^{\pm x_{ij}} \rangle \in \mathcal{F}$.

We define $\xi_k : X \to G$ by taking ξ_k trivial on $\langle x_i, x_j \rangle \cap X$ for $k \neq i, j$, $\xi_k = \xi_{ki}$ on $\langle x_i, x_k \rangle$, and decreeing that ξ_k interchanges x_{ijk} and y_{ijk}, x_{kij} and x_{ikj}, and y_{kij} and y_{ikj}. Take ξ_k to fix s. Observe that remarks in the previous two paragraphs and easy calculations show

(19.2) ξ_k *preserves the Hall relations and hence preserves valid statements about* X. *If* (G, S) *is a Hall system of rank 4*, ξ_k *induces an automorphism of* G.

(19.3)

     (1) $\langle x_{ij} \rangle = Z(\langle s, x_i, x_j \rangle)$.

     (2) $\langle x_i, x_j \rangle$ *is of exponent 3.*

     (3) $xyx = y^{-1}x^{-1}y^{-1}$ *for all* $x, y \in \langle x_i, x_j \rangle$.

     (4) $x_{ij}^{-1} = x_{ji} = [x_i^{-1}, x_j] = [x_i, x_j^{-1}]$.

     (5) $[x_i^{-1}, x_j^{-1}] = x_{ij}$.

Proof: This is a consequence of 19.1, 5.5, and our earlier remark that $\langle s, x_i, x_j \rangle$ is in $\mathcal{F}$.

(19.4) $x_{ijk}^{x_k^{-1}} = y_{ijk}^{-1}$ *and* $y_{ijk}^{x_k^{-1}} = y_{ijk}^{-1} x_{ijk}.$

Proof: Calculate.

(19.5)

    (1) $F = \langle s, x_k, x_k^{-x_{ij}} \rangle \in \mathcal{F}.$

    (2) $y_{ijk} x_{ijk}$ *generates* $Z(F).$

    (3) $x_{ijk} \in F.$

    (4) $[y_{ijk}, x_{ijk}] = 1.$

Proof: We have already observed that (1) holds. By (1), $Z(F)$ is generated by $[x_k^{-x_{ij}}, x_k] = y_{ijk} x_{ijk}$. As $x_{ijk} = x_k^{-x_{ij}} x_k$, (3) holds. Then (2) and (3) imply (4).

(19.6)

    (1) $x_{ijk}^{x_i} = x_{ji} x_{ik} x_{ij} x_{ijk} x_{ki}.$

    (2) $x_{ijk}^{x_i^{-1}} = x_{ji} x_{ki} x_{ij} x_{ijk} x_{ik}.$

    (3) $y_{ijk}^{x_i} = x_{ji} x_{ki} x_{ij} y_{ijk} x_{ik}.$

    (4) $y_{ijk}^{x_i^{-1}} = x_{ji} x_{ik} x_{ij} y_{ijk} x_{ki}.$

Proof: By 19.3, $x_{ij}^{-1} = x_{ji}$ centralizes x_i, so $x_i^{-1} x_{ij}^{-1} = x_i^{-1} x_{ji} = x_{ji} x_i^{-1}$ and similarly $x_i^{-1} x_{ij} = x_{ji} x_i^{-1}$. Therefore

$$x_{ijk}^{x_i} = x_i^{-1} x_{ij}^{-1} x_k^{-1} x_{ij} x_k x_i = x_{ji} x_i^{-1} x_k^{-1} x_{ij} x_k x_i$$
$$= x_{ji} x_{ik} x_k^{-1} x_i^{-1} x_{ij} x_k x_i = x_{ji} x_{ik} x_k^{-1} x_{ij} x_i^{-1} x_k x_i$$
$$= x_{ji} x_{ik} (x_{ij} x_{ij}^{-1}) x_k^{-1} x_{ij} (x_k x_k^{-1}) x_i^{-1} x_k x_i = x_{ji} x_{ik} x_{ij} x_{ijk} x_{ki}$$

establishing (1). Then by (1),

$$x_{ijk}^{x_i^{-1}} = (x_{ijk}^{x_i})^{x_i} = (x_{ji} x_{ik} x_{ij} x_{ijk} x_{ki})^{x_i}$$
$$= x_{ji} x_{ik} x_{ij} x_{ijk}^{x_i} x_{ki} = x_{ji} x_{ik} x_{ij} (x_{ji} x_{ik} x_{ij} x_{ijk} x_{ki}) x_{ki}$$
$$= x_{ji} x_{ik}^2 x_{ij} x_{ijk} x_{ki}^2 = x_{ji} x_{ki} x_{ij} x_{ijk} x_{ik}$$

proving (2). Finally apply ξ_k to (1) and (2) to get (3) and (4), respectively.

(19.7) $y_{ijk}x_{kj}x_{ki}y_{kij}x_{jk}x_{kij}^{-1}x_{ik} = 1.$

Proof: By the Hall relation $(H4)$,

(1) $$1 = (x_k x_i x_j x_i)^3.$$

Next by 19.3.4, $x_{ik} = [x_k, x_i^{-1}]$, so $x_i x_k x_i = x_k x_{ik} x_i^{-1}$. Similarly $x_{ki} = [x_k, x_i]$ so $x_k x_i = x_i x_k x_{ki}$. Then the equalities $x_i x_k x_i = x_k x_{ik} x_i^{-1}$ and $x_k x_i = x_i x_k x_{ki}$ transform (1) into

(2) $$1 = x_k x_i x_j x_k x_{ik} x_i^{-1} x_j x_i x_i x_k x_{ki} x_j x_i.$$

Now $x_i^2 = x_i^{-1}$ and $x_{ij} = [x_j, x_i^{-1}]$, so $x_i^{-1} x_j x_i^{-1} = x_i x_j x_{ij}$. Substituting these equations into (2), we obtain

(3) $$1 = x_k x_i x_j x_k x_{ik} x_i x_j x_{ij} x_k x_{ki} x_j x_i.$$

Similarly $x_k x_{ik} x_i = x_i x_k$ and $x_i x_j x_i = x_j x_{ij} x_i^{-1}$ so (3) can be replaced by

(4) $$1 = x_k x_j x_{ij} x_i^{-1} x_k x_j x_{ij} x_k x_{ki} x_j x_i.$$

As $x_i^{-1} x_k = x_k x_{ki} x_i^{-1}$, (4) becomes

(5) $$1 = x_k x_j x_{ij} x_k x_{ki} x_i^{-1} x_j x_{ij} x_k x_{ki} x_j x_i.$$

Then making the replacements $x_i^{-1} x_j x_{ij} = x_j x_i^{-1}$ and $x_j x_i = x_i x_j x_{ji}$ we get

(6) $$1 = x_k x_j x_{ij} x_k x_{ki} x_j x_i^{-1} x_k x_{ki} x_i x_j x_{ji}.$$

Next
$$x_{ik} = x_{ki}^2 = x_k^{-1} x_i^{-1} x_k x_i x_{ki} = x_k^{-1} x_i^{-1} x_k x_{ki} x_i$$
so $x_i^{-1} x_k x_{ki} x_i = x_k x_{ik}$, and substituting this into (6) we obtain

(7) $$1 = x_k x_j x_{ij} x_k x_{ki} x_j x_k x_{ik} x_j x_{ji}.$$

Also $x_{ki} x_j = x_j x_{ki} x_{kij}$, so we have

(8) $$1 = x_k x_j x_{ij} x_k x_j x_{ki} x_{kij} x_k x_{ik} x_j x_{ji}.$$

Now $x_{kij} x_k = x_k (x_{kij})^{x_k}$, and we use 19.6 to evaluate $x_{kij}^{x_k}$ and obtain

(9) $$1 = x_k x_j x_{ij} x_k x_j x_{ki} x_k x_{ik} x_{kj} x_{ki} x_{kij} x_{jk} x_{ik} x_j x_{ji}.$$

Next x_k commutes with $x_{ik} = x_{ki}^{-1}$, x_j commutes with x_{ij}, and $x_j x_k$ is of order 3 so $(x_j x_k)^2 = x_k^{-1} x_j^{-1}$. Hence we get

(10) $$1 = x_k x_{ij} x_k^{-1} x_j^{-1} x_{kj} x_{ki} x_{kij} x_{jk} x_{ik} x_j x_{ji}.$$

Similarly x_j commutes with x_{kj} and $x_j^{-1}x_{ki} = x_{ki}x_j^{-1}y_{kij}^{-1}$, so (10) becomes

(11) $$1 = x_k x_{ij} x_k^{-1} x_{kj} x_{ki} x_j^{-1} y_{kij}^{-1} x_{kij} x_{jk} x_{ik} x_j x_{ji}.$$

Now $x_{ik}x_j = x_{ki}^{-1}x_j = x_j(x_{ki}^{x_j})^{-1} = x_j(x_{ki}x_{kij})^{-1} = x_j x_{kij}^{-1} x_{ik}$, so we replace (11) by

(12) $$1 = x_k x_{ij} x_k^{-1} x_{kj} x_{ki} x_j^{-1} y_{kij}^{-1} x_{kij} x_{jk} x_j x_{kij}^{-1} x_{ik} x_{ji}.$$

Then by 19.4, $y_{kij}^{x_j^{-1}} = y_{kij}^{-1}x_{kij}$ whereas by 19.3.1, $[x_{jk}, x_j] = 1$, so (12) becomes

(13) $$1 = x_k x_{ij} x_k^{-1} x_{kj} x_{ki} y_{kij} x_{jk} x_{kij}^{-1} x_{ik} x_{ji}.$$

Finally we conjugate (13) by x_{ij} to complete the proof of the lemma.

(19.8) $\quad y_{ijk}x_{ijk}^{-1} = (y_{kij}x_{kij}^{-1})^{x_{ik}x_{kj}}.$

Proof: Conjugate the equation of 19.7 by x_k^{-1} and use the fact that x_k commutes with x_{kr} to obtain

(a) $$1 = y_{ijk}^{x_k^{-1}} x_{kj} x_{ki} y_{kij}^{x_k^{-1}} x_{jk} x_{kij}^{-x_k^{-1}} x_{ik}.$$

Now apply 19.4 and 19.6 to (a) and recall $[y_{ijk}, x_{ijk}] = 1$ by 19.5 to obtain the lemma.

(19.9) $\quad [y_{kij}x_{kij}^{-1}, x_{ik}x_{jk}] = 1.$

Proof: Applying ξ_k to 19.8 we get

(b) $$x_{ijk}y_{ijk}^{-1} = (y_{ikj}x_{ikj}^{-1})^{x_{ki}x_{jk}}.$$

Also by 19.5.4, $x_{ikj}y_{ijk}^{-1} = y_{ikj}^{-1}x_{ikj}$. Hence we conclude from 19.8 and (b) that $(y_{kij}x_{kij}^{-1})^{x_{ik}x_{kj}} = (y_{ikj}^{-1}x_{ikj})^{x_{ki}x_{jk}}$ and conjugating this equality by $x_{jk}^{-1} = x_{kj}$, we get

$$(y_{kij}x_{kij}^{-1})^{x_{ik}x_{jk}} = (y_{ikj}^{-1}x_{ikj})^{x_{ki}}.$$

But also $x_{ikj}^{x_{ki}} = x_{kij}^{-1}$ and $y_{ikj}^{-x_{ki}} = y_{kij}$, so $(y_{kij}x_{kij}^{-1})^{x_{ik}x_{jk}} = y_{kij}x_{kij}^{-1}$, completing the proof.

(19.10) $\quad x_{ijk}$ *and* y_{ijk} *commute with* x_{rs} *for all* $i, j, k, r, s.$

Proof: Conjugating 19.9 by x_j and using 19.4, we get

(c) $$[y_{kij}, x_{kij}x_{ik}x_{jk}] = 1.$$

We conclude from (c) and 19.5.4 that $[y_{kij}, x_{ik}x_{jk}] = 1$, and then conclude from this equality and 19.9 that

(d) $$[y_{kij}, x_{ik}x_{jk}] = [x_{kij}, x_{ik}x_{jk}] = 1.$$

Then applying ξ_j to (d) we conclude $x_{ik}x_{kj}$ also centralizes $\langle y_{kij}, x_{kij} \rangle$, so that $(x_{ik}x_{kj})^{-1}(x_{ik}x_{jk}) = x_{kj}$ does too. Then by (d), x_{ik} centralizes $\langle y_{kij}, x_{kij} \rangle$. Finally $x_{jk} = x_{kj}\xi_k$ centralizes $y_{kij}\xi_k = y_{ikj}$ and $x_{kij}\xi_k = x_{ikj}$, so that x_{ij} centralizes $\langle y_{kij}, x_{kij} \rangle$.

(19.11) $y_{ijk}x_{ijk}^{-1} = y_{i\sigma j\sigma k\sigma}x_{i\sigma j\sigma k\sigma}^{-1}$ *for all even permutations σ of $\{1, 2, 3\}$.*

Proof: This follows from 19.8 and 19.10.

(19.12) $[x_{jk}, x_{ik}] = y_{ijk}x_{ijk}$ *is in the center of G.*

Proof: By 19.7 and 19.10, we have

(e) $$x_{kj}x_{ki}x_{jk}x_{ik} = [x_{jk}, x_{ik}] = x_{kij}y_{kij}^{-1}y_{ijk}^{-1}.$$

But by 19.11, $x_{kij}y_{kij}^{-1} = x_{ijk}y_{ijk}^{-1}$, so we get

(f) $$[x_{jk}, x_{ik}] = x_{ijk}y_{ijk}.$$

Now

$$[x_{jk}, x_{ik}]^{x_i} = [x_{jk}^{x_i}, x_{ik}] = [x_{jk}x_{jki}, x_{ik}] = [x_{jk}, x_{ik}]^{x_{jki}}[x_{jki}, x_{ik}] = [x_{jk}, x_{ik}]$$

by 19.10. Similarly $[x_{jk}, x_{ik}]$ centralizes x_j and x_k; whereas by 19.5, $z = x_{ijk}y_{ijk}$ centralizes s. So $z \in Z(G)$.

Let $z_{ijk} = x_{ijk}y_{ijk}$, $G_0 = G$, $G_1 = \langle x_i : i \rangle$, $G_2 = \langle x_{ij} : i, j \rangle$,

$G_3 = \langle x_{ijk}, y_{ijk} : i, j, k \rangle$, $\qquad G_4 = \langle z_{ijk} : i, j, k \rangle$, and $G_5 = 1$.

Theorem G (M. Hall)

(1) $G = G_1\langle s \rangle$ with G_1 a normal 3-subgroup of G of order at most 3^{10}.

(2) $G_1 > \ldots > G_5$ is the descending central series for G_1.

(3) s inverts, centralizes G_i/G_{i+1} for i odd, even, respectively.

(4) G_i/G_{i+1} is elementary abelian of order at most 27 and $|G_3/G_4| \leq 3$.

Proof: Recall from Section 9 of [FGT] that the descending central series of G_1 is $L_i(G_1)$, $i = 1, 2, \ldots$, where $L_i(G)$ is defined recursively by $L_1(G_1) = G_1$ and $L_n(G_1) = [L_{n-1}(G_1), G_1]$ for $n > 1$. By definition of the members of X, $G_i \leq L_i(G_1)$ for $1 \leq i \leq 3$. By 19.5, $z_{ijk} = [x_{ijk}, x_k]$, so $G_4 \leq L_4(G_1)$. Thus $G_i \leq L_i(G_1)$ for $1 \leq i \leq 5$.

Suppose $G_i \trianglelefteq G$ for each i. As G_2 contains all commutators $x_{ij} = [x_i, x_j]$ of generators of G_1, G_1/G_2 is abelian, so $L_2(G_1) = G_2$. Similarly by 19.3, $[x_{ij}, x_r] = 1$ or x_{ijk} for $r \neq k$, $r = k$, respectively, so $L_3(G_1) = G_3$. So to establish (2) we need only show for each $1 \leq i \leq 4$, $G_i \trianglelefteq G$ and $[G_1, G_j] \leq G_{j+1}$ for $j = 3$ and 4.

By 19.12, $G_4 \leq Z(G)$. By 19.5, G_4 is of exponent 3. As $[a, b]^{-1} = [b, a]$ and $z_{ijk} = [x_{jk}, x_{ik}]$ by 19.12, we conclude $z_{jik} = z_{ijk}^{-1}$. Thus $|G_4| \leq 27$. So passing to G/G_4, we may assume $G_4 = 1$, and it remains to show $G_3 \leq Z(G_1)$ and $G_i \trianglelefteq G$.

As $G_4 = 1$, $x_{ijk} = y_{ijk}^{-1}$. Then

$$x_{ijk}^s = [x_{ij}, x_k]^s = [x_{ij}^s, x_k^s] = [x_{ij}, x_k^{-1}] = y_{ijk} = x_{ijk}^{-1}.$$

As $x_{ijk} = y_{ijk}^{-1}$, x_k centralizes x_{ijk} by 19.4, and by 19.11, $x_{ijk} = x_{i\sigma j\sigma k\sigma}$ for all even permutations σ. Hence $G_3 \leq Z(G)$ and as s inverts generators for G_3, s inverts G_3. Indeed $G_2/\langle x_{123}\rangle \leq Z(G_1/\langle x_{123}\rangle)$ as $[x_{ij}, x_r]$ or $[x_{ji}, x_r] \in \langle x_{123}\rangle$ for all i, j, r. Thus $L_3(G_1) = \langle x_{123}\rangle$, so $x_{213} \in \langle x_{123}\rangle$ and hence $|G_3| \leq 3$. So passing to G/G_3, we may assume $G_3 = 1$.

Then $[x_{ij}, x_r] \in G_3 = 1$ for all i, j, r, so $G_2 \leq Z(G)$ and as $G_2 = \langle x_{12}, x_{13}, x_{23}\rangle$, G_2 is of exponent 3 and order at most 27. By 19.3, s centralizes G_2. Then passing to G/G_2, we may take $G_2 = 1$. Therefore $G_1 = \langle x_1, x_2, x_3\rangle$ is abelian of exponent 3 and order at most 27 and inverted by s, completing the proof.

(19.13) *Let D be a set of 3-transpositions of a group M, $T = \{t_1, t_2, t_3, t_4\} \subseteq D$, $M = \langle T\rangle$, and assume M is of width 1. Let (G, S) be a Hall system of rank 4. Then*

(1) T satisfies the Hall relations in M.

(2) There is a surjective homomorphism of G onto M mapping s_i to t_i for all i.

Proof: Let $t = t_4$, $y_i = tt_i$, and so on. The Hall relations $(H1)$ and $(H2)$ follow immediately from the hypothesis that D is a set of 3-transpositions of M of width 1 (cf. 5.5). In particular $\langle t, y_i, y_j\rangle \in \mathcal{F}$ by 5.5, so t centralizes y_{ij} and then $ty_k^{-y_{ij}} \in D$. Hence $\langle t, y_k, y_k^{y_{ij}}\rangle \in \mathcal{F}$, so that the Hall relation $(H3)$ is satisfied. Similarly $t y_i^\alpha y_j^\beta y_i^\alpha \in D$ for $\alpha, \beta \in \{\pm 1\}$, so $\langle t, y_k, y_i^\alpha y_j^\beta y_i^\alpha \rangle \in \mathcal{F}$ and hence the Hall relation $(H4)$ is satisfied.

Theorem 19.14. *Let (G, S) be the Hall system of rank 4. Then*

(1) s^G *is a set of 3-transpositions of G.*

(2) $|G| = 2 \cdot 3^{10}$.

(3) $\text{Aut}(G)/\text{Inn}(G) \cong SL_3(3)$.

(4) *The split extension of G by $SL_3(3)$ is a 3-local of the Fischer group $M(23)$.*

Proof: We will see in 40.13 that there is a subgroup K of M of type $M(23)$ such that $O_3(K) = P$ is of order 3^{10}, $K/P \cong GL_3(3)$, K/P is faithful on $P/\Phi(P)$ of order 27, and an involution $t \in K$ inverting $P/\Phi(P)$ is a 3-transposition in M. It follows that $\langle t \rangle P = \langle t^P \rangle = \langle t, t_1, t_2, t_3 \rangle$, where $t_i = t y_i$ and $P = \langle y_1, y_2, y_3 \rangle$. Therefore by 19.12, there is a surjective homomorphism $\pi : G \to P\langle t \rangle$ with $s_i \pi = t_i$. Further by Theorem G, $|G| \le 2 \cdot 3^{10} = |P\langle s \rangle|$, so π is an isomorphism. Therefore (1) and (2) are established.

Identify P with G_1 and t with s via π. Then as K/P is faithful on $P/\Phi(P)$, $\text{Aut}(G)/U \cong GL_3(3)$, where $U = C_{\text{Aut}}(P/\Phi(P))$ and it remains to show $U = V$, where $V = P/G_4$. By a Frattini argument, $U = VC_U(s)$, so it suffices to show $u \in C_U(s)$ is in V.

If u centralizes P/G_4 then u acts on $x_i G_4$ for each i. Hence as x_i is the unique element in $x_i G_4$ inverted by s, $x_i^u = x_i$. But then as $P = \langle x_1, x_2, x_3 \rangle$, $u = 1$. So we may assume u is faithful on P/G_4. On the other hand u centralizes $P/\Phi(P) = P/G_2$ and $G_2/G_3 = C_{P/G_3}(s)$, so the same argument shows $[u, P] \le G_3$.

Let $\bar{P} = P/G_4$. Then $\bar{G}_3 \cong Z_3$ and u centralizes $\bar{P}/\bar{G}_3$. Hence $[\bar{x}_i, u] = \bar{x}^{a_i}$ for $a_i = 0, 1, 2$ and $x = x_{123}$. But there exists $v \in C_P(s)$ with $[\bar{x}_i, v] = \bar{x}^{a_i}$, so $v = u$ by an earlier remark, completing the proof.

20. Beyond Fischer's Theorem

Fischer's Theorem gives us a complete description of finite almost simple 3-transposition groups. Further we have many elementary results about finite 3-transposition groups that are not almost simple. In this section we survey some papers in the literature that give deeper results about finite 3-transposition groups that are not almost simple, and results that discuss *infinite* 3-transposition groups. The best of these results are due to H. Cuypers and J. Hall [CH1], [CH2], building on earlier work of Hall [H2], [H3], Weiss [W3], and Zara [Z1], [Z2],

[Z3]. We also discuss generalizations of the 3-transposition hypothesis that are important in the study of groups of Lie type.

We saw in Section 18 that the study of centerless 3-transposition groups is equivalent to the study of Fischer spaces via the correspondences

$$(G, D) \mapsto (D, \mathcal{L}(D)) \quad \text{and} \quad X \mapsto (\langle T(X) \rangle, T(X))$$

between sets D of 3-transpositions in a group G and Fischer spaces $X = (X, \mathcal{L})$. The papers of Cuypers and Hall make use of these correspondences and some of their theorems can be stated in the language of Fischer spaces as well as in the language of groups. For that matter Fischer's Theorem can also be stated in the language of Fischer spaces. To state the Cuypers–Hall results we need more notation and terminology.

Let $X = (X, \mathcal{L})$ be a partial linear space. The *collinearity graph* Γ of X is the graph with vertex set X where vertices x and y are adjacent if x and y are collinear, that is if x and y are incident with a common line.

Example:

(1) If D is a set of 3-transpositions of a group G then the collinearity graph of the Fischer space $(D, \mathcal{L}(D))$ is the complementary graph of the commuting graph $\mathcal{D}(D)$.

With this example in mind we write A_x for the set of vertices $y \in X$ distinct from x and collinear with x and write D_x for the set of vertices distinct from x and not collinear with x. Let $V_x = \{v \in X : A_x = A_v\}$, and $W_x = \{w \in X : D_x = D_w\}$. Define X to be *2-reduced* if $\{x\} = V_x$ for all $x \in X$ and define X to be *3-reduced* if $W_x = \{x\}$ for all $x \in X$. Define X to be *reduced* if X is both 2-reduced and 3-reduced. Observe that X is either 2-reduced or 3-reduced.

Define X to be *connected* if the collinearity graph Γ is connected and define X to be *coconnected* if the complementary graph Γ^c of the collinearity graph is connected.

Example:

(2) Let $X = (X, \mathcal{F})$ be a Fischer space. Let $D = T(X)$ and $G = \langle D \rangle$ so that D is a set of 3-transpositions of G by 18.1. By 18.1, $X \cong (D, \mathcal{L}(D))$ so by Example 1, $\Gamma \cong \mathcal{D}(D)^c$ and $\Gamma^c \cong \mathcal{D}(D)$. In particular X is coconnected if and only if $\mathcal{D}(D)$ is connected. Also by 8.1, D is locally conjugate in D, so by 7.3, G is transitive on D if and only if Γ is connected. If X is finite then by Exercise 3.1, X is 3-reduced if and only if $O_3(G) \leq Z(G)$, and if X is nondegenerate then X is 2-reduced if and only if $O_2(G) \leq Z(G)$.

Further as $G \le \mathrm{Aut}(X)$, G is faithful on X, so as the map $x \mapsto t(x)$ is G-equivariant, $Z(G) = 1$. Therefore if X is finite and nondegenerate, then X is reduced if and only if $O_2(G) = O_3(G) = 1$.

To illustrate these notions further we state one of our results about finite 3-transposition groups in the language of Fischer spaces.

(20.1) *Let $X = (X, \mathcal{L})$ be a finite reduced Fischer space that is connected and coconnected. Then*

  (1) *The collinearity graph Γ of X is simple; that is Γ possesses no nontrivial contractions.*
  (2) *For each $x \in X$, D_x is a connected subgraph of Γ and A_x is a connected subgraph of Γ^c.*

Proof: Proceeding as in Example 2, let $D = T(X)$ and $G = \langle D \rangle$. As X is connected and coconnected, the discussion in Example 2 says G is transitive on D and $\mathcal{D}(D)$ is connected. As X is reduced the discussion in Example 2 says $O_3(G) = O_2(G) = 1$. Therefore 9.4 says G is primitive on D, and then 6.5 completes the proof.

We next consider some examples.

Examples:

  (3) Let Ω and Ω' be disjoint sets with $|\Omega| \ge 2$. Define $T(\Omega, \Omega')$ to be the partial linear space whose point set X is the set of all subsets x of $\Omega \cup \Omega'$ such that $x \cap \Omega$ is of order 2, and whose lines are the 3-subsets $\{x, y, z\}$ of X such that $x + y + z = 0$, where the sum of two subsets is their symmetric difference. Write $T(\Omega)$ for $T(\Omega, \emptyset)$ and $T(n)$ for $T(\Omega)$ when Ω is of order n. Notice that if Ω is of order n then $D = T(T(\Omega))$ is the set of transpositions on Ω and so $\langle D \rangle$ is the symmetric group on Ω. That is $T(n)$ is the Fischer space of the symmetric group S_n.
  Observe next that we have a morphism $f : T(\Omega, \Omega') \to T(\Omega)$ of partial linear spaces defined by $f(x) = x \cap \Omega$. The fibers of the map f are the sets V_x, $x \in X$, as long as $|\Omega| \ne 4$. We leave as an exercise the check that the subspace generated by any pair of distinct intersecting lines of $T(\Omega, \Omega')$ is isomorphic to the dual affine space of order 2. Therefore $T(\Omega, \Omega')$ is a Fischer space. Also $T(\Omega, \Omega')$ is connected if $|\Omega| > 2$ and coconnected if $|\Omega| > 4$. We have seen that if $|\Omega| \ne 4$ then for x a 2-subset of Ω, the fiber $f^{-1}(x) = V_x$, so in particular if $\Omega' \ne \emptyset$ then

$V_x \neq \{x\}$, so from Example 2, $T(\Omega, \Omega')$ is not 2-reduced and therefore *is* 3-reduced.

The transposition (a, b) on Ω is the map $t(x)$ on $T(\Omega)$, for $x = \{a, b\}$, and the set $T(T(\Omega))$ of transpositions generates the subgroup $F \operatorname{Sym}(\Omega)$ of the symmetric group $\operatorname{Sym}(\Omega)$ on Ω consisting of all finitary g, where $g \in \operatorname{Sym}(\Omega)$ is *finitary* if g moves only a finite number of points of Ω. Define the *natural module* V for $F \operatorname{Sym}(\Omega)$ to be the submodule V_0 of the $GF(2)$-permutation module U consisting of all vectors of even weight, unless Ω is finite of even order, when $V = V_0/\langle\Omega\rangle$.

(4) Let V be a vector space over the field of order 2; we allow V to be infinite-dimensional. Let f be a nontrivial symmetric bilinear form on V such that $f(v, v) = 0$ for all $v \in V$. Define $Sp(V, f)$ to be the partial linear space whose point set is the set of all points of V not contained in $\operatorname{Rad}(V)$ and whose lines are the triples of points in nondegenerate lines of V. Thus $x \in A_y$ if and only if $f(x, y) \neq 0$. Again we leave it as an exercise to prove that the subspace generated by distinct intersecting lines is a dual affine plane of order 2, so $Sp(V, f)$ is a Fischer space. Further $Sp(V, f)$ is connected and $Sp(V, f)$ is coconnected if $\dim(V/\operatorname{Rad}(V)) > 2$. Indeed if $Sp(V, f)$ is coconnected then for each point x, V_x is the set of points in $\operatorname{Rad}(V) + x$, so if $\operatorname{Rad}(V) \neq 0$ then $Sp(V, f)$ is not 2-reduced and hence is 3-reduced.

Define a linear map g on V to be *finitary* if $V/C_V(g)$ is of finite dimension. For x a point of $Sp(V, f)$, $t(x)$ is the transvection with center x and axis $x^\perp$ and if f is nondegenerate then $T(Sp(V, f))$ generates the group $FSp(V, f)$ of all finitary isometries of $Sp(V, f)$.

(5) Let V be a vector space over the field of order 2; again we allow V to be infinite-dimensional. Let q be a quadratic form on V with associated bilinear form f satisfying the hypotheses of Example (4). Let $\mathcal{N}(V, q)$ be the subspace of $Sp(V, f)$ consisting of all points x with $q(x) = 1$ and all lines of $Sp(V, f)$ through pairs of such points. As subspaces of a Fischer space are Fischer spaces, $\mathcal{N}(V, q)$ is a Fischer space. If q is nondegenerate, the group generated by the transvections $T(\mathcal{N}(V, q))$ is the group $FO(V, q)$ of all finitary isometries of (V, q), unless (V, q) is of dimension 4 and sign $+1$.

(6) Let (V, q) be an orthogonal space over the field F of order 3 and denote by $\mathcal{N}(V, q)$ the partial linear space with points the points Fv of V with $Q(v) = 1$ and lines the points in a line of V that is degenerate but not totally singular. Then $\mathcal{N}(V, q)$ is a Fischer space with $t(Fv)$ the reflection through Fv. If (V, q) is nondegenerate we write $^+\Omega(V, q)$ for the group generated by the reflections $T(\mathcal{N}(V, q))$.

(7) Let (V, h) be a unitary space over the field of order 4 and $\mathcal{U}(V, h)$ the partial linear space whose points are the singular points of V and whose lines are the points in nondegenerate lines of V. Then $\mathcal{U}(V, h)$ is a Fischer space with $t(x)$ the transvection with center x and axis $x^{\perp}$ and as h is nondegenerate, $T(\mathcal{U}(V, h))$ generates the group $FU(V, h)$ of all finitary isometries of (V, h).

(8) Denote by $\mathcal{M}(n)$ the Fischer space of the Fischer group $M(n)$, for $n = 22, 23$, and 24.

We can now state the theorem of Cuypers and Hall extending Fischer's Theorem to infinite groups. The result can be stated either as a result about 3-transposition groups or as a result about Fischer spaces. Here is the Fischer space version; the spaces in (2)–(5) are all nondegenerate.

Theorem 20.2 (Cuypers–Hall [CH1], [CH2]). *Let X be a reduced, connected, and coconnected Fischer space. Then X is isomorphic to one of the following:*

(1) $T(\Omega)$, $|\Omega| \geq 5$.

(2) $Sp(V, f)$, for some symplectic space (V, f) over $GF(2)$ of dimension at least 3.

(3) $\mathcal{U}(V, h)$, for some unitary space (V, h) over $GF(4)$ of dimension at least 4.

(4) $\mathcal{N}(V, q)$ for some orthogonal space over $GF(2)$ of dimension at least 6.

(5) $\mathcal{N}(V, q)$ for some orthogonal space over $GF(3)$ of dimension at least 5.

(6) $\mathcal{M}(22)$, $\mathcal{M}(23)$, or $\mathcal{M}(24)$.

Recall from Example 2 that if $G = \langle T(X) \rangle$ is finite then X is reduced if and only if $O_3(G) = O_2(G) = 1$. Thus Theorem 20.2 is the generalization of Fischer's Theorem obtained by removing the constraint that G be finite.

The case where the Fischer space X is not reduced has also been treated by Cuypers and Hall. Here the result is best stated in the language of groups, but even then there is a fairly complicated list of examples. Thus we only state a theorem of Hall handling a special case of the Cuypers–Hall Theorem, and follow that with a discussion of the types of examples that occur in the unrestricted version.

Theorem 20.3 (J. Hall [H2]). *Let X be a connected partial linear space with at least two lines such that the subspace generated by any pair of distinct intersecting lines is a dual affine plane of order 2. Then X is isomorphic to a Fischer space $Sp(V, f)$ or $\mathcal{N}(V, q)$ for some vector space V over the field of order 2 and some form f or q on V, or $X \cong T(\Omega, \Omega')$ for some pair of disjoint sets Ω, Ω'.*

Notice Hall requires that the subspace generated by any pair of distinct intersecting lines be the dual affine plane of order 2; thus Hall is excluding the possibility of affine subplanes. However he does not require his partial linear space to be reduced. The hypothesis that X has no affine subplanes plus the nondegeneracy conditions requiring connectivity and at least two lines force X to be 3-reduced, but leave (as we have seen) plenty of examples that are not 2-reduced.

The Cuypers–Hall Theorem says that if D is a conjugacy class of 3-transpositions of a group G with $Z(G) = 1$ then generically G is the split extension of an abelian subgroup A by a D-subgroup H isomorphic to $FSp(V, f)$, $FO(V, q)$, $FU(V, h)$, or $^+\Omega(V, Q)$ for some symplectic space (V, f) or orthogonal space (V, q) over $GF(2)$, some unitary space (V, h) over $GF(4)$, or some orthogonal space (V, Q) over $GF(3)$. Moreover as an H-module, A is the direct sum of copies of V. There is however a long list of exceptional examples that are more complicated. We direct the reader to the paper [CH1] for the complete list of examples.

The theory of 3-transposition groups can be embedded in a larger theory in a different way. Recall the notion of a *group of Lie type*; for example there is a discussion of these groups in Section 47 of [FGT]. In particular each finite group G of Lie type comes equipped with a Tits system (G, B, N, S) with a finite Weyl group $W = N/(B \cap N)$ and a root system Φ. Further there exists an injection $\alpha \mapsto U_\alpha$ of Φ into the set of subgroups of G, and we call U_α the *root subgroup* of G determined by α. The conjugates of the subgroups U_α under G are called root subgroups of G. Now W has one or two orbits on the set Φ of roots, with the conjugates U_α^g, $g \in G$, of the root subgroup of a long root α called the *long root subgroups* of G. It develops:

(20.4)   *Let G be a finite group of Lie type over the field of order q distinct from $L_2(q)$, $^2B_2(q)$, $^2G_2(q)$, and $^2F_4(q)$. Let X be the set of subgroups $Z(U)$ as U varies over the long root subgroups of G. Then*

  (1) $X \cong E_q$ for each $X \in X$.
  (2) For distinct $X, Y \in X$, $\langle X, Y \rangle$ is isomorphic to E_{q^2}, $SL_2(q)$, or a special group of order q^3 with $Z(\langle X, Y \rangle) \in X$.
  (3) If G is $Sp_n(q)$ or $U_n(q)$ then the last case does not occur.

Here is a sketch of a proof. Let $X \in X$. Then $P = N_G(X)$ is a parabolic subgroup of G corresponding to some set J of simple roots, so by Exercise 14.6 in [FGT], there is a bijection $W_J w W_J \mapsto X^{wP}$ between the orbits of W_J on W/W_J and the orbits of P on X. Thus if $w_1, \ldots, w_r$ are double coset representatives for W_J in W, we get a set of representatives $(X, X^{w_i}), 1 \le i \le r$, for the orbits of G on $\mathcal{X} \times \mathcal{X}$. It is then relatively easy to inspect this set of representatives to complete the proof.

Example:

(9) The nontrivial elements of a group G of Lie type fused into long root subgroups are called *long root elements* of G. Suppose $q = 2$ and let D be the set of long root elements of G. Then each long root subgroup is generated by an involution in D. Hence by 20.4, if $a, b \in D$ then $|ab| \leq 4$ with $[a, b] \in D$ if $|ab| = 4$.

Following Timmesfeld in [T1], define D to be a set of $\{3, 4\}^+$-*transpositions* of a finite group G if D is a union of conjugacy classes of involutions of G that generates G and such that for all $a, b \in D$, $|ab| \leq 4$ with $[a, b] \in D$ in case of equality. This last part of the definition is the "+" condition. Then 20.2 says the set of long root elements in a finite group of Lie type over the field of order 2 is a class of $\{3, 4\}^+$-transpositions. Moreover if G is $Sp_n(2)$ or $U_n(2)$ then D is even a set of 3-transpositions of G.

Example:

(10) Let G be a finite group of Lie type over a field of even order q and D the set of long root elements of G. Then the elements of D are involutions and for $a, b \in D$, $a \in X \in \mathcal{X}$ and $b \in Y \in \mathcal{X}$, so $\langle a, b \rangle \leq \langle X, Y \rangle$. If $\langle X, Y \rangle$ is abelian then $|ab| \leq 2$. If $\langle X, Y \rangle$ is special of order q^3 with $Z(\langle X, Y \rangle) \in \mathcal{X}$, then $|ab| = 4$ with $[a, b] \in D$. Finally if $\langle X, Y \rangle \cong SL_2(q)$ then $|ab|$ divides $q^2 - 1$, so in particular $|ab|$ is odd. Indeed if G is $U_n(q)$ or $Sp_n(q)$ then the case where $\langle X, Y \rangle$ is special of order q^3 does not occur, so D is a set of odd transpositions of G in the sense of Exercise 2.2 or Section 24. That is for all $a, b \in D$, either $|ab| = 2$ or $|ab|$ is odd.

Again following Timmesfeld, this time in [T2], define D to be a set of *root involutions* of a finite group G if D is a union of conjugacy classes of involutions of G that generates G and such that for all $a, b \in D$, $|ab|$ is odd, $|ab| = 2$, or $|ab| = 4$ and $[a, b] \in D$. Thus the long root elements in a finite group of Lie type over a field of even order form a class of root elements in the sense of Timmesfeld. Papers of the author [A1] and Timmesfeld [T2] classify the almost simple finite groups generated by a class of root involutions. Examples consist of the 3-transposition groups, the groups of Lie type and even characteristic, plus a few extra classes of examples. Cuypers, Hall, and Timmesfeld have investigated infinite groups containing a conjugacy class $\mathcal{X}$ of subgroups satisfying hypotheses like those in 20.4.2.

Exercises

1. A *Shult space* is a partial linear space $(X, \mathcal{L})$ such that there exist noncollinear points whereas for each $l \in \mathcal{L}$ and each $x \in X$, x is collinear with exactly one point of l or all points of l. Let $G = Sp_n(2)$ or $U_n(2)$, $n \geq 4$, D the set of transvections in G, and S the set of singular lines $d * b$, $d \in D$, $b \in D_d$. Also let V be the space defining G, X the set of singular points of V, and $\mathcal{L}$ the set of totally singular lines in V. Prove
   (1) (D, S) is a Shult space.
   (2) $(X, \mathcal{L})$ is a Shult space.
   (3) $(D, S) \cong (X, \mathcal{L})$.
2. Prove $\mathcal{T}(\Omega, \Omega')$ and $\mathcal{S}p(V, f)$ are Fischer spaces.

PART II

THE EXISTENCE AND UNIQUENESS OF THE FISCHER GROUPS

Introduction

In Part II of *3-Transposition groups* we prove the existence of the three Fischer groups and show each Fischer group is the unique finite group with a certain involution centralizer. Establishing the existence of the Fischer groups supplies a coda to Fischer's Theorem by demonstrating that the sporadic 3-transposition groups of type $M(22)$, $M(23)$, and $M(24)$ appearing in the statement of Fischer's Theorem do indeed exist. It also establishes the existence of these sporadic groups for purposes of the classification of the finite simple groups.

Fischer's Theorem says the Fischer groups are unique as 3-transposition groups with suitable properties, but this is not the right uniqueness result for purposes of the classification. The centralizer of involution characterizations of Part II supplies the appropriate uniqueness results.

The finite simple groups are classified in terms of properties of their local subgroups and particularly centralizers of suitable involutions. See Section 48 in [FGT] and [SG] for a more detailed discussion of the classification and the place of characterizations by involution centralizers in the classification. In particular in the Introduction to [SG] the author's preferred hypothesis, Hypothesis $\mathcal{H}(w, L)$, for characterizing sporadic groups is discussed. Let w be a positive integer and L a group; consider the following hypothesis:

Hypothesis $\mathcal{H}(w, L)$. *G is a finite group containing an involution z such that $F^*(C_G(z)) = Q$ is an extraspecial 2-subgroup of order 2^{2w+1}, $C_G(z)/Q \cong L$, and z is not weakly closed in Q with respect to G.*

The majority of the sporadic groups satisfy Hypothesis $\mathcal{H}(w, L)$ for suitable w and L, and it seems that the right uniqueness result characterizing those sporadics says: Up to isomorphism there exists at most one finite group G satisfying Hypothesis $\mathcal{H}(w, L)$. In [SG], 5 of the 26 sporadics are characterized by this hypothesis. For example it is shown in [SG] that Conway's largest group Co_1 is the unique finite group satisfying Hypothesis $\mathcal{H}(4, \Omega_8^+(2))$.

In Lemma 32.4 in [SG] it is shown that there exists a subgroup X of the largest sporadic group the Monster, such that $E(X)$ is quasisimple with center Z of order 3, X is the split extension of $E(X)$ by an involution inverting Z, and $E(X)/Z$ satisfies Hypothesis $\mathcal{H}(6, \mathbf{Z}_2/U_4(3)/\mathbf{Z}_3)$. We say that a group satisfying Hypothesis $\mathcal{H}(6, \mathbf{Z}_2/U_4(3)/\mathbf{Z}_3)$ is of *type* F_{24}. Further we say a group $\hat{G}$ is of *type* $\mathrm{Aut}(F_{24})$ if $\hat{G}$ possesses a subgroup G of index 2 and type F_{24} such that if z is a 2-central involution in G then $O_2(C_G(z)) = O_2(C_{\hat{G}}(z))$. In Chapter 11 we show that up to isomorphism there are unique groups of type F_{24} and type $\mathrm{Aut}(F_{24})$ and that each group of type $\mathrm{Aut}(F_{24})$ is a 3-transposition group of type $M(24)$. This proves the existence of the Fischer groups and supplies the necessary centralizer of involution characterization of the derived group of $M(24)$.

Although the majority of the sporadic groups satisfy Hypothesis $\mathcal{H}(w, L)$ for some w and L, some do not. In particular $M(22)$ and $M(23)$ have no such involutions (although $\mathrm{Aut}(M(22))$ does; cf. Lemma 37.8). When a simple group G has no involution z such that $F^*(C_G(z))$ is an extraspecial 2-group, the next best choice is an involution t such that $C_G(t)/\langle t \rangle$ is a nonabelian simple group. Here two cases are possible: $C_G(t)$ splits over $\langle t \rangle$ or does not split. The first case is the less attractive since we are always assured of at least one nontrivial example that is not simple. Namely if L is a nonabelian simple group then the wreath product $G = L \,\mathrm{wr}\, \mathbf{Z}_2$ possesses an involution t such that $C_G(t) \cong \mathbf{Z}_2 \times L$. The existence of this example virtually assures that the analysis of groups with such an involution centralizer will be fairly complicated.

On the other hand if $C_G(t)$ does *not* split over $\langle t \rangle$ then $C_G(t)$ is quasisimple, the troublesome wreath product example does not arise, and usually the corresponding centralizer of involution characterization goes smoothly. In particular this is the case for $M(22)$ and $M(23)$, which are of types $\tilde{M}(22)$ and $\tilde{M}(23)$, respectively. Thus in Chapter 11 we prove that up to isomorphism there exists at most one group of type $\tilde{M}$ for $M = M(22)$ and $M(23)$. This seems to be the best uniqueness result for the smaller Fischer groups for the purposes of the classification.

The first four chapters in Part II cover preliminary material necessary to the proof of the existence and uniqueness of the Fischer groups in Chapter 11. Section 21 contains a discussion of 2-cohomology that we use to establish that certain subgroups of the Fischer groups are determined up to isomorphism. Section 22 contains a discussion of the Todd modules for the Mathieu groups. We find in Lemma 25.7 that if G is of type $M(n)$ with 3-transposition set D, $T \in \mathrm{Syl}_2(G)$, and $A = \langle T \cap D \rangle$, then A is the Todd module for $N_G(A)/A \cong M_n$.

This subgroup plays a big role in our analysis, so we need detailed facts about the Todd modules that appear in Section 22. Section 23 contains information about $U_6(2)$ and its covering group.

We prove the uniqueness of groups of type $\tilde{M}$ by demonstrating each such group is a 3-transposition group of type M, and then appealing to Fischer's Theorem. To do so we need some techniques for showing a suitable set D of involutions in a group G is a set of 3-transpositions of G. Results of this flavor appear in Section 24. Then Section 25 contains some specialized results about 3-transposition groups that we will need to prove the uniqueness of the Fischer groups.

We also use some techniques with the flavor of combinatorial topology in our uniqueness proofs. This machinery was developed by Yoav Segev and the author in [AS], is discussed in Chapters 12 and 13 of [SG], and is recalled in Chapter 9 of this book. With these techniques, we can say that a group G is determined up to isomorphism by a pair of subgroups G_1 and G_3 and by $G_1 \cap G_1^t$ for some involution $t \in G_3 - G_1$. Such a characterization of $M(23)$ and $M(24)$ appears in Lemma 27.8.

In Chapter 10 we consider a certain 6-dimensional lattice over $\mathbf{Z}[e^{2\pi i/2}]$ and use it to establish the embedding of $U_4(3)$ in $U_6(2)$ and to prove numerous properties of this embedding. For example we used one fact about this embedding to prove Lemma 15.10.

Finally Chapter 15 brings together all this machinery in a proof of the existence and uniqueness of the Fischer groups. We begin in Section 30 with a centralizer of involution characterization of $U_6(2)$. We then go on in Sections 31 and 32 to prove the uniqueness of groups of type $\tilde{M}(22)$ and $\tilde{M}(23)$, respectively. These three uniqueness results are established by using the machinery from Section 24 to show our group is generated by a class of 3-transpositions, and then appealing to Fischer's Theorem.

In Section 33 we provide a weak characterization of $M(24)$ as a group G with an involution d such that $C_G(d) = \langle d \rangle \times E(C_G(d))$ with $E(C_G(d))$ of type $M(23)$ and $\tilde{M}(23)$, and satisfying two other conditions. As mentioned, without the extra conditions, the wreath product $M(23) \, \mathrm{wr} \, \mathbf{Z}_2$ would appear here and require a much more difficult proof. But this weak result is sufficient for our purposes. Again we prove the theorem by showing our group is generated by 3-transpositions and appealing to Fischer's Theorem.

In Section 34 we begin our analysis of groups of type F_{24}. We produce a subgroup M that is a nonsplit extension of $A \cong E_{2^{11}}$ by M_{24} with A the Todd module for M/A. The results in Chapter 7 on cohomology and the Todd modules allow us to conclude M is determined up to isomorphism. Then we use our earlier characterizations of $U_6(2)$ and $M(22)$ from Sections 30 and 31 to prove there is an involution t in G with $F^*(C_G(t))$ quasisimple and $C_G(t)/\langle t \rangle \cong$ Aut($M(22)$). In particular at this point we have established the existence of groups of type $M(22)$ and shown such a group is of type $\tilde{M}(22)$.

Then in Section 35 we prove the uniqueness of groups of type Aut(F_{24}) by achieving the hypotheses of Section 33 and appealing to the weak characterization appearing there. Since Section 33 shows our group is generated by 3-transpositions, and since we know that groups of type Aut(F_{24}) exist as sections of the Monster, at this point we have established the existence of the Fischer groups. The proof uses the results on groups of type F_{24} from the previous section.

Finally in Section 36, we complete our proof of the uniqueness of groups of type F_{24}. This proof uses the existence of the Fischer groups. This is why we first proved that groups of type Aut(F_{24}) are of type $M(24)$. The proof proceeds by constructing a subgroup of type $M(23)$ using the combinatorial topological result of Chapter 9, and then using this subgroup together with the subgroup M alluded to previously and the combinatorial topology of Chapter 9 to prove G is unique. Notice we can't prove G is a 3-transposition group as the Fischer group $M(24)$ is not simple, so the 3-transpositions do not live in the simple derived subgroup of type F_{24}.

The Fischer groups $M(22)$ and $M(23)$ were first characterized as groups of type $\tilde{M}$ by Hunt in [Hu2] and [Hu3]. Hunt assumes the existence of the groups $M(22)$ and $M(23)$; this assumption supplies a significant technical advantage. He also uses the character tables of several groups (including $M(22)$), which he calculates in [Hu1]. His initial approach is similar to the one given here, but the latter part of his argument diverges from ours, using the existence of the Fischer groups heavily, analyzing the fusion of 2-elements, and using the Thompson order formula.

Parrott was the first to characterize $M(24)'$ as a group of type F_{24} [P]. His approach is much different from ours, again involving detailed analysis of fusion of 2-elements and the Thompson order formula. He does construct the $M(23)$-subgroup and analyze the graph on the cosets of this subgroup corresponding to the 3-transpositions of $M(24)$, but identifies this graph as the $M(24)$ graph in a different way.

See [vBW] for another proof by van Bon and Weiss of the existence of the Fischer groups.

7

Some group extensions

In this chapter we investigate certain extensions of groups. We begin in Section 21 with a general discussion of the 2-cohomology group $H^2(G, V)$ of the representation of a group G on an abelian group V. The 2-cohomology group keeps track of the number of isomorphism types of extensions $\hat{G}$ of V by G in which the induced representation of $\hat{G}/V$ on V is equivalent to that of G on V. We recall the standard theory of 2-cohomology in Section 21. The result we need is 21.8, which guarantees that if G_1 and G_2 are two such extensions with V an abelian p-group, G faithful on V, and H_i a subgroup of G_i with $|G_i : H_i|$ prime to p, then under suitable conditions an isomorphism $\varphi : H_1 \to H_2$ extends to an isomorphism of G_1 with G_2.

In Section 22 we recall facts about the Todd modules V for the Mathieu groups G found in [SG]. Then we go on to determine the 1-cohomology group $H^1(G, V)$ of the Todd modules. The 1-cohomology group has at least three important group theoretic interpretations; see for example Section 17 in [FGT]. We will use the results on the 1-cohomology of the Todd modules in proving the uniqueness of the Fischer groups subject to suitable hypotheses. In particular we use such results in Section 23 to determine the perfect central extensions of certain sections of the Fischer groups. We also determine the conjugacy classes of elements of order 2 and 3 in $U_6(2)$ in Section 23.

21. Some 2-cohomology

In this section G is a group, V is an abelian group written additively, and $\pi : G \to \text{Aut}(V)$ is a group homomorphism. We record some facts about the 2-cohomology group $H^2(G, V)$ of the representation π of G on V. Most of the discussion is standard.

Write $\text{Ext}(G, V) = \text{Ext}(G, V, \pi)$ for the set of triples $(\hat{G}, \alpha, \beta)$ such that $\hat{G}$ is a group, $\alpha : \hat{G} \to G$ is a surjective homomorphism, and $\beta : \ker(\alpha) = \hat{V} \to V$

is an isomorphism such that $u\beta^{x\alpha\pi} = (u^x)\beta$ for all $x \in \hat{G}$ and $u \in \hat{V}$. Equivalently the following diagram commutes for all $x \in \hat{G}$,

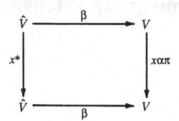

where $x^* : \hat{v} \mapsto \hat{v}^x$ is the homomorphism induced on $\hat{V}$ by x via conjugation. Thus $\mathrm{Ext}(G, V, \pi)$ is in essence the set of extensions of V by G with respect to the representation π.

REMARK 21.1: Let G_1 be a group, $\alpha : G_1 \to G$ a surjective homomorphism, $V_1 = \ker(\alpha)$, and $\beta : V_1 \to V$ an isomorphism. Let $\tilde{G}_1 = G_1/V_1, \tilde{\alpha} : \tilde{G}_1 \to G$ the isomorphism induced by α, and $\pi_1 : \tilde{G} \to \mathrm{Aut}(V_1)$ the representation induced by the action of G_1 on V_1 via conjugation. That is $\tilde{x}\tilde{\pi}_1 : u \mapsto u^x$ for $x \in G_1$ and $u \in V_1$. Observe that $(G_1, \alpha, \beta) \in \mathrm{Ext}(G, V, \pi)$ if and only if $\tilde{\alpha}, \beta$ is a quasiequivalence of the representations π_1 and π, in the sense of Section 1.

Define an *isomorphism* $\varphi : (\hat{G}_1, \alpha_1, \beta_1) \to (\hat{G}_2, \alpha_2, \beta_2)$ of extensions to be a group isomorphism $\varphi : \hat{G}_1 \to \hat{G}_2$ such that $\alpha_1 = \varphi\alpha_2$ and $\beta_1 = \varphi\beta_2$. That is the following diagrams commute:

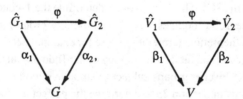

Let $Z^2(G, V) = Z^2(G, V, \pi)$ be the set of functions

$$f : G \times G \to V$$

such that

(B0) $f(g, 1) = f(1, g) = 0$ for all $g \in G$.

(B1) $f(g, h)^{k\pi} + f(gh, k) = f(h, k) + f(g, hk)$ for all $g, h, k \in G$.

And let $B^2(G, V) = B^2(G, V, \pi)$ be the set of functions $f_\theta : G \times G \to V$ determined by maps $\theta : G \to V$ with $\theta(1) = 0$ via

$$f_\theta(g, h) = \theta(g)^{h\pi} + \theta(h) - \theta(gh).$$

Make $Z^2(G, V)$ into an abelian group by defining addition componentwise and observe that $B^2(G, V)$ forms a subgroup of $Z^2(G, V)$. $Z^2(G, V)$ is the group of *2-cocycles* of the representation, $B^2(G, V)$ is the subgroup of *2-coboundaries* and $H^2(G, V) = Z^2(G, V)/B^2(G, V)$ is the *second cohomology group* of the representation. We say two cocycles $f, f' \in Z^2(G, V)$ are *cohomologous* if $f + B^2(G, V) = f' + B^2(G, V)$.

We see in the next few lemmas that the second cohomology group counts the number of isomorphism types of extensions of V by G with respect to π.

(21.2) *Let* $(\hat{G}, \alpha, \beta) \in \text{Ext}(G, V)$, $\mathcal{X}$ *the set of all sets of coset representatives for* $\hat{V}$ *in* $\hat{G}$ *containing* 1, *and for* $X \in \mathcal{X}$ *define* $f = f^X : G \times G \to V$ *by*

$$f(g, h) = (x_{gh}^{-1} x_g x_h)\beta, \quad \text{where } x_g = X \cap g\alpha^{-1}.$$

Then

(1) $f^X \in Z^2(G, V)$.

(2) *There is a bijection* $\theta \mapsto X_\theta$ *between the set of maps* $\theta : G \to V$ *and* $\mathcal{X}$ *given by* $X_\theta = \{x \cdot \theta(x\alpha)\beta^{-1} : x \in X\}$.

(3) $f^{X_\theta} = f^X + f_\theta$.

(4) X *is a subgroup of* $\hat{G}$ *if and only if* $f^X = 0$. *Thus* $\hat{G}$ *splits over* $\hat{V}$ *if and only if* f^X *is homologous to* 0.

Proof: As $1 \in X$, f^X satisfies condition (B0). Associativity in $\hat{G}$ implies condition (B1). Thus (1) holds. Part (2) is trivial and part (3) is a straightforward calculation. The first statement of (4) is trivial and together with (2) and (3) implies the second statement in (4).

(21.3) *Let* $(\hat{G}_i, \alpha_i, \beta_i) \in \text{Ext}(G, V)$, $i = 1, 2$, *and let* X_i *be a set of coset representatives for* $\hat{V}_i$ *in* $\hat{G}_i$. *Then*

(1) *A map* $\varphi : \hat{G}_1 \to \hat{G}_2$ *defines an isomorphism* $\varphi : (\hat{G}_1, \alpha_1, \beta_1) \to (\hat{G}_2, \alpha_2, \beta_2)$ *if and only if* $\alpha_1 = \varphi\alpha_2$, $\beta_1 = \varphi\beta_2$, $(xu)\varphi = x\varphi u\varphi$ *for each* $x \in X_1$ *and* $u \in \hat{V}_1$, *and* $f^{X_1} = f^{X_1\varphi}$.

(2) *The extension* $(\hat{G}_1, \alpha_1, \beta_1)$ *is isomorphic to* $(\hat{G}_2, \alpha_2, \beta_2)$ *if and only if* f^{X_1} *is cohomologous to* f^{X_2}.

Proof: Part (1) is straightforward. For example if $\varphi : \hat{G}_1 \to \hat{G}_2$ satisfies the conditions of (1) then $\varphi = \beta_1\beta_2^{-1} : \hat{V}_1 \to \hat{V}_2$ is an isomorphism. Also φ induces $\tilde{\varphi} : \tilde{G}_1 = \hat{G}_1/\hat{V}_1 \to \hat{G}_2/\hat{V}_2$ via $\tilde{\varphi} : x\hat{V}_1 \to x\varphi\hat{V}_2$ for $x \in X_1$. Then as $\alpha_1 = \varphi\alpha_2$, $\tilde{\varphi} = \tilde{\alpha}_1\tilde{\alpha}_2^{-1} : \tilde{G}_1 \to \tilde{G}_2$ is an isomorphism. Therefore $X_1\varphi$ is a set of coset representatives for $\hat{V}_2$ in $\hat{G}_2$. Then as $(xu)\varphi = x\varphi u\varphi$ for $x \in X_1$ and $u \in \hat{V}_1$, $(x_g\hat{V}_1)\varphi = y_g\hat{V}_2$ for each $g \in G$, where x_g, y_g are the members of

X_1 and $X_1\varphi$ mapping onto $g \in G$ under α_1 and α_2, respectively. In particular $\varphi : \hat{G}_1 \to \hat{G}_2$ is a bijection. Also for $x \in X_1$ and $u \in \hat{V}_1$,

$$(u^x)\varphi\beta_2 = (u^x)\beta_1 = u\beta_1^{x\alpha_1\pi} = (u\varphi\beta_2)^{x\varphi\alpha_2\pi} = (u\varphi^{x\varphi})\beta_2, \qquad (u^x)\varphi = u\varphi^{x\varphi}.$$

Finally use this observation and $f^{X_1} = f^{X_1\varphi}$ to check φ is a group homomorphism. We leave the other direction of (1) as an exercise.

So (1) holds and it remains to prove (2). Suppose $\varphi : (\hat{G}_1, \alpha_1, \beta_1) \to (\hat{G}_2, \alpha_2, \beta_2)$ is an isomorphism. Then by (1), $f^{X_1\varphi} = f^{X_1}$ and by 21.2, $f^{X_1\varphi}$ is cohomologous to f^{X_2}. Conversely if f^{X_1} is cohomologous to f^{X_2} then by 21.2 we may assume the cocycles are equal and then (1) says we can define an isomorphism $\varphi : (\hat{G}_1, \alpha_1, \beta_1) \to (\hat{G}_2, \alpha_2, \beta_2)$ via $(x_g u)\varphi = y_g u\beta_1\beta_2^{-1}$ using the notation of the previous paragraph.

(21.4) Let $f \in Z^2(G, V)$ and define $\hat{G}(f)$ to be the set product $G \times V$ with multiplication defined by

$$(g, v)(h, u) = (gh, v^{h\pi} + u + f(g, h)).$$

Then $(\hat{G}(f), p_G, p_V) \in \text{Ext}(G, V)$, where $p_G : (g, v) \mapsto g$ and $p_V : (g, v) \mapsto v$ are the projections on G and V, respectively, and $X = \{(g, 0) : g \in G\}$ is a set of coset representatives for $\hat{V}$ in $\hat{G}$ with $f^X = f$.

Proof: This is an easy calculation.

Theorem 21.5. The map $f + B^2(G, V) \mapsto [\hat{G}(f), p_G, p_V]$ is a bijection between $H^2(G, V)$ and isomorphism classes of extensions of V by G with respect to π.

Proof: By 21.4, $(\hat{G}(f), p_G, p_V)$ is an extension of V by G with respect to π. If $f_\theta \in B^2(G, V)$ then by 21.3.2,

$$[\hat{G}(f), p_G, p_V] = [\hat{G}(f + f_\theta), p_G, p_V]$$

so the map $\varphi : f \mapsto [\hat{G}(f), p_G, p_V]$ is well defined. Similarly φ is a bijection by 21.3.2.

In the remainder of this section assume G and V are finite.

(21.6) Let p be a prime and assume V is a p-group and $H \leq G$ with $|G : H|$ prime to p. Then the restriction map $f + B^2(G, V) \mapsto f_{|H} + B^2(H, V)$ induces an injection of $H^2(G, V)$ into $H^2(H, V)$.

Proof: First if $f \in Z^2(G, V)$ then $f_{|H} \in Z^2(H, V)$ and if $\theta : G \to V$ with $\theta(1) = 0$ then $(f_\theta)_{|H} = f_{\theta_{|H}}$, so the map ψ of the lemma is well defined. Visibly ψ preserves addition, so it remains to show $\ker(\psi) = 0$.

Let $f + B^2(G, V) \in \ker(\psi)$, $\hat{G} = \hat{G}(f)$, and $\hat{H} = H \times V \leq \hat{G}$. Then $\phi = f_{|H}$ is a coboundary, so $\hat{H} = \hat{H}(\phi)$ splits over $\hat{V}$ by 21.2.4. Therefore as $\hat{V}$ is an abelian p-group and $|\hat{G} : \hat{H}|$ is prime to p, $\hat{G}$ splits over $\hat{H}$ by Gaschütz's Theorem (cf. 2.6). Then by 21.2.4, $f + B^2(G, V) = 0$.

(21.7) *Let p be a prime, V a p-group, $(G_i, \alpha_i, \beta_i) \in \mathrm{Ext}(G, V)$, $i = 1, 2$, and $H_i \leq G_i$ with $|G_i : H_i|$ prime to p. Assume $H_1\alpha_1 = H = H_2\alpha_2$ and $\varphi : (H_1, (\alpha_1)_{|H_1}, \beta_1) \to (H_2, (\alpha_2)_{|H_2}, \beta_2)$ is an isomorphism of extensions of V by H with respect to $\pi_{|H}$. Then φ extends to an isomorphism $\psi : (G_1, \alpha_1, \beta_1) \to (G_2, \alpha_2, \beta_2)$ of extensions.*

Proof: Pick a set X_H of coset representatives for V_1 in H_1 and extend X_H to a set X of coset representatives for V_1 in G_1. Let $f = f^X$ be the cocycle determined by X and $Y_H = X_H\varphi$. As φ is an isomorphism of extensions, 21.3.1 says $f_{|H} = f^{X_H} = f^{Y_H}$.

Let Y be a set of coset representatives for V_2 in G_2 containing Y_H and $\phi = f^Y$. Then $f_{|H} = f^{Y_H} = \phi_{|H}$ so by 21.6, f and ϕ are cohomologous. Thus $f = \phi + f_\theta$ for some $\theta : G \to V$ with $\theta(1) = 0$, so by 21.2.3, $f = f^{Y_\theta}$. As $f^{Y_H} = f_{|H} = f^{Y_\theta}_{|H}$, replacing Y by Y_θ, we may assume $f = f^Y$. Then by the proof of 21.3.2, the map $\psi : (G_1, \alpha_1, \beta_1) \to (G_2, \alpha_2, \beta_2)$ is an isomorphism, where ψ is defined by $(x_g u)\psi = y_g(u\beta_1\beta_2^{-1})$, for $x_g \in X$ and $u \in V_1$, and $g = x_g\alpha_1 = y_g\alpha_2$. In particular for $g \in H$, $y_g = x_g\varphi$ and as $\varphi : (H_1, (\alpha_1)_{|H_1}, \beta_1) \to (H_2, (\alpha_2)_{|H_2}, \beta_2)$ is an isomorphism, $\beta_1\beta_2^{-1} = \varphi$ on V_1, so $(x_g u)\psi = x_g\varphi u\varphi = (x_g u)\varphi$. That is ψ extends φ.

(21.8) *Let $\pi_i : G_i \to \mathrm{Aut}(V_i)$, $i = 1, 2$, be faithful representations of a group G_i on an object V_i, and $\alpha : H_1 \to H_2$, $\beta : V_1 \to V_2$, a quasiequivalence of the restricted representations $\pi_i : H_i \to \mathrm{Aut}(V_i)$. Assume*

    (1) There is a quasiequivalence $\psi : G_1 \to G_2$, $\varphi : V_1 \to V_2$ of the representations π_1 and π_2 with $H_1\psi = H_2$.

    (2) $H_2\pi_2$ is contained in a unique $\mathrm{Aut}(V_2)$ conjugate of $G_2\pi_2$.

Then α extends to $\gamma : G_1 \to G_2$ such that γ, β is a quasiequivalence of π_1 and π_2.

Proof: As π_1 and π_2 are faithful, we may take them to be inclusions. Then the statement that α, β and ψ, φ are quasiequivalences reduces to $\alpha = \beta^*$ on H_1 and $\psi = \varphi^*$. Now $H_2 \leq G_2 = G_1\psi = G_1\varphi^*$ and $H_2 = H_1\alpha = H_1\beta^* \leq G_1\beta^*$. But π_1 is quasiequivalent to $\pi_1\varphi^*$ and $\pi_1\beta^*$, so $\pi_1\varphi^*$ is quasiequivalent to $\pi_1\beta^*$. Therefore $G_2 = G_1\varphi^*$ and $G_2\beta^*$ are conjugate in $\mathrm{Aut}(V_2)$, so as $H_2 \leq G_2 \cap G_1\beta^*$, condition (2) says that $G_2 = G_1\beta^*$. Thus the pair β^*, β is a quasiequivalence of π_1 and π_2 with $\beta^* = \alpha$ on H_1. That is we may take γ to be β^*.

REMARK 21.9: Let V be an object in some category and $H \leq G \leq \text{Aut}(V)$. Then H is contained in a unique $\text{Aut}(V)$-conjugate of G if and only if

(1) $N_{\text{Aut}(V)}(H) \leq N_{\text{Aut}(V)}(G)$, and
(2) $H^{\text{Aut}(V)} \cap G = H^{N_{\text{Aut}(V)}(G)}$.

(21.10) *Let V_i be a normal self-centralizing abelian p-subgroup of the finite group G_i and $H_i \leq G_i$ with $|G_i : H_i|$ prime to p for $i = 1, 2$. Let $\tilde{G}_i = G_i/V_i$, $\pi_i : \tilde{G}_i \to \text{Aut}(V_i)$ the representation induced by the action of G_i on V_i via conjugation, and assume*

(1) There exists an isomorphism $\varphi : H_1 \to H_2$ with $V_1\varphi = V_2$.
(2) There exists a quasiequivalence $\xi : \tilde{G}_1 \to \tilde{G}_2$, $\zeta : V_1 \to V_2$ of the representations π_1 and π_2 with $\tilde{H}_1\xi = \tilde{H}_2$.
(3) $\tilde{H}_2\pi_2$ is contained in a unique $\text{Aut}(V_2)$-conjugate of $\tilde{G}_2\pi_2$.

Then φ extends to an isomorphism $\psi : G_1 \to G_2$.

Proof: Let $V = V_1$, $G = \tilde{G}_1$, $\pi = \pi_1$, $\alpha_1 : G_1 \to \tilde{G}_1$ the natural map, and $\beta_1 : V_1 \to V_1$ the identity map. Then $(G_1, \alpha_1, \beta_1) \in \text{Ext}(G, V, \pi)$. Next define $\alpha_2 = q\xi^{-1} : G_2 \to G$, where $q : G_2 \to \tilde{G}_2$ is the natural map, and define $\beta_2 : V_2 \to V$ by $\beta_2 = \zeta^{-1}$. As ξ, ζ is a quasiequivalence of the representations π_1 and π_2, $(G_2, \alpha_2, \beta_2) \in \text{Ext}(G, V, \pi)$ by Remark 21.1.

Let $\tilde{\varphi} : \tilde{H}_1 \to \tilde{H}_2$ be the map induced by φ. By (3), $\tilde{H}_2\pi_2$ is contained in a unique $\text{Aut}(V_2)$-conjugate of $\tilde{G}_2\pi_2$. Therefore by (2) and 21.8, $\tilde{\varphi}$ extends to $\gamma : \tilde{G}_1 \to \tilde{G}_2$ such that γ and $\varphi : V_1 \to V_2$ define a quasiequivalence of π_1 and π_2. So replacing ξ, ζ by γ, φ, we may assume ξ extends $\tilde{\varphi}$ and $\zeta = \varphi$ on V_1. Therefore

$$\varphi : (H_1, (\alpha_1)_{|H_1}, \beta_1) \to (H_2, (\alpha_2)_{|H_2}, \beta_2)$$

is an isomorphism. Then we appeal to 21.7.

(21.11) *Let V_i be a normal abelian p-subgroup of the finite group G_i, $i = 1, 2$, and $H_i \leq G_i$, with $|G_i : H_i|$ prime to p. Let $\tilde{G}_i = G_i/V_i$ and $\pi_i : \tilde{G}_i \to \text{Aut}(V_i)$ the representation induced by conjugation, and assume*

(1) There exists an automorphism $\varphi : H_i \to H_2$ with $V_1\varphi = V_2$.
(2) There exists an isomorphism $\xi : \tilde{G}_1 \to \tilde{G}_2$ extending the induced map $\tilde{\varphi} : \tilde{H}_1 \to \tilde{H}_2$.

(3) There exists a quasiequivalence

$$\eta : \hat{G}_1 = G_1/C_{G_1}(V_1) \to G_2/C_{G_2}(V_2) = \hat{G}_2, \qquad \mu : V_1 \to V_2$$

of the representations $\hat{\pi}_1$ and $\hat{\pi}_2$ induced by π_1 and π_2 with $\hat{H}_1\eta = \hat{H}_2$.

(4) $\hat{H}_2\hat{\pi}_2$ is contained in a unique $\mathrm{Aut}(V_2)$-conjugate of $\hat{G}_2\hat{\pi}_2$.

(5) $C_{\mathrm{Aut}(\hat{G}_1)}(\hat{H}_1) = 1$.

(6) $C_{G_i}(V_i) = \langle C_{H_i}(V_i)^{G_i}\rangle$.

Then there exists an isomorphism $\psi : G_1 \to G_2$ extending φ with $\tilde{\psi} = \xi$.

Proof: Let $\hat{\varphi} : \hat{H}_1 \to \hat{H}_2$ be the map induced by φ. By (3), (4), and 21.8, $\hat{\varphi}$ extends to $\eta_0 : \hat{G}_1 \to \hat{G}_2$ such that η_0 and $\varphi : V_1 \to V_2$ define a quasiequivalence of $\hat{\pi}_1$ and $\hat{\pi}_2$, so we may assume $\eta = \eta_0$ extends $\hat{\varphi}$ and $\mu = \varphi$ on V_1. Then the induced map $\hat{\xi} : \hat{G}_1 \to \hat{G}_2$ and η extend $\hat{\varphi}$, so $\hat{\xi}\eta^{-1} \in C_{\mathrm{Aut}(\hat{G}_1)}(\hat{H}_1) = 1$ by (5), and hence $\eta = \hat{\xi}$. This says that ξ, φ define a quasiequivalence of π_1 and π_2. Namely

$$C_{\tilde{H}_1}(V_1)\xi = C_{H_1}(V_1)\varphi/V_2 = C_{H_2}(V_2)/V_2 = C_{\tilde{H}_2}(V_2)$$

so

$$C_{\tilde{G}_1}(V_1)\xi = \langle C_{\tilde{H}_1}(V_1)^{\tilde{G}_1}\rangle\xi = \langle(C_{\tilde{H}_1}(V_1)\xi)^{\tilde{G}_2}\rangle = \langle C_{\tilde{H}_2}(V_2)^{\tilde{G}_2}\rangle = C_{\tilde{G}_2}(V_2),$$

and hence, as $\hat{\xi}, \varphi$ is a quasiequivalence of $\hat{\pi}_1$ and $\hat{\pi}_2$, the observation is established.

Let $V = V_1$, $G = \tilde{G}_1$, $\pi = \pi_1$, $\alpha_1 : G_1 \to \tilde{G}_1$ the natural map, and $\beta_1 : V_1 \to V_1$ the identity map. Then $(G_1, \alpha_1, \beta_1) \in \mathrm{Ext}(G, V, \pi)$. Next define $\alpha_2 = q\xi^{-1} : G_2 \to G$, where $q : G_2 \to \tilde{G}_2$ is the natural map, and define $\beta_2 : V_2 \to V$ by $\beta_2 = \varphi^{-1}$. As ξ, φ is a quasiequivalence of the representations π_1 and π_2, $(G_2, \alpha_2, \beta_2) \in \mathrm{Ext}(G, V, \pi)$ by Remark 21.1, and as also ξ extends $\tilde{\varphi} : \tilde{H}_1 \to \tilde{H}_2$,

$$\varphi : (H_1, (\alpha_1)_{|H_1}, \beta_1) \to (H_2, (\alpha_2)_{|H_2}, \beta_2)$$

is an isomorphism. Then we appeal to 21.7.

(21.12) *Let G_i, $i = 1, 2$, be finite groups and $H_i \le G_i$ with $|G_i : H_i|$ prime to p. Assume*

(1) There is a G_i-chief series

$$1 = H_{i,0} \trianglelefteq \cdots \trianglelefteq H_{i,m} = O_p(G_i)$$

and an isomorphism $\varphi : H_1 \to H_2$ with $H_{1,j}\varphi = H_{2,j}$ for each j.

(2) Let $V_{i,j} = H_{i,j}/H_{i,j-1}$. Assume $C_{G_i}(V_{i,j}) = \langle C_{H_i}(V_{i,j})^{G_i} \rangle$ for $i = 1, 2$ and each $1 \le j \le m$.

(3) Assume for each $1 \le j \le m$, there exists a quasiequivalence

$$\eta_j : G_1^{V_{1,j}} = G_1/C_G(V_{1,j}) \to G_2/C_{G_2}(V_{2,j}) = G_2^{V_{2,j}},$$

$$\mu_j : V_{1,j} \to V_{2,j}$$

of the representations via conjugation with $H_1^{V_{1,j}} \eta_j = H_2^{V_{2,j}}$.

(4) For each $1 \le j \le m$, $H_2^{V_{2,j}}$ is contained in a unique $\mathrm{Aut}(V_{2,j})$-conjugate of $G_2^{V_{2,j}}$.

(5) For each $1 \le j \le m$, $C_{\mathrm{Aut}(G_1^{V_{1,j}})}(H_1^{V_{1,j}}) = 1$.

(6) There exists an isomorphism $\xi_0 : \bar{G}_1 = G_1/O_p(G_1) \to G_2/O_p(G_2) = \bar{G}_2$ extending $\bar{\varphi}$.

Then there exists an isomorphism $\psi : G_1 \to G_2$ extending φ.

Proof: Induct on the length m of the series in (1). The case $m = 0$ satisfies the conclusion of the lemma by (6). Let $V_i = V_{i,1}$ and $\tilde{G}_i = G_i/V_i$. By induction on m there is an isomorphism $\xi : \tilde{G}_1 \to \tilde{G}_2$ extending $\tilde{\varphi}$. We now have the hypotheses of 21.11, and that lemma completes the proof.

(21.13) Let V be an elementary abelian p-group and $H \le K \le GL(V)$ such that

(1) $\mathrm{Aut}_{GL(V)}(H) = \mathrm{Aut}_K(H)$, and

(2) Each H-chief factor on V is an absolutely irreducible $GF(p)H$-module, distinct chief factors are not isomorphic, and V is an indecomposable $GF(p)H$-module.

Then $N_{GL(V)}(H) \le N_{GL(V)}(K)$.

Proof: By (1), $N_{GL(V)}(H) = (N_{GL(V)}(H) \cap N_{GL(V)}(K))C_{GL(V)}(H)$, so it suffices to show $C_{GL(V)}(H) = Z(GL(V))$. But as distinct H-chief factors are not isomorphic, $C_{GL(V)}(H)$ acts on each chief factor, and then as each chief factor is absolutely irreducible, $C_{GL(V)}(H)$ induces scalar action on each chief factor. As V is an indecomposable H-module, each semisimple element in $C_{GL(V)}(H)$ is a scalar in $Z(GL(V))$, so $C_{GL(V)}(H) = Z(GL(V))R$, where R is the unipotent radical of $C_{GL(V)}(H)$ and is trivial on each H-chief factor. Finally if $r \in R^{\#}$ and U is the maximal H-submodule of V then $C_V(r) \le U$ and indeed by induction on the number of chief factors of H even $U = C_V(r)$, so the map $\theta : V/U \to U$ defined by $\theta : v + U \mapsto [v, r]$ is nontrivial with

$\theta \in \text{Hom}_{GF(p)H}(V/U, U)$. This is impossible since V/U is not isomorphic to any chief factor in U.

22. The Todd modules for the Mathieu groups

In this section we prove a few facts about Todd modules for the Mathieu groups. In particular we calculate orbits of each Mathieu group on the vectors in its Todd module, and calculate the 1-cohomology of the Mathieu groups on their irreducible Todd modules.

Let V be an elementary abelian p-group and G a group of automorphisms of V. We can regard V as a $GF(p)G$-module. Then the 1-cohomology group $H^1(G, V)$ is a $GF(p)$-space in one to one correspondence with the conjugacy classes of complements to G in the semidirect product GV of V by G with respect to the representation of G on V. Moreover if $C_V(G) = 0$ there exists a largest $GF(p)G$-module U containing V such that $C_U(G) = 0$ and $[G, U] \leq V$, and it turns out that $U/V \cong H^1(G, V)$. The dual of this statement is that if $V = [V, G]$ then there exists a largest $GF(p)G$-module W such that $W = [W, G]$ and there exists $Z \leq C_W(G)$ with $W/Z \cong V$, and it turns out that $H^1(G, V^*) \cong Z$. See Section 17 in [SG] for a discussion of these facts.

Let $G = M_{24}$ and (X, C) the Steiner system for G as described in Chapter 7 of [SG]. Thus X is a set of order 24, C is a collection of 8-subsets of X called *octads* such that each 5-subset of X is contained in a unique octad, and G is a 5-transitive subgroup of the symmetric group on X that permutes the octads.

Proceeding as in Section 19 of [SG], let V_0 be the binary permutation module for G on $X = \{1, \ldots, 24\}$ and identify V_0 with the power set of X by identifying $v \in V_0$ with its support. Under this identification, addition is symmetric difference. Let V_C be the *12-dimensional Golay code submodule* of V_0 generated by C and form $\tilde{V}_0 = V_0/V_C$ as in Section 19 of [SG]. Thus $\tilde{V}_0$ is the *12-dimensional Todd module* for G.

The *weight* of a vector $v \in V_0$ is just its order. We also speak of the weight of $\tilde{v} \in \tilde{V}_0$, which we define to be the minimum weight of a vector u in the coset $\tilde{v}$.

Let V be the core of V_0, consisting of all $v \in V_0$ of even weight. Then $\tilde{V}$ is an 11-dimensional irreducible $GF(2)G$-module called the *11-dimensional Todd module* for G. The restriction of $\tilde{V}$ to M_{23} is the *11-dimensional Todd module* for M_{23}.

(22.1)

(1) *G has 4 orbits V_m, $1 \leq m \leq 4$, on $\tilde{V}_0^{\#}$ where V_m consists of the vectors $\tilde{v}$ of weight m.*

(2) *If v is of weight $m \leq 3$ then v is the unique member of the coset $\tilde{v}$ of weight m, whereas if v is of weight 4 then there are six members of $\tilde{v}$ of weight 4 and they form a partition of X called a sextet.*

(3) *The stabilizer $G_{\tilde{v}}$ of $\tilde{v} \in \tilde{V}$ of weight 4 (or equivalently the stabilizer of the corresponding sextet) is the split extension of E_{16} by $\hat{S}_6$, where $\hat{S}_6$ is the extension of $\hat{A}_6$ with $\hat{S}_6/Z(\hat{A}_6) \cong S_6$ and $\hat{A}_6$ is quasisimple with center of order 3.*

Proof: See 19.9 and 19.10 in [SG].

For $Y \subseteq X$ write e_Y for the subset Y regarded as a vector of V_0. Then $e_{1,2}$ is of weight 2 and $G_{\tilde{e}_{1,2}}$ is the global stabilizer $G(\{1, 2\})$ of the 2-subset $\{1, 2\}$ of X and hence isomorphic to $\text{Aut}(M_{22})$, an extension of $G_{1,2} \cong M_{22}$ by $\mathbf{Z}_2$. Therefore $\hat{V} = \tilde{V}/\langle \tilde{e}_{1,2} \rangle$ is a 10-dimensional module for M_{22} called the *10-dimensional Todd module* for M_{22}.

Next the stabilizer in M_{22} of the image $\hat{e}_{1,3}$ of $\tilde{e}_{1,3}$ in $\hat{V}$ is the pointwise stabilizer $G_{1,2,3} \cong L_3(4)$ of the 3-subset $\{1, 2, 3\}$ and $G_{1,2,3}$ acts on $X - \{1, 2, 3\}$ as on the 21 points of the projective plane of order 4. Further $\bar{V} = \tilde{V}/\langle \tilde{e}_{1,2}, \tilde{e}_{1,3} \rangle$ is a 9-dimensional $GF(2)L_3(4)$-module that we call the *9-dimensional Todd module* for $L_3(4)$.

(22.2) *Let $L = G_{1,2,3} \cong L_3(4)$. Then L has three orbits I_i, $i = 1, 2, 3$, on the nonzero vectors of the 9-dimensional Todd module $\bar{V}$, where*

(1) *I_1 of order 21 consists of the vectors $\bar{e}_Y$ with Y a 4-subset containing $\{1, 2, 3\}$, and L acts 2-transitively on I_1 as the points of the projective plane $\mathcal{P}$ of L.*

(2) *$|I_2| = \binom{21}{2} = 210$, $I_2 = \{a + b : a, b \in I_1, a \neq b\}$, and $L_{a+b} = N_L(\{a, b\})$.*

(3) *$|I_3| = 280$ with I_3 consisting of the elements $a + b + c$, where $\{a, b, c\} \subseteq I_1$ is a triple of independent points of $\mathcal{P}$. Further $L_{a+b+c} = N_L(P)$ where $P = O_3(L\{a, b, c\}) \in \text{Syl}_3(L)$.*

Proof: Let $\mathcal{I}_1$ be the set of 4-subsets of X containing $\{1, 2, 3\}$. Then L acts 2-transitively on $\mathcal{I}_1$ and we may identify $\mathcal{I}_1$ with the points of the projective plane $\mathcal{P}$ of L. Further for $Y \in \mathcal{I}_1$, e_Y is the unique member of $\tilde{e}_Y$ in $\mathcal{I}_1$ since the 4-subsets of $\tilde{e}_Y$ form a sextet. Therefore $I_1 = \{\bar{e}_Y : Y \in \mathcal{I}_1\}$ can also be identified with the points of $\mathcal{P}$ and is an orbit of length 21 under L.

Let $\mathcal{I}_2$ be the set of subsets $R + S$, with $R, S \in \mathcal{I}_1$ distinct. Then $R + S = \{x, y\}$ for some $x, y \in X - \{1, 2, 3\}$ and $I_2 = \{\bar{I} : I \in \mathcal{I}_2\}$ satisfies the conclusions of part (2) of the lemma.

Finally let $R, S, T \in \mathcal{I}_1$ with $R \cup S \cup T$ not contained in an octad; equivalently $\{\bar{R}, \bar{S}, \bar{T}\}$ is a triple of independent points of $I_1 \cong \mathcal{P}$. Let $v = \bar{R} + \bar{S} + \bar{T}$ and observe that the global stabilizer of $\{\bar{R}, \bar{S}, \bar{T}\}$ in L is $P \in \mathrm{Syl}_3(L)$ extended by an involution. Then $v \in C_{\bar{V}}(P)$ so, as $|C_{\bar{V}}(P)| = 2$, $\langle v \rangle = C_{\bar{V}}(P)$ is $N_L(P)$-invariant. Then as $N_L(P)$ is maximal in L, $N_L(P) = L_v$ and $I_3 = v^L$ is of order $|L : N_L(P)| = 280$. As $2^9 = 1 + 21 + 210 + 280$, our proof is complete.

(22.3) *Let $L = G_{1,2} \cong M_{22}$. Then L has three orbits I_i, $i = 1, 2, 3$, on the nonzero vectors of the 10-dimensional Todd module $\hat{V}$, where*

(1) I_1 *of order 22 consists of the vectors $\hat{e}_{1i}$, $3 \leq i \leq 24$, and L acts 3-transitively on I_1 as on $X - \{1, 2\}$.*

(2) $I_2 = \{a + b : a, b \in I_1, a \neq b\}$ *is of order $\binom{22}{2} = 231$. If $a = \hat{e}_{1i}$ and $b = \hat{e}_{1j}$ then $L_{a+b} = N_L(\{a, b\}) = N_L(\{i, j\})$.*

(3) $|I_3| = \binom{22}{3/2} = 770$ *with I_3 consisting of the elements $a + b + c$, where $\{a, b, c\}$ is a 3-subset of I_1. Further*

$$a_1 + a_2 + a_3 = a_4 + a_5 + a_6$$

for distinct triples if and only if $a_i = \hat{e}_{1, j(i)}$ with $\{1, 2, j(i) : 1 \leq i \leq 6\}$ the unique octad containing $\{1, 2, j(i) : 1 \leq i \leq 3\}$.

(4) *If a, b, c, d are distinct members of I_1 with c not in the block of the Steiner system for L on I_1 containing $\{a, b, d\}$, then $a + b + c + d \in I_3$.*

Proof: By 3-transitivity of L on $Y = X - \{1, 2\}$, I_i, $i = 1, 2, 3$, are orbits of L on $\hat{V}$. Also $\hat{e}_{1i} = \{\tilde{e}_{1i}, \tilde{e}_{2i}\}$ and $\hat{e}_{ij} = \{\tilde{e}_{ij}, \tilde{e}_{1,2,i,j}\}$ contain a unique vector $\tilde{e}_{1,r}, \tilde{e}_{rs}$, respectively, so $|I_1| = |Y| = 22$ and $|I_2| = \binom{22}{2} = 770$. On the other hand if $I = \{i, j, k\}$ is a 3-subset of Y and $v(I) = \hat{e}_{1,i} + \hat{e}_{1j} + \hat{e}_{1k}$ then $v(I) = \{\tilde{e}_{I \cup \{1\}}, \tilde{e}_{I \cup \{2\}}\}$, so $v(I) = v(J)$ if and only if $I \cup J \cup \{1, 2\}$ is the unique octad containing $I \cup \{1, 2\}$. Thus $|I_3| = \binom{22}{3/2} = 770$. Then as $2^{10} - 1 = |I_1| + |I_2| + |I_3|$, $V^{\#} = \bigcup_i I_i$, completing the proof of (1)–(3).

For $i \in I_1$, let $i = \hat{e}_{1, y(i)}$ with $y(i) \in Y$. Then $a + b + c + d = \hat{e}_J$, where $J = \{y(a), y(b), y(c), y(d)\}$. Let $K = \{1, r, s, t\}$ be the member of the sextet containing J such that $1 \in K$. Then $2 \notin K$ as the block of the Steiner system on I_1 containing a, b, d is $\{\hat{e}_{1,i} : i \in B - \{1, 2\}\}$, where B is the octad of X containing $\{1, 2, y(a), y(b), y(c)\}$, and c is not in that block. Hence

$$\hat{e}_J = \hat{e}_K = \hat{e}_{1,r} + \hat{e}_{1,s} + \hat{e}_{1,t} \in I_3$$

establishing (4).

(22.4) *Let* $L = G_1 \cong M_{23}$. *Then* L *has three orbits* I_i, $i = 1, 2, 3$, *on the nonzero vectors of the* 11-*dimensional Todd module* $\tilde{V}$, *where*

(1) I_1 *of order* 23 *consists of the vectors* $\tilde{e}_{1i}$, $2 \leq i \leq 24$, *and* L *acts* 4-*transitively on* I_1 *as on* $X - \{1\}$.

(2) $I_2 = \{a + b : a, b \in I_1,\ a \neq b\}$ *is of order* $\binom{23}{2} = 253$. *If* $a = \hat{e}_{1i}$ *and* $b = \hat{e}_{1j}$ *then* $L_{a+b} = N_L(\{a, b\}) = N_L(\{i, j\})$.

(3) $|I_3| = \binom{23}{3} = 1{,}771$ *with* I_3 *consisting of the elements* $a + b + c$, *where* $\{a, b, c\}$ *is a* 3-*subset of* I_1 *and* $\tilde{e}_{1i} + \tilde{e}_{1j} + \tilde{e}_{1k} = \tilde{e}_{1,i,j,k}$ *has stabilizer* $N_L(\{\tilde{e}_{1i}, \tilde{e}_{1j}, \tilde{e}_{1k}\}) = N_L(\{i, j, k\})$.

Proof: The proof is quite similar to that of 22.3, except easier.

(22.5) *The Todd modules of dimension* 9, 10, 11, 11 *for* $L = L_3(4)$, M_{22}, M_{23}, *and* M_{24} *are absolutely irreducible* $GF(2)L$-*modules.*

Proof: A sum of a proper subset of orbits does not add up to 2^m, so the Todd module M for L is an irreducible $GF(2)L$-module. Further for $x \in M^{\#}$, $C_M(C_L(x)) = \langle x \rangle$, so $\mathrm{End}_{GF(2)L}(M) = GF(2)$ and hence M is absolutely irreducible.

(22.6) *Let* $B \in C$, $M = N_G(B)$, $x \in X - B$, $y \in B$, $H = G_{x,y} \cong M_{22}$, *and* $t \in G$ *with cycle* (x, y). *Then*

(1) $M_{x,y} \cong A_7$.
(2) $M_{x,y}^t \notin M_{x,y}^H$.
(3) $M_{x,y}$ *is a maximal subgroup of* H.
(4) $M_{x,y}$ *has chief factors of dimension* 1, 4, 6 *on the Todd module* $\tilde{V}$ *for* G.
(5) $H^1(M_{x,y}, \hat{V}) = 0$.

Proof: From 19.1 in [SG], $O_2(M) = Q \cong E_{16}$ is the pointwise stabilizer G_B of B in G and is regular on $X - B$, so M_x is a complement to Q in M and $M_x \cong A_8$ acts faithfully as A_8 on B. Hence $M_{x,y} = K$ acts faithfully as A_7 on B and as Q is regular on $X - B$, the representation of M_x on $X - B$ is equivalent to its representation via conjugation on Q (cf. 15.11 in [FGT]). However M_x acts as $GL(Q) \cong GL_4(2)$ on Q so its A_7-subgroup K is transitive on $Q^{\#}$ by Exercise 7.1, and hence also on $X - B - \{x\}$. Therefore K has orbits $Y_1 = B \cap Y$ and $Y_2 = Y - B$ on $Y = X - \{x, y\}$, and (1) is established.

If $K^t \in K^H$ then by a Frattini argument we may take $t \in N_G(K)$. Then t acts on the orbit Y_1 of K of length 7 so as B is the unique octad containing Y_1, t acts on B, contradicting $yt = x \notin B$. So (2) is established.

Next from 20.6 in [SG], M has a chief series

$$0 < W_1 < W_2 < W$$

on the 11-dimensional Golay code module $W = V_C/\langle e_X \rangle$ with W_1 a point, and $W_2/W_1 = U_1$ and $W/W_2 = U_2$ the natural modules of dimension 4, 6 for M_x regarded as $L_4(2)$ and A_8, respectively. Then the restriction of K to these chief factors remains irreducible with $H^1(K, U_1) = H^1(K, U_2) = 0$ by Exercise 7.1 and by Exercise 6.3 in [FGT], respectively. As W is the dual of the Todd module $\tilde{V}$, it follows that (4) holds and that K has chief factors U_1 and U_2 on $\hat{V}$. Thus as $H^1(K, U) = 0$ for each chief factor U of K on $\hat{V}$, (5) holds.

It remains to prove (3), so assume $K < J < H$. As K is the stabilizer of B in H, K is the stabilizer of Y_1 in H, so J is transitive on Y. Thus $11 \cdot |K|$ divides the order of J, so $|H : J| = 2^a$ with $1 \le a \le 4$. Let $S_p \le K$ be of order $p = 5, 7$. Then by 21.5 in [SG], $N_H(S_p) = N_K(S_p)$, so $N_H(S_p) = N_J(S_p)$. Further by Sylow's Theorem, $|H : N_H(S_p)| \equiv |J : N_J(S_p)| \equiv 1 \bmod p$, so $1 \equiv |H : J| = 2^a \bmod p$, with $1 \le a \le 4$. This is a contradiction.

(22.7) *Let $H = G_{1,2} \cong M_{22}$ and $\hat{V}$ the 10-dimensional Todd module for H. Then $H^1(H, \hat{V}) = 0$.*

Proof: From 17.11 in [FGT] it suffices to assume W is a $GF(2)H$-module with $\hat{V} = [W, M]$ of codimension 1 in W and to prove W splits over $\hat{V}$. So assume the extension does not split.

By 22.6, H has a maximal subgroup K isomorphic to A_7, and $H^1(K, \hat{V}) = 0$, so K fixes a point $w \in W - \hat{V}$. As K is maximal in H, $H_w = K$.

Now K contains a Sylow 3-subgroup P of H. Then $\mathrm{Fix}_X(P) = \{1, 2, 3\}$ so $N_H(P) \le G_{1,2,3} = L \cong L_3(4)$ and hence $N_H(P)$ is the split extension of P by Q_8. Let $\bar{W} = W/\langle \hat{e}_{13} \rangle$, so that the image $\bar{V}$ of $\hat{V}$ in $\bar{W}$ is the 9-dimensional Todd module for L. Then

$$|L : L_{\bar{w}}| = |\bar{w}^L| \le |\bar{W} - \bar{V}| = 2^9,$$

so as $|L : N_K(P)| > 2^9$, $N_K(P) < L \cap K$. Hence (cf. Exercise 1.5) as A_6 is the unique maximal subgroup of K containing $N_K(P)$ and $N_K(P)$ is maximal in A_6, $L \cap K \cong A_6$.

Let $\langle g, k \rangle \cong Q_8$ be a Sylow 2-subgroup of $N_H(P)$ with g, k of order 4 and $k \in K$. From 22.2, $C_W(P) = \langle \hat{e}_{1,3}, v, w \rangle$ with $C_W(L \cap K) = \langle \hat{e}_{1,3}, w \rangle$ and $v \in \hat{V}$. Therefore $C_{\bar{W}}(P) = \langle \bar{v}, \bar{w} \rangle$, so as $g \notin K = G_{\bar{w}}$, $\bar{w}^g = \bar{w} + \bar{v}$. Then as g inverts k and k centralizes $\langle \hat{e}_{1,3}, w \rangle$, we conclude k centralizes $C_W(P)$.

Now by 22.6, K has a chief factor U that is the natural 6-dimensional module for K. Thus U is the 6-dimensional permutation module for $L \cap K$, say with basis $u_1, \ldots, u_6$ permuted by $L \cap K$. But then if $P = \langle (1, 2, 3), (4, 5, 6) \rangle$ in its action on this basis and $k = (1, 4, 2, 5)(3, 6)$ then $C_U(P) = \langle u_1 + u_2 + u_3, u_4 + u_5 + u_6 \rangle$ is not centralized by k, contradicting $C_W(P) \leq C_W(k)$.

(22.8)

(1) $H^1(M_{23}, \tilde{V}) = 0$, where $\tilde{V}$ is the 11-dimensional Todd module for M_{23}.

(2) $H^1(M_{24}, \tilde{V}) \cong \mathbf{Z}_2$, where $\tilde{V}$ is the 11-dimensional Todd module for $G \cong M_{24}$. Indeed if W is a 12-dimensional indecomposable $GF(2)G$-module with $[W, G] \cong \tilde{V}$ then W is isomorphic to the 12-dimensional Todd module.

Proof: Let $D = G$ or $G_1 \cong M_{23}$. By 17.11 in [FGT] it suffices to assume W is an indecomposable $GF(2)D$-module such that $[W, D] = \tilde{V}$ and if $D \cong M_{23}$ to take $\tilde{V}$ to be a hyperplane of W and derive a contradiction, wheareas if $D \cong M_{24}$ to take $\dim(W/\tilde{V}) = 1$ or 2 and show W is isomorphic to the 12-dimensional Todd module.

Suppose first $D \cong M_{23}$ and let $H = G_{1,2}$. Then $H = D_v$, where $v = \tilde{e}_{12}$ and $\tilde{V}/\langle v \rangle$ is the 10-dimensional Todd module for H. Hence by 22.7, $W/\langle v \rangle$ splits over $\tilde{V}/\langle v \rangle$ as an H-module, so H fixes $w \in W - \tilde{V}$. Thus W is an image of the 23-dimensional permutation module U for D on D/H. But U is a summand of $V_0 = \langle e_1 \rangle \oplus U$ with $U = U_1 \oplus U_2$ where $U_1 = C_U(D)$ and U_2 has chief factors the 11-dimensional Todd modules and Golay code modules for D. In particular the unique 1-dimensional chief factor U_1 for D on U is a direct summand, so W splits over $\tilde{V}$, completing the proof of (1).

So assume $D = G \cong M_{24}$. By (1), $W = C_W(G_1) \oplus \tilde{V}$, so if $w \in C_W(G_1)$ then $W(w) = \langle w^G \rangle$ is an image of the permutation module V_0 for G on G/G_1. Hence from the structure of V_0, $W(w)$ is isomorphic to the 12-dimensional Todd module. Thus we may assume $C_W(G_1) = \langle w_1, w_2 \rangle$ is of dimension 2. Let $Y \in \mathrm{Syl}_{11}(G_1)$; then $\dim(C_W(Y)) = 3$ and there is $g \in N_G(Y) - G_1$. Then $C_W(G_1)$ and $C_W(G_1^g)$ are hyperplanes of $C_W(Y)$ so

$$0 \neq C_W(G_1) \cap C_W(G_1^g) = C_W(\langle G_1, G_1^g \rangle) = C_W(G),$$

completing the proof.

23. Central extensions, Schur multipliers, and $U_6(2)$

In this section we recall some facts about central extensions and Schur multipliers and prove various facts about central extensions and 1-cohomology connected with the group $U_6(2)$. We also determine the conjugacy classes of involutions in unitary groups over fields of even order and classes of elements of order 3 in $U_6(2)$.

Our reference on central extensions is Section 33 of [FGT]. Recall a *central extension* of a group G is a pair (H, π) where H is a group and $\pi : H \to G$ is a surjective group homomorphism with $\ker(\pi) \leq Z(H)$. Further if G is perfect then G possesses a universal central extension $(\text{Cov}(G), \pi)$ (cf. 33.4 in [FGT]). The group $\text{Cov}(G)$ is the *universal covering group* of G and $\ker(\pi)$ is the *Schur multiplier* of G, which we denote by $\text{Schur}(G)$.

If G is finite and perfect then so is $\text{Cov}(G)$ (cf. 33.10 in [FGT]) and hence $\text{Schur}(G)$ is a finite abelian group. Given a prime p let

$$\text{Cov}_p(G) = \text{Cov}(G)/O^p(\text{Schur}(G))\Phi(O_p(\text{Schur}(G)))$$

and

$$\text{Schur}_p(G) = \text{Schur}(G)/O^p(\text{Schur}(G))\Phi(O_p(\text{Schur}(G))).$$

Therefore $\text{Cov}_p(G)$ is the largest perfect central extension of an elementary abelian p-group by G.

(23.1) Let G be a perfect group with $F^*(G) = Q$ an extraspecial 2-group and G/Q quasisimple. Let $\hat{G} = \text{Cov}_2(G)$, $\hat{Z} = \text{Schur}_2(G)$, and $\hat{Q} = O_2(\hat{G})$. Assume G is irreducible on $\tilde{Q} = Q/Z(G)$, $Z(G/Q)$ is of order 3, and $\text{End}_{G/Q}(\tilde{Q})$ is of order 4. Then

  (1) $\hat{Q} = [\hat{Q}, G] \times \hat{Z}$, with $[\hat{Q}, \hat{G}] \cong Q$.
  (2) $\hat{G}/[\hat{Q}, \hat{G}] \cong \text{Cov}_2(G/Q)$ and $\hat{Z} \cong \text{Schur}_2(G/Q)$.
  (3) H is a finite group with $F^*(H) \cong Q$ and $H/Z(H) \cong G/Z(G)$ if and only if $H \cong \hat{G}/V$ for some complement V to $[\hat{Q}, \hat{G}]$ in $\hat{Q}$.

Proof: The proof is essentially the same as that of 8.17 in [SG] modulo two changes. Let $G^* = G/Q$ and $X^* = Z(G^*)$. As $Q = F^*(G)$, G^* is faithful on $\tilde{Q}$. As G is irreducible on $\tilde{Q}$, $\tilde{Q} = [\tilde{Q}, X^*]$, so as X^* is of odd order, $H^1(G^*, \tilde{Q}) = 0$ by a Frattini argument. This establishes Hypothesis (b) of 8.17 in [SG]. It remains to observe that Hypothesis (a) of 8.17 in [SG] can also be dispensed with as it was used only to prove there is a unique nontrivial G^*-invariant symmetric bilinear form on $\tilde{Q}$, and we now supply an alternate proof of that fact under our hypotheses.

As X^* is of order 3 and $F = \text{End}_{G^*}(\tilde{Q})$ is of order 4, $X^* = F^\#$. Now by Exercise 9.1 in [FGT]), we have an isomorphism $\psi : \text{Hom}_{FG^*}(\tilde{Q}, \tilde{Q}^*) \to L^2_{G^*}(\tilde{Q})$ defined by $\psi(\alpha)(u, v) = v(u\alpha)$, where $L^2_{G^*}(\tilde{Q})$ is the set of bilinear forms on $\tilde{Q}$ preserved by G^*. From 8.3 in [SG], the commutator map on Q defines a nondegenerate symmetric bilinear form f on $\tilde{Q}$ preserved by G^*; let $\alpha = \psi^{-1}(f)$. As G^* is irreducible on $\tilde{Q}$, $\text{Hom}_{FG^*}(\tilde{Q}, \tilde{Q}^*) = F\alpha$, so if $0 \neq f' \in L^2_{G^*}(\tilde{Q})$ then

$f' = \psi(x\alpha)$ for some $x \in X^*$ and hence

$$f'(u, v) = v(u(x\alpha)) = v((ux)\alpha) = f(ux, v).$$

Therefore as X^* preserves f, f' is symmetric if and only if $f' = f$. That is f is the unique nontrivial G^*-invariant symmetric bilinear form on $\tilde{Q}$, so that the proof of 8.17 in [SG] now applies in our situation.

(23.2) *Let V be an n-dimensional unitary space over a field F of even order, N the greatest integer less than or equal to $n/2$, and $G = SU(V)$. Then*

(1) *G has N classes of involutions $\mathcal{J}_m$, $1 \leq m \leq N$, where $\mathcal{J}_m$ is the set of involutions of G with m Jordon blocks of size 2 on V.*

(2) *For $j \in \mathcal{J}_m$, $V = V_0 \oplus V_1 \oplus \cdots V_m$ is the orthogonal direct sum of subspaces V_i such that $V_0 \leq C_V(j)$ and j induces a transvection on the 2-dimensional subspace V_i for $1 \leq i \leq m$.*

(3) *$j = t_1 \cdots t_m$ where t_i is the transvection in G with $(t_i)_{|V_i} = j_{|V_i}$.*

Proof: Let $g \in G$ be an involution. By 22.1 in [FGT], $C_V(g) = [V, g]^{\perp}$, so as $[V, g] \leq C_V(g)$, $[V, g]$ is totally singular of dimension $m \leq N$. Thus $g \in \mathcal{J}_m$ and of course G acts on $\mathcal{J}_m$ by conjugation, so to prove (1) it remains to show G is transitive on $\mathcal{J}_m$.

Let $\theta = 1 - g \in \text{End}(V)$, $\tilde{V} = V/C_V(g)$, and define $f : \tilde{V} \times \tilde{V} \to F$ by $f(\tilde{u}, \tilde{v}) = (u, v\theta)$, well defined as $C_V(g) = [V, g]^{\perp}$. Then f is linear in its first variable. Also

$$(u, v) = (ug, vg) = (u + u\theta, v + v\theta) = (u, v) + (u, v\theta) + (u\theta, v),$$

so $(u, v\theta) = (u\theta, v) = (v, u\theta)^q$, where F is of order q^2. Hence f is hermitian symmetric. Finally as $C_V(g) = [V, g]^{\perp}$, f is nondegenerate, so f is a unitary form.

In particular there is $u \in V$ with $0 \neq f(u, u)$. Then $V_1 = \langle u, u\theta \rangle$ is a nondegenerate 2-subspace and g induces a transvection on V_1 with center $Fu\theta$. Further by induction on n, $V_1^{\perp} = V_2 \oplus \cdots V_m \oplus V_0$ has the required orthogonal decomposition, so V does too and (2) and (3) are established. Finally (2) and Witt's Lemma 10.9 imply transitivity of G on $\mathcal{J}_m$ and complete the proof of (1).

In the remainder of this section let V be a 6-dimensional unitary space over the field F of order 4, $H_0 = SU(V) \cong SU_6(2)$ the special unitary group of V, and $H = H_0/Z(H_0) \cong U_6(2)$ the projective special unitary group. Thus V is a projective space for H. We abuse notation and identify each involution j of H with the unique involution in H_0 mapping onto j. By 23.2, H has three classes $\mathcal{J}_m$, $1 \leq m \leq 3$, of involutions, where $j_m \in \mathcal{J}_m$ has m Jordon blocks of size 2 on V. In particular $\mathcal{J}_1$ is the set of transvections of H and a set of 3-transpositions of H.

Let U be a maximal totally singular subspace of V, so that $\dim_F(U) = 3$ and $K = N_H(U)$ is a maximal parabolic subgroup of H with K the split extension of $A = O_2(K) \cong E_{2^9}$ by $L \cong L_3(4)$. Let $\mathcal{I}_m = \mathcal{J}_m \cap A$. Then $\mathcal{I}_m$ consists of those $j \in \mathcal{J}_m$ with $[V, j] \leq U \leq C_V(j)$, so $\mathcal{I}_m \neq \emptyset$. Indeed we see in the next lemma that A is the 9-dimensional Todd module for L, so the orbits $\mathcal{I}_m$ are described in 22.2.

(23.3)

    (1) A is the 9-dimensional Todd module for L.

    (2) The sets $\mathcal{I}_m$, $1 \leq m \leq 3$, are the orbits of L on $A^\#$ of length 21, 210, 280, respectively, and described in Lemma 22.2.

Proof: First U has 21 points and the map $j \mapsto [V, j]$ is a bijection of $\mathcal{I}_1$ with the set of points of U such that $C_L(j) = N_L([V, j])$, so $\mathcal{I}_1$ is of length 21. Thus the subspace $A_1 = \langle \mathcal{I}_1 \rangle$ of A is an image of the permutation module W for L on $\mathcal{I}_1$ and we can view L as the stabilizer in $G = M_{24}$ of 3 points 1, 2, 3 in the set X of order 24 permuted by G, and W as the subspace of the permutation module V_0 for G on X spanned by $Y = X - \{1, 2, 3\}$ described in Section 22. As G is 3-transitive on X, for each $x \in \{1, 2, 3\}$ there is an octad containing only x from $\{1, 2, 3\}$, so $V_0 = W + V_C$ and thus $W_C = V_C \cap W$ is of dimension 9. Now L is irreducible on the Todd module by 22.5, and hence also on its dual W_C. Therefore $[W, L]$ has two composition factors W_C and $[W, L]/W_C$, with the latter the Todd module for L. As $W = \langle e_Y \rangle \oplus [W, L]$ and L has no orbit of length 21 on W_C it follows that the Todd module is the unique 9-dimensional image of W, so (1) holds.

By (1) and 22.2, L has three orbits $\mathcal{I}'_m$ on $A^\#$ with $\mathcal{I}'_1 = \mathcal{I}_1$ of length 21, $|\mathcal{I}'_2| = 210$, and $|\mathcal{I}'_3| = 280$. For $a, b, c \in \mathcal{I}_1$ distinct, $ab \in \mathcal{I}_2$ and $abc \in \mathcal{I}_3$, so $\mathcal{I}_2 = \mathcal{I}'_2$ and $\mathcal{I}_3 = \mathcal{I}'_3$, from the description of $\mathcal{I}'_k$ in 22.2. Thus (2) holds.

Recall that if S is a 2-group then $J(S) = \langle A(S) \rangle$ is the *Thompson subgroup* of S, where $A(S)$ is the set of elementary abelian subgroups of S of maximal order.

(23.4)

    (1) L is irreducible on A.

    (2) $A = J(S)$ for $A \leq S \in \mathrm{Syl}_2(H)$.

Proof: Part (1) follows from 23.3 and 22.5.

Next let $j \in \mathcal{J}_3$. By 23.3 and 22.2, $C_L(j) = N_L(P)$, where $E_9 \cong P \in$ $\mathrm{Syl}_3(L)$, $N_L(P)/P \cong Q_8$, and $\langle j \rangle = C_A(P)$. Then $A = C_A(P) \oplus [A, P]$ with P inverted by an involution $t \in N_L(P)$, so

$$m([A, t]) = m([A, P])/2 = 4.$$

But $L \cong L_3(4)$ has one class of involutions so if $t \in B \leq L$ with $E_{2^k} \cong B$ and $k \geq m(A/C_A(B))$ then as $m(A/C_A(t)) = m(L) = 4$, we conclude $B \cong$ E_{16} and $C_A(t) = C_A(B)$. This is impossible as t is the unique involution in $C_L(t) \cap C_L(j)$. Therefore $A = J(S)$ and (2) holds.

(23.5) *Assume $\hat{H}$ is quasisimple with $Z(\hat{H}) = \langle z \rangle \cong \mathbf{Z}_2$ and $\hat{H}/Z(\hat{H}) = H$. For $X \leq H$ write $\hat{X}$ for the preimage of X in $\hat{H}$ under the natural map. Then*

(1) $\hat{A} \cong E_{2^{10}}$.

(2) Each involution in H lifts to an involution in $\hat{H}$.

(3) z is not a square in $\hat{H}$.

(4) If $a \in \hat{A}$ with $a\langle z \rangle \in \mathcal{J}_1$ then $az \notin a^{\hat{H}}$, and $a^{\hat{H}}$ is a set of 3-transpositions of $\hat{H}$.

(5) $[\hat{K}, \hat{A}] = \hat{A}$.

Proof: By 23.4.1 either $\hat{A} \cong E_{2^{10}}$ or $\hat{A}$ is extraspecial; the latter case is impossible as $|\hat{A}| = 2^{10}$ and 10 is even. Hence (1) holds. As each involution in H is fused into A, (2) follows from (1). Then (2) implies (3).

For $Y \subseteq \hat{H}$ write $\tilde{Y}$ for the image of Y in $H = \hat{H}/\langle z \rangle$ under the natural map. If $az \in a^{\hat{H}}$ then by 23.4.2 and 2.8.1, $az \in a^{\hat{K}}$. Thus $az = a^k$ for some $k \in C_{\hat{L}}(\tilde{a})$. This is impossible since by 23.3 and 22.2, $C_L(\tilde{a}) = C_L(\tilde{a})^{\infty}$, so $C_{\hat{L}}(\tilde{a}) = C_{\hat{L}}(a)$. Thus $az \notin a^{\hat{H}}$. Then as $\tilde{a}^H = \mathcal{J}_1$ is a set of 3-transpositions of H, 24.5 completes the proof of (4).

Assume $[\hat{K}, \hat{A}] \neq \hat{A}$. Then $B = [\hat{K}, \hat{A}]$ is a $\hat{K}$-invariant complement to $\langle z \rangle$ in $\hat{A}$. As $\hat{H}$ is quasisimple, $\hat{K}$ does not split over $\langle z \rangle$ by Gaschütz's Theorem, 2.6. Therefore $\hat{K}/B$ is quasisimple. On the other hand as K splits over $A = \tilde{B}$, $\hat{K}$ has a complement K_0 to B with $\tilde{K}_0 = L$.

Let $S \in \mathrm{Syl}_2(K)$, $\langle b \rangle = C_B(\hat{S})$, and $\hat{X} = C_{\hat{H}}(b)$, so that $\langle \tilde{b} \rangle = Z(S)$, $O_2(X) = Q \cong Q_8^4$, and $X/Q \cong U_4(2)$. Then QA/A is the natural module for $(K \cap X)/QA \cong L_2(4)$, so $\hat{Q}\hat{A}/B \cong E_{32}$. Also $L \cap X$ has a complement X_0 to $Q \cap L$ and each involution in L lifts to an involution in K_0, so $\hat{X}_0$ has a complement $\hat{X}_1$ to $\langle z \rangle$ in $\hat{X}_0$.

By Gaschütz, $\hat{X} \cap \hat{K}$ does not split over $\langle z \rangle$, so as $\hat{X}_1$ is a complement to $\hat{Q}\hat{A}$ in $\hat{X} \cap \hat{K}$, we conclude $\hat{Q}\hat{A}/B \cong \hat{Q}/(\hat{Q} \cap B)$ is an indecomposable module for $\hat{X}_1$ and hence $\hat{Q}/\langle b \rangle$ is an indecomposable module for $\hat{X}/\hat{Q} = X^* \cong U_4(2)$ with $\hat{Q}/\langle b, z \rangle$ the natural module for X^*.

Let W be the dual of $\hat{Q}/\langle b \rangle$, so that W is indecomposable with $[W, X^*] = W_0$ the natural module for X^*. Now $[B, \hat{Q}] \leq \hat{Q} \cap B$, so $B^* \cong E_{16}$ centralizes the annihilator W_1 of $(\hat{Q} \cap B)/\langle b \rangle$ in W, $W_1 \not\leq W_0$, and W_1 is isomorphic as an $N_{X^*}(B^*)$-module to the dual of the indecomposable module $\hat{Q}/(\hat{Q} \cap B)$. In particular $N_{X^*}(B^*)$ has an orbit $u^{N_{X^*}(B^*)}$ of length 6 on $W_1 - W_0$ with stabilizer I such that $I/B^* \cong D_{10}$. Let $Y \in \text{Syl}_5(I)$; then $C_{W_0}(Y) = 0$ so $\langle u \rangle = C_W(Y)$ and then $X^* = \langle I, N_{X^*}(Y) \rangle \leq C_{X^*}(u)$, a contradiction. Here we use Exercise 7.2.

REMARK: In the next 3 lemmas we determine $\text{Schur}_2(G)$ and $H^1(G, U)$ for certain groups G and $GF(2)G$-modules U. On the one hand, we assume the existence of an appropriate Fischer group to obtain lower bounds on the order of these cohomology groups, and on the other, prove an upper bound on the order and use that upper bound later to establish the existence and uniqueness of the Fischer groups. Once the Fischer groups are shown to exist in Chapter 11, our lower bound, and hence the full strength of the lemmas, become valid. Since we do not use the lower bound in our proof of existence, and since the proof of the upper bound does not use existence, our proof is not circular. This is a good example of the subtleties involved in studying the existence, uniqueness, and structure of the Fischer groups simultaneously.

(23.6)   *Let A^* be the dual of the 9-dimensional Todd module. Then*

$$\dim_{GF(2)}(H^1(L, A^*)) = 2$$

and an outer automorphism of L of order 3 is faithful on $H^1(L, A^*)$.

Proof: Let $\tilde{H}$ be the centralizer in a group G of type $M(23)$ and $\tilde{M}(23)$ of a 4-group $\langle a, b \rangle$ generated by a pair a, b of distinct commuting 3-transpositions of G. As G is of type $\tilde{M}(23)$, $C_G(a)$ is quasisimple with $C_G(a)/\langle a \rangle$ of type $\tilde{M}(22)$, so $\tilde{H}$ is quasisimple with $Z(\tilde{H}) = \langle a, b \rangle \cong E_4$ and $\tilde{H}/Z(\tilde{H}) \cong H$. Let $\tilde{A}$ and $\tilde{K}$ be the preimage in $\tilde{H}$ of A and K, respectively. By 23.5.1 and 23.5.5, $\tilde{A} \cong E_{2^{11}}$ with $[\tilde{K}, \tilde{A}] = \tilde{A}$, so that $\dim_{GF(2)}(H^1(L, A^*)) \geq \dim(C_{\tilde{A}}(\tilde{K})) = 2$ (cf. 17.12 in [FGT]). Hence it remains to show $\dim(H^1(L, A^*)) \leq 2$ to prove the first part of the lemma.

Let W be an indecomposable L-module with $A^* = [W, L]$. We saw during the proof of 22.2 that for $P \in \text{Syl}_3(L)$, $\dim(C_A(P)) = 1$, so

$$\dim(C_{A^*}(P)) = 1 + \dim(W/A^*) \quad \text{with } C_{A^*}(P) = \langle v \rangle \text{ 1-dimensional.}$$

Now $Q \in \text{Syl}_2(N_L(P))$ is isomorphic to Q_8 and induces a group of transvections on $C_{A^*}(P)$ with center v. Therefore $[Z(Q), C_{A^*}(P)] = 0$ and if

$\dim(W/A^*) > 2$ then there is $w \in C_W(PQ) - A^*$. To complete the proof that $\dim(W/A^*) \leq 2$, we derive a contradiction from the existence of such a vector w. Then taking $W = U(L, A^*)$, in the language of Section 17 of [FGT], $W/A^* \cong H^1(L, A^*)$ and a group B of outer automorphisms of L of order 3 acts on W by the universal property of $U(L, A^*)$. We may take B to act on P and faithfully on Q, so as $Q/Z(Q)$ induces the full group of transvections with center $\langle v \rangle$ on the 3-dimensional space $C_W(P)$, and as B is faithful on $Q/Z(Q)$, B is faithful on $C_W(P)/\langle v \rangle$, and hence on $W/A^* \cong H^1(L, A^*)$.

Thus it remains to show w does not exist, and we may take $W = \langle w, A^* \rangle$ of dimension 10. As PQ is maximal in L, $PQ = C_L(w) = C_L(v + w)$, so

$$|w^L| = |(w + v)^L| = |L : PQ| = 280.$$

But $v + w \notin w^L$ since $\{w\}$ is the unique point of w^L fixed by PQ (since $PQ = N_L(PQ) = C_L(w)$) so $|w^L \cup (v + w)^L| = 560 > 2^9 = |W - A^*|$, a contradiction.

(23.7) $\mathrm{Schur}_2(U_6(2)) \cong E_4$ *and a group of outer automorphisms of* $U_6(2)$ *of order 3 is faithful on* $\mathrm{Schur}_2(U_6(2))$.

Proof: Let $\tilde{H} = \mathrm{Cov}_2(H)$ and $Z = Z(\tilde{H})$, so that $Z \cong \mathrm{Schur}_2(H) \cong E_{2^k}$. By 23.5, for each hyperplane U of Z, $\tilde{A}/U = [\tilde{A}/U, \tilde{K}] \cong E_{2^{10}}$, so $\tilde{A} \cong E_{2^{k+9}}$ and $[\tilde{K}, \tilde{A}] = \tilde{A}$. We saw during the proof of the previous lemma that $k \geq 2$, so it remains to show $k \leq 2$. But (cf. 17.12 in [FGT]) $\dim(H^1(L, A^*)) \geq k$, so 23.6 says $k = 2$. Further as a group B of outer automorphisms of order 3 of H acts on $\tilde{H}$ by the universal property of $\mathrm{Cov}_2(H)$, and as we may take B to act on $\tilde{K}$ and induce outer automorphisms on $\tilde{K}/\tilde{A}$, B is faithful on $\mathrm{Schur}_2(H) = Z(\tilde{K})$ by 23.6.

(23.8)   *Let G be of type $M(22)$ and type $\tilde{M}(22)$, D the set of 3-transpositions of G, $T \in \mathrm{Syl}_2(G)$, $B = \langle T \cap D \rangle$, and $\tilde{G} = \mathrm{Cov}_2(M(22))$. Then*

(1) $\mathrm{Schur}_2(G) \cong Z(\tilde{G}) \cong \langle z \rangle \cong \mathbf{Z}_2$.

(2) $\tilde{B} \cong E_{2^{11}}$.

(3) z is not a square in $\tilde{G}$ and each involution in G lifts to an involution in $\tilde{G}$.

(4) If $\tilde{a} \in \tilde{G}$ is a lift of $a \in D$ then $\tilde{a}z \notin \tilde{a}^{\tilde{G}}$ and $\tilde{a}^{\tilde{G}}$ is a set of 3-transpositions of $\tilde{G}$.

(5) $\tilde{H} = C_{\tilde{G}}(\tilde{a})$ is quasisimple with $\tilde{H}/\langle \tilde{a}, z \rangle \cong U_6(2)$.

(6) $[N_{\tilde{G}}(\tilde{B}), \tilde{B}] = \tilde{B}$.

Proof: The centralizer of a 3-transposition in a group of type $\tilde{M}(23)$ is a perfect central extension of $\mathbf{Z}_2$ by a group of type $M(22)$ and $\tilde{M}(22)$, so $\mathrm{Schur}_2(G) \neq 1$. We prove that (2)–(6) hold for each quasisimple group $\tilde{G}$ with center $\langle z \rangle$ of order 2 and $\tilde{G}/\langle z \rangle = G$ of type $M(22)$ and $\tilde{M}(22)$. It follows that in $\hat{G} = \mathrm{Cov}_2(G)$, $\hat{H} = C_{\hat{G}}(\hat{a})$ is quasisimple with center isomorphic to $\mathbf{Z}_2 \times \mathrm{Schur}_2(G)$, so $\mathrm{Schur}_2(G) \cong \mathbf{Z}_2$ by 23.7.

Thus it remains to assume $\tilde{G}$ is quasisimple with center $\langle z \rangle$ of order 2 and $\tilde{H}/\langle z \rangle = G$, and to prove (2)–(6). Let $a \in B$ and $M = N_G(B)$, so that $B \cong E_{2^{10}}$ and B is the 10-dimensional Todd module for $M/B \cong M_{22}$ (cf. 25.6). In particular M is irreducible on B, so either (2) holds or $\tilde{B}$ is extraspecial. The latter is impossible as B is not a self-dual M/B-module; for example $B = [B, M_a]$ so M_a has no invariant hyperplane.

As each involution in G is fused into B (cf. 37.4), part (2) implies part (3). The proof of (4) is the same as that of 23.5.4.

As $\tilde{G}$ is quasisimple, $\tilde{M}$ does not split over z by Gaschütz's Theorem 2.6. Let $a \neq b \in T \cap D$. Then $N_M(\langle a, b \rangle)$ contains a Sylow 2-subgroup of M, so by another application of Gaschütz, $N_{\tilde{M}}(\langle \tilde{a}, \tilde{b} \rangle)$ does not split over $\langle z \rangle$. Then as $C_{\tilde{M}}(\langle \tilde{a}, \tilde{b} \rangle) = O^2(N_{\tilde{M}}(\langle \tilde{a}, \tilde{b} \rangle))$ is of index 2 in $N_{\tilde{M}}(\langle \tilde{a}, \tilde{b} \rangle)$ and the extension splits, $C_{\tilde{M}}(\langle \tilde{a}, \tilde{b} \rangle)$ does not split over $\langle z \rangle$ and hence $\tilde{H} = C_{\tilde{M}}(\tilde{a})$ does not split over $\langle z \rangle$. That is (5) holds.

Further $\tilde{B} \leq \tilde{H}$ and as we observed during the proof of 23.7, $[\tilde{H} \cap \tilde{M}, \tilde{B}] = \tilde{B}$, so (6) holds.

If A and B are groups then we denote the *wreath product* of A by B by $A \ wr \ B$. See page 33 in [FGT] for a definition and discussion of the wreath product.

(23.9) *Let D be the set of transvections in $H = U_6(2)$. Then*

(1) H has three conjugacy classes of elements of order 3 with representatives x_i, $1 \leq i \leq 3$.

(2) Let $\omega \in F$ be of order 3. Then the multiplicity of $(\omega, \omega^{-1}, 1)$ as an eigenvalue of a suitable lift $\tilde{x}_i$ of x_i in $H_0 = SU(V)$ on V is $(3, 0, 3)$, $(1, 1, 4)$, $(2, 2, 2)$, respectively.

(3) $w_D(C_H(x_1)) = 2$ and $C_H(x_1)$ is of index 3 in $GU_3(2) * GU_3(2)$.

(4) x_2 is inverted by a member of D, $w_D(C_H(x_2)) = 5$, $C_G(x_2) = \langle x_2 \rangle \times \langle C_D(x_2) \rangle$, and $\langle C_D(x_2) \rangle \cong U_4(2)$.

(5) $w_D(C_G(x_3)) = 3$, $C_G(x_3) \cong \mathbf{Z}_3 \times (S_3 \ wr \ \mathbf{Z}_2)$, and $\langle C_D(x_3) \rangle \cong S_3^3$.

Proof: First if $x \in H_0$ is of order 3 then there is an orthonormal basis $(V_i : 1 \leq i \leq 6)$ of eigenvectors for x. Let ω^{n_i} be the eigenvalue of x at v_i. Multiplying x

by a power of $z = \omega I$, we can assume $n_5 = n_6 = 0$. Now $\sum_{i \leq 4} n_i \equiv 0 \bmod 3$ as $\det(x) = 1$, so we determine the multiplicities of the eigenvalues of x are one of the three triples listed in (2).

Thus to prove (1) and (2) it remains to show that if $y \in H_0$ has an image in H of order 3 then y is of order 3. If not, $y^3 = z$. Let U be a 6-dimensional unitary space over $K = GF(64)$ and $G = GU(U)$. Then $GU(V) = C_G(\sigma)$ for some field automorphism σ of G determined by a generator of $Gal(K/F)$. Now $y \in C_G(\sigma)$ has an orthonormal basis $(u_i : 1 \leq i \leq 6)$ of eigenvectors of U. Let a_i be the eigenvalue of y at u_i. As $y^3 = z$, $a_i^3 = \omega$ for each i, so $a_i = a\omega^{m_i}$ for some $a \in K$ of order 9 with $a^3 = \omega$. Further as $y \in C_G(\sigma)$, the multiplicity of $a\omega^m$ as an eigenvalue of y is independent of m, so $\det(y) = a^6 = \omega^{-1} \neq 1$. Thus $y \notin O^3(G)$, so $y \notin O^3(C_G(\sigma)) = O^3(GU(V)) = H_0$, a contradiction.

So (1) and (2) are established. Now an easy calculation in linear algebra establishes (3)–(5).

Exercises

1. Let V be a 4-dimensional vector space over $GF(2)$ and $A_7 \cong G \leq GL(V)$. Prove
   (1) G is transitive on $V^{\#}$, and
   (2) $H^1(G, V) = 0$.
   [*Hint:* For (2), use the Gorenstein–Alperin argument of Exercise 6.4 in [FGT] applied to the class of subgroups of G of order 3 moving three points in the permutation representation of G of degree 7.]

2. Let $G = U_4(2)$, D the set of transvections of G, $T \in \mathrm{Syl}_2(G)$, and $B = \langle T \cap D \rangle$. Prove the maximal subgroups of G containing B are $N_G(B)$ and $C_G(d)$, $d \in T \cap D$.
   [*Hint:* Let V be the natural module for G and $B \leq M < G$. Observe that for $d \in T \cap D$ we have $\langle d^{\perp} \rangle = \langle T \cap D, a \rangle$ for each $a \in d^{\perp} - T$, so either $d^{\perp} \subseteq M$ or $T \cap D = d^{\perp} \cap M$. In the former case conclude $M \leq C_G(d)$ by 7.8.5. On the other hand if $T \cap D = d^{\perp}$ for each $d \in T \cap D$ then by 9.2.2, $db \in O_2(M)$ for each $b \in T \cap D$, so $U = C_V(O_2(M)) \leq C_V(db) = C_V(B)$. Therefore $U = C_V(B)$ or $[V, c]$ for some $c \in T \cap D$ and hence $M \leq N_G(U) = N_G(B)$ or $C_G(c)$.]

3. Prove $\mathrm{Schur}_2(U_4(3)) \cong \mathbf{Z}_2$.
   [*Hint:* Argue as in 23.7 with respect to a subgroup M of $G = U_4(3)$ with $A = O_2(M) \cong E_{16}$ and $M/A \cong A_6$. First observe that as $U_4(3)$ has one class of involutions, each involution in $U_4(3)$ lifts to an involution in $\mathrm{Cov}_2(U_4(3)) = \tilde{G}$. Then using Gaschütz's Theorem 2.6 and 33.15 in

[FGT], conclude $\tilde{A} = [\tilde{A}, \tilde{M}]$. Finally use Exercise 6.3 in [FGT] to conclude $\dim(\mathrm{Schur}_2(G)) \leq \dim(H^1(M/A, A)) = 1$ and appeal to the existence of $SU_4(3)$ for the opposite inequality.]

4. Let $\tilde{V}_0$ be the 11-dimensional Todd module for $G \cong M_{24}$, $\tilde{v} \in \tilde{V}_0$ of weight 4, Z of order 3 in $O_{2,3}(G_{\tilde{v}})$, and $K = C_G(Z)$. Prove $\tilde{V}_0 = [\tilde{V}_0, Z] \oplus C_{\tilde{V}_0}(Z)$ with $[\tilde{V}_0, Z]$ a 6-dimensional irreducible for K and $C_{\tilde{V}_0}(Z)$ a 5-dimensional indecomposable for K isomorphic to the core of the 6-dimensional permutation module for $K/Z \cong A_6$.

   [*Hint:* Let Δ be the sextet determining $\tilde{V}$. From 20.4.6 in [SG], K acts on Δ and hence on $\tilde{v}$ and $K \cong \hat{A}_6$ is quasisimple with $Z = Z(K)$ and $K/Z \cong A_6$. Prove $\mathrm{Fix}_X(Z) = Y$ is of order 6 and $C_{\tilde{V}_0}(Z)$ consists of the 15 vectors of weight 2 in Y, the 15 vectors of weight 4 in Y, and $\tilde{v} = \tilde{e}_Y = \tilde{e}_I + \tilde{e}_{Y+I}$ for I a 2-subset of Y. Use 19.11.5 in [SG] to prove this last fact.]

5. Let $\tilde{V}_0$ be the 11-dimensional Todd module for $G \cong M_{24}$ and $\tilde{v} \in \tilde{V}_0$ of weight 4. Prove
   (1) $\mathrm{Aut}(G_{\tilde{v}}) = G_{\tilde{v}}$.
   (2) $N_{GL(\tilde{V}_0)}(G_{\tilde{v}}) \leq G$.
      [*Hint:* For (1), use Exercise 16.3 in [FGT] to see $\mathrm{Aut}(A_6) = P\Gamma L_2(9)$. Then observe the two subgroups of $G_{\tilde{v}}$ of index 15 act differently on $O_2(G_{\tilde{v}})$, whereas their images are conjugate under $\alpha \in P\Gamma L_2(8) - S_6$. For (2) use Lemma 21.13. Use Exercise 7.4 to verify the second hypothesis of 21.13.]

6. Let (X, C) be the Steiner system for $G = M_{24}$, $M = G(\{1, 2\}) \cong \mathrm{Aut}(M_{22})$ the global stabilizer of the 2-subset $\{1, 2\}$ of X, $\hat{V}$ the 10-dimensional Todd module for M, and I_1 the orbit of M on $\hat{V}$ of length 22 supplied by Lemma 22.3. Prove
   (1) $u = \hat{e}_{1234} = \hat{e}_{34} = \hat{e}_{12} + \hat{e}_{14}$ with $\hat{e}_{12}, \hat{e}_{14} \in I_1$.
   (2) M_u is the stabilizer in M of the sextet through $\{1, 2, 3, 4\}$ and the stabilizer in G of $\{1, 2\}$ and $\{1, 2, 3, 4\}$.
   (3) M_u is the split extension of $O_2(M_u) \cong E_{32}$ by S_5 with $O_2(M_u)$ the image of the 6-dimensional permutation module for S_5.
   (4) u is centralized by some $T \in \mathrm{Syl}_2(M)$.
   (5) M is transitive on the set T of involutions of M with no fixed points on I_1 and there is $t \in O_2(M) \cap T$ with $C_M(t) = C_{M_u}(t)$ the extension of $O_2(M_u)$ by a Frobenius group of order 20.
   (6) There is $s \in T \cap M$ with $\hat{e}_{13}^s = \hat{e}_{14}$ and $t \in M_u^\infty O_2(M_u) - O_2(M_u)$.
      [*Hint:* Use 21.1 in [SG] to obtain the class of involutions of G and t as in (5).]

7. Let $G = U_4(2)$, D the set of transvections in G, and t an involution in G. Prove there is $d \in D$ with $d^t \in A_d$.

8. Assume the hypotheses of Exercise 7.6. Let C be an octad with $1, 2 \in C$, $J = \{\hat{e}_{1c} : c \in C - \{1, 2\}\}$, and B the subspace of $\hat{V}$ generated by J. Prove
   (1) $N_M(C)$ is the split extension of $E = O_2(N_M(C)) \cong E_{16}$ by S_6 acting as $Sp_4(2)$ on E.
   (2) $\dim(B) = 5$, $J = I_1 \cap B$, and $N_M(C)$ acts 6-transitively as S_6 on J.
   (3) $[\hat{V}/B, N_M(C)]$ is $N_M(C)/E$-isomorphic to $[B, N_M(C)]^\sigma$, where σ is an outer automorphism of $N_M(C)/E$.
   [*Hint:* See 19.1 in [SG] for $N_G(C)$.]

9. Assume the hypotheses of Exercise 7.6 and let $L = G_{12} = M_{22}$. Prove
   (1) L has one class of involutions t^L and $\text{Fix}_X(t)$ is an octad containing $1, 2$.
   (2) $\dim(C_{\hat{V}}(t)) = 6$.
   (3) $m(A) < m(\hat{V}/C_{\hat{V}}(A))$ for each nontrivial elementary abelian 2-subgroup A of L.
   [*Hint:* Use 2.1 in [SG] to prove (1). Then by (1) we can take t in the subgroup E of Exercise 7.6.8. Now calculate $|C_{\hat{V}}(t) : B| = 2$ to prove (2). Then prove $m_2(L) = 4$, so if A is a nontrivial elementary abelian 2-subgroup of L with $m(A) \geq m(\hat{V}/C_{\hat{V}}(A))$ and $t \in A$ then $C_{\hat{V}}(A) = C_{\hat{V}}(t)$. Conclude $A = E$, whereas $C_{\hat{V}}(E) = B \neq C_{\hat{V}}(t)$.]

10. Let $(X, \mathcal{C})$ be the Steiner system for $G = M_{24}$, $B \in \mathcal{C}$, and $M = N_G(B)$. Prove
    (1) $M = M_x E$, where $E_{16} \cong E = G_B$, $x \in X - B$, and M_x acts faithfully as $GL(E)$ on E.
    (2) $H^1(M_x, E) = 0$.
    (3) $\text{Schur}_2(G) = 1$.
    [*Hint:* Part (1) is 19.1 in [SG]. See the Hint for Exercise 7.1 for (2). If (3) fails there is quasisimple $\hat{G}$ with $Z(\hat{G}) = Z \cong \mathbf{Z}_2$ and $\hat{G}/Z = G$. Let $\hat{E}/Z = E$ and $\hat{M}_x/Z = M_x$. Use (1) and (2) to show $\hat{E} = Z \times [\hat{E}, \hat{M}_x] \cong E_{32}$, so $\hat{M}_x$ is quasisimple. By 33.15 in [FGT], some involution in M_x does not lift to an involution in $\hat{M}_x$. But by 21.1 in [SG], all involutions in G fixing a point of X are fused.]

8

Almost 3-transposition groups

In this short chapter we prove some more results about general 3-transposition groups, prove some facts about the Fischer groups, and also establish some sufficient conditions for a group to be a 3-transposition group.

24. Sufficient conditions to be a 3-transposition group

In this section G is a finite group, D is a G-invariant set of involutions of G, $\mathcal{D} = \mathcal{D}(D)$ is the commuting graph on D, and for $d \in D$, $D_d = C_D(d) - \{d\}$ and $A_d = D - C_D(d)$. We seek sufficient conditions to ensure D is a set of 3-transpositions of $\langle D \rangle$.

We say D is a set of *odd transpositions* of $\langle D \rangle$ if $|ab|$ is odd or 2 for all $a, b \in D$.

(24.1) D is a set of odd transpositions of $\langle D \rangle$ if and only if $C_D(ab) = C_D(a) \cap C_D(b)$ for all $a \in D$ and $b \in D_a$.

Proof: If D is a set of odd transpositions then the proof of 8.4 shows

$$C_D(a) \cap C_D(b) = C_D(ab) \quad \text{for } a \in D \quad \text{and} \quad b \in D_a.$$

Conversely assume this condition and let $a, d \in D$ with ad of even order. Let z be the involution in $\langle ad \rangle$. Then dz is conjugate in $\langle a, d \rangle$ to a or d, so $dz \in D$. Then $a \in C_D(z) = C_D(d) \cap C_D(dz) \subseteq C_D(d)$, so $|ad| = 2$.

(24.2) If D is a set of odd transpositions of $\langle D \rangle$, $H \leq G$, and $\alpha : G \to G\alpha$ is a group homomorphism then

(1) $H \cap D$ is a set of odd transpositions of $\langle H \cap D \rangle$.
(2) $D\alpha$ is a set of odd transpositions of $\langle D\alpha \rangle$.
(3) $\langle T \cap D \rangle$ is abelian for $T \in \mathrm{Syl}_2(G)$.

137

(4) D is locally conjugate in G.

(5) If $\langle D \rangle$ is not transitive on D then $\mathcal{D}$ is connected.

Proof: Parts (1)–(3) are straightforward. Part (4) is Exercise 1.2.1; (4) and 7.3 imply (5).

(24.3) *Assume D is a set of odd transpositions of $\langle D \rangle$, let $T \in \mathrm{Syl}_2(G)$, and assume*

(1) $\langle T \cap D \rangle^{\#} \neq T \cap D$.

(2) $N_G(T \cap D) \cap C_G(abc)$ is not transitive on $T \cap D$ for $a, b, c \in T \cap D$.

(3) *If Δ is a connected component of $\mathcal{D}$ and x is an involution in $N_G(\Delta)$ then x fixes a member of Δ.*

Then $\mathcal{D}$ is connected.

Proof: Let $A = \langle T \cap D \rangle$. By 24.2.3, A is abelian so $T \cap D$ is a complete subgraph of $\mathcal{D}$ and hence $T \cap D$ is contained in some connected component Δ of $\mathcal{D}$. Assume $D \neq \Delta$ and let $H = N_G(\Delta)$. Notice that by hypothesis (3), $\Delta = H \cap D$.

Let $d \in \Delta$ and $a \in D_d$. Then $C_G(ad)$ acts on $C_D(ad) = C_D(a) \cap C_D(d) \subseteq \Delta$ by 24.1, so $C_G(ad) \leq H$.

Let $a, b, c \in T \cap D$. Then by hypothesis (2), $C_G(abc)$ is not transitive on $C_D(abc)$ since $N_G(T \cap D) \cap C_G(abc)$ controls $C_G(abc)$-fusion in $T \cap D$ by 2.8. Therefore $\mathcal{D}(C_D(abc))$ is connected by 24.2.5, so $C_D(abc) \subseteq \Delta$ and hence $C_G(abc) \leq H$.

By hypothesis (1) there exists $d, e \in T \cap D$ with $x = de \notin D$. By a previous reduction, $C_G(x) \leq H$. Let $a \in D - \Delta$. As $x \notin D \supseteq a^G$, xa has even order and hence contains an involution z. Then za is conjugate to x or a in $\langle x, a \rangle$. Further if $za \in x^G$ then $za = bc$ for $b \in D$ and $c \in D_b$ so $a \in C_D(za) = C_D(b) \cap C_D(c)$ by 24.1. Thus in any event $z = za \cdot a$ is the product of two or three commuting involutions in the connected component Δ_a of $\mathcal{D}$ containing a, so by previous reductions, $x \in C_G(z) \leq N_G(\Delta_a)$. Then by hypothesis (3), there is $f \in \Delta_a$ fixed by x. Now $f \in C_D(x) \leq H \cap D = \Delta$, so $\Delta = \Delta_a$, contrary to the choice of a.

(24.4) *Assume G is transitive on D, let $d \in D$, and let Δ be the connected component of $\mathcal{D}$ containing d. Assume*

(i) *D_d is a conjugacy class of 3-transpositions of $\langle D_d \rangle$ and $|D_d| > 1$.*

(ii) *For $b \in D_d$ and $a \in A_d \cap D_b$, $c^{\perp} \cap D_{a,d} \neq \emptyset$ for all $c \in D_d$.*

Then

(1) Δ is a set of 3-transpositions of $\langle \Delta \rangle$.

(2) Let $d \in T \in \mathrm{Syl}_2(G)$, $A = \langle T \cap D \rangle$, and assume

(iii) A is abelian,

(iv) $A = \langle ab : a, b \in T \cap D \rangle$,

(v) The map $\{a, d\} \mapsto ad$ is an injection on 2-subsets of $T \cap D$, and

(vi) Hypotheses (1), (2), and (3) of 24.3 hold.

Then $\mathcal{D}$ is connected so D is a set of 3-transpositions of $\langle D \rangle$.

Proof: If Δ is of diameter 2 then hypothesis (i) implies Δ is a set of 3-transpositions of $\langle \Delta \rangle$. So assume $cdba$ is a path in $\mathcal{D}$ with c at distance 3 from a. Then by hypothesis (ii), $c^\perp \cap D_{a,d} \neq \emptyset$, so c is at distance at most 2 from a, a contradiction.

Therefore (1) is established so we assume the hypotheses of (2). By 24.3 it suffices to show D is a set of odd transpositions of $\langle D \rangle$, and thus by 24.1 it suffices to show $C_D(ad) \subseteq C_D(d)$ for $a \in D_d$. Assume not and let $H = C_G(ad)$ and $H^* = H/\langle ad \rangle$.

Claim d^* is weakly closed in $C_{H^*}(d^*)$ with respect to H^*. Without loss $a \in T$ and $h \in H$ with $a, d \neq d^h \in T$. Then $d^h \in T \cap D \subseteq A$, and by hypothesis (iii) and 2.8, $N_H(A)$ controls fusion in A, so we may take $h \in N_H(A)$. Then $ad = (ad)^h = a^h d^h$. But by hypothesis (v) the map $\{a, d\} \mapsto ad$ is an injection on 2-subsets of $T \cap D$, so $\{a, d\} = \{a^h, d^h\}$, contrary to the choice of d^h.

Thus the claim is established so by Glauberman's Z^*-Theorem [G], $d^* \in Z^*(J^*)$. If $d^* \in Z(H^*)$ then $\langle a, d \rangle \trianglelefteq H$ so as A is abelian, $H \cap D \subseteq C_D(d)$, completing the proof.

Therefore we may assume $X = [d, O(H)] \neq 1$ and it remains to produce a contradiction. Let $a, d \neq b \in T \cap D$. Then D_b is a set of 3-transpositions of $\langle D_b \rangle$ and $a, d \in D_b$, so $d^X \cap D_b \subseteq C_{D_b}(ad) \subseteq D_{b,d}$ and hence $d^X \cap D_b = \{d\}$. Therefore $[d, C_X(b)] = 1$, so $ef \in C_G(X)$ for all $e, f \in T \cap D$ by Exercise 8.1 in [FGT]. But then by hypothesis (iv), $d \in A = \langle ef : e, f \in T \cap D \rangle \leq C_G(X)$, a contradiction. Hence the proof is at last complete.

(24.5) *Let t be an involution in $Z(G)$, $\tilde{G} = G/\langle t \rangle$, and assume $\tilde{D}$ is a set of 3-transpositions of $\langle \tilde{D} \rangle$ and $dt \notin D$ for $d \in D$. Then D is a set of 3-transpositions of $\langle D \rangle$.*

Proof: Let $a, b \in D$. Then $|\tilde{a}\tilde{b}| \leq 3$. If $|\tilde{a}\tilde{b}| \leq 2$ then $a^b \in a\langle t \rangle \cap D = \{a\}$, so $|ab| \leq 2$. If $|\tilde{a}\tilde{b}| = 3$ then either $|ab| = 3$ or $\langle a, b \rangle \cong D_{12}$ with $\langle t \rangle = Z(\langle a, b \rangle)$. But in the latter case $at \in b^{\langle a, b \rangle} \subseteq D$, a contradiction.

25. Some results on 3-transposition groups

In this section D is a union of conjugacy classes in a finite group G such that D is a set of 3-transpositions of $\langle D \rangle$. Let $d \in D$.

(25.1) *Assume $d^{\perp} \subseteq M \leq G$. Then*

(1) $d^{O_3(M)} \subseteq d^{O_3(G)} = W_d$.

(2) *If $d^M \cap D_d \neq \emptyset$ then $d^{O_3(M)} = W_d$.*

(3) *If $O_3(M) = \langle [O_3(M), d]^M \rangle$ then $O_3(M) \leq O_3(G)$.*

Proof: Let $U_d = \{e \in M \cap D : D_e \cap M = D_d \cap M\}$. By Exercise 3.1.2, $U_d = d^{O_3(M)}$ and $W_d = d^{O_3(G)}$. Now if $e \in U_d$ then as $e \in d^M$, $D_e \subseteq M$, so

$$D_d = D_d \cap M = D_e \cap M = D_e,$$

and hence $e \in W_d$. That is $U_d \subseteq W_d$, so $d^{O_3(M)} = U_d \subseteq W_d = d^{O_3(G)}$, establishing (1). In particular

$$[O_3(M), d] = \langle de : e \in U_d \rangle \leq \langle dw : w \in W_d \rangle = [O_3(G), d] \leq O_3(G),$$

so (3) holds.

Assume $b \in d^M \cap D_d$. Then $W_d \subseteq D_b \subseteq M$, so $W_d \subseteq U_d$, so as we have seen that $U_d \subseteq W_d$, we conclude (2) holds.

(25.2) *Assume $H = \langle D_d \rangle$ is transitive on D_d and $d^G \cap D_d \neq \emptyset$. Then*

(1) *G is transitive on D.*

(2) *Either $\langle D \rangle$ is transitive on D or G is of width 2, H is of width 1, and $\langle D \rangle = H * \langle d^G \rangle$.*

(3) *If $O_3(H) \leq Z(H)$ and $|D_d| > 1$ then setting $X = \langle D \rangle$, one of the following holds:*

(a) *X is primitive on D.*

(b) *$H/Z(H) \cong \mathrm{Sp}_6(2)$ and $X/Z(X) \cong S_3/\Omega_8^+(2)$.*

(c) *$H/Z(H) \cong PO_7^{-\pi,\pi}(3)$ and $X/Z(X) \cong S_3/P\Omega_8^+(3)$.*

Proof: Let $X = \langle D \rangle$. If $a \in A_d$ then a is conjugate to d in $\langle a, d \rangle$ by 8.2. On the other hand by hypothesis there is $b \in D_d \cap d^G$ and if $a \in D_d$ then a is conjugate to b in H by transitivity of H on D_d. Therefore (1) is established.

Assume X is not transitive on D. Then by 8.2, there is a nontrivial partition $D = D_1 \cup D_2$ of D such that $X = X_1 * X_2$ where $X_i = \langle D_i \rangle$. Take $d \in D_1$. Then $D_2 \subseteq A_d$ and $H \leq X$, so by transitivity of H on D_d, we have $D_d = D_2$ and $H = X_2$. Therefore X_1 is of width 1. But by (1), d is fused into D_2 under

G, so by symmetry, H is also of width 1, and hence G is of width 2. Therefore (2) is established.

We next prove (3), so assume $O_3(H) \leq Z(H)$, $|D_d| > 1$, and X is not primitive on D. If X is not transitive on D then H is of width 1 by (2), so $|H : O_3(H)| = 2$ by 8.6. Then as $O_3(H) \leq Z(H)$, $|D_d| = 1$, contrary to our hypotheses. Therefore X is transitive on D. Now by Exercise 3.3, $O_2(G) \leq Z(G) \geq O_3(G)$. Hence by 9.4, $\mathcal{D}(D)$ is disconnected, and then by 17.2, (b) or (c) holds. So the proof of (3) is complete.

(25.3) *Assume* $\langle D_d \rangle \cong S_3$ *and* $O_3(C_G(d)) \cong \mathbf{Z}_3$ *for each* $d \in D$. *Then* $\langle D \rangle \cong S_5$ *or* $S_3 \times S_3$.

Proof: Without loss, $G = \langle D \rangle$. If G is not transitive on D then by 8.2 there is a nontrivial G-invariant partition $D = D_1 \cup D_2$ of D such that $G = G_1 * G_2$ for $G_i = \langle D_i \rangle$. Let $d_i \in D_i$; as $\langle D_{d_i} \rangle \cong S_3$ we conclude $D_{d_i} = D_{3-i}$, and hence $G \cong S_3^2$.

Therefore we may assume G is transitive on D. Suppose next that $X = O_3(G) \not\leq Z(G)$. By 9.2, $W_d = d^{O_3(G)}$. Let $b \in D_d$, so that $W_d \subseteq D_b$. Then as D_b is of order 3, $W_d = D_b$. Therefore $X = [X, d] \times C_X(d)$ with $[X, d] = O_3(\langle D_b \rangle) \cong \mathbf{Z}_3$. Then as $O_3(C_G(d)) \cong \mathbf{Z}_3$, $C_X(d) = O_3(\langle D_d \rangle) = [X, b]$, and then $X = [X, d] \times [X, b] \cong E_9$. Now $G^* = G/C_G(X) \leq GL_2(3)$ with D^* a class of 3-transpositions of G^* and $D_{d^*} = \{b^*\}$. As G^* is transitive on D^*, there is $a^* \in A_{d^*}$. As $D_{b^*} = \{d^*\}$, $a^* \in A_{b^*}$, so $\langle d^*, a^*, b^* \rangle \cong S_4$ by 8.3. This is impossible as $GL_2(3)$ contains no such subgroup.

So $O_3(G) \leq Z(G)$, and then as $O_3(C_G(d)) = O_3(\langle D_d \rangle)$, $O_3(G) = 1$. By Exercise 3.2, $O_2(G) \leq Z(G)$. By 17.2, $\mathcal{D}(D)$ is connected and then by 9.4, G is primitive on D. Finally by 13.4, $G \cong S_5$.

We prove the next two lemmas together.

(25.4) *Let* L *be a D-subgroup of* G *and assume one of the following holds:*

   (a) $G \cong U_6(2)$ *and* $L/Z(L) \cong PO_6^{+,\pi}(3)$.
   (b) G *is of type* $M(22)$ *and* $L/Z(L) \cong PO_7^{-\pi,\pi}(3)$.
   (c) G *is of type* $M(23)$ *and* $L/Z(L) \cong S_3/P\Omega_8^+(3)$.

Then L *is transitive on* $D - L$ *and for* $b \in D - L$, $L_b \cong S_6/E_{16}$, $Sp_6(2)$, $S_3/\Omega_8^+(2)$ *in cases* (a),(b),(c), *respectively, and* L_b *is a D-subgroup of* G.

(25.5) *Assume* G *is of type* $M(23)$ *and* L *is a D-subgroup of* G *with* $L/Z(L) \cong P\Omega_8^{+,\pi}(3)$. *Then* $N_G(E(L)) \cong S_3/P\Omega_8^+(3)$ *is a D-subgroup of* G.

Now the proof of 25.4 and 25.5. So assume the hypotheses of 25.4 or 25.5 and let $\Gamma = L \cap D$ and pick $d \in \Gamma, b \in D_d - \Gamma$, set $H = \langle D_d \rangle$, $K = \langle D_d \cap \Gamma \rangle$, $H^* = H\langle d \rangle / \langle d \rangle$, and $I = \langle D_b \cap \Gamma \rangle$.

We first deal with case (a) of 25.4. There G is $U_6(2)$ and L is $PO_6^{+,\pi}(3)$, so by 29.9.3, L is determined up to conjugacy under $PGU_6(2)$, so in particular the embedding of L in G is described in Chapter 10. Now the action of G on D is equivalent to its action on the singular points of the natural projective module V for G, and by 28.20, 29.1, and 29.5, L has two orbits on these points with stabilizers D-subgroups isomorphic to $O_5^{-\pi,\pi}(3)$ and S_6/E_{16}. Hence L is transitive on Γ and on $D - \Gamma$ with $L_b \cong S_6/E_{16}$.

So case (a) of 25.4 is handled and hence we may assume G is of type $M(22)$ or $M(23)$. Therefore $H^* \cong U_6(2)$ or H^* is of type $M(22)$. Also $K^* \cong PO_6^{+,\pi}(3)$ or $PO_7^{\pi,\pi}(3)$ in the respective case by 11.11. Notice K^* is maximal in H^* by 16.1 and 16.3, so $K\langle d \rangle = L \cap H\langle d \rangle$. Next by induction on the order of G, $K_b \leq I$, K is transitive on $D_d - \Gamma$, and $K_b^* \cong S_6/E_{16}$ or $Sp_6(2)$, respectively. Moreover the corresponding remarks hold for $d' \in D_{d,b} \cap \Gamma$, so by Exercise 8.2, $I/Z(I) \cong Sp_6(2)$ if G is of type $M(22)$ and $I/Z(I)$ is isomorphic to $O_8^\epsilon(2)$ or $S_3/\Omega_8^+(2)$ if G is of type $M(23)$.

By a Frattini argument, $L_b = IC_{L_b}(d)$, while $C_{L_b}(d) \leq L_b \cap H\langle d \rangle = K_b\langle d \rangle \leq I$, so $L_b = I$. Similarly $C_L(I) \leq C_{L \cap H\langle d \rangle}(K_b) = C_{K\langle d \rangle}(K_b) \leq \langle d \rangle$, so $Z(I) = 1$. Now if $I \cong Sp_6(2)$ or $S_3/\Omega_8^+(2)$, respectively, then $|b^L| = |L : I| = |D - \Gamma|$ by direct calculation (cf. 16.7–16.10 for $|D|$ and Exercise 4.13.2 for $|\Gamma|$). Therefore $b^L = D - \Gamma$ and the lemmas are established in this case.

Thus we may assume G is of type $M(23)$ and $I \cong O_8^\epsilon(2)$. As $|O_8^-(2)|$ does not divide $|L|$, $I \cong O_8^+(2)$. But then if $L \cong S_3/P\Omega_8^+(3)$, we have $|b^L| = |L : I| = 3|D - \Gamma|$, contradicting $b^L \subseteq D - \Gamma$. Therefore $L \cong PO_8^{+,\pi}(3)$ and we are reduced to proving 25.5.

Now by the previous calculation, $|D-\Gamma| = |b^L|+2|\Gamma|$. Also as K is transitive on $D_d - \Gamma$ and L is transitive on Γ, $b^L = \bigcup_{d \in \Gamma} D_d - \Gamma$. So $S = \bigcap_{d \in \Gamma} A_d$ is of order $2|\Gamma|$. Let $T = \Gamma \cup S$, $M = \langle T \rangle$, $s \in S$, and $r = s^d$. Claim that T acts on T and that Γ, Γ^s, and Γ^r are the connected components of $\mathcal{D}(T)$. It then follows that M is a D-subgroup of G with $T = M \cap D$ and $E(L) \cong P\Omega_8^+(3)$ is the kernel of the action of M on $\{\Gamma, \Gamma^s, \Gamma^r\}$ with $M/E(L) \cong S_3$. Finally $M = N_G(E(L))$ by 16.11.2.

So it remains to establish the claim. By definition of S, L acts on S and hence on T. Let $a \in \Gamma$. By 8.7, $a^s = s^a \in S$, so $\Gamma^s \subseteq S$. Similarly $\Gamma^r \subseteq S$ and if $c \in \Gamma$ with $c^r = a^s$ then $c = a^{sr} = a^{ds} \in \Gamma^s \cap \Gamma \subseteq S \cap \Gamma = \emptyset$, a contradiction. Therefore $|\Gamma^s \cup \Gamma^r| = 2|\Gamma| = |S|$, so $T = \Gamma \cup \Gamma^s \cup \Gamma^r$ is a partition of T. Finally r interchanges Γ and Γ^r and $a^{sr} \in \Gamma^s$ for $a^s \in \Gamma^s$, so r acts on T. Thus T acts on T and as Γ is connected with $S \subseteq A_d$ for each $d \in \Gamma$, we conclude Γ, Γ^s, and Γ^r are the connected components of $\mathcal{D}(T)$, completing the proof.

(25.6) *Let G be of type $M(22)$, $M(23)$, or $M(24)$; $a \in A_d$; $b \in D_d \cap A_a$; and $S = \langle a, b, d \rangle$. Then*

    (1) G is transitive on 3-strings.
    (2) $E(\langle D_{a,d} \rangle)$ is transitive on $D_d \cap A_a$.
    (3) $\langle C_D(S) \rangle \cong S_6/E_{16}$, $\mathrm{Sp}_6(2)$, or $S_3/\Omega_8^+(2)$ for G of type $M(22)$, $M(23)$, or $M(24)$, respectively.
    (4) $\langle C_D(S) \rangle = C_G(S)$ if G is of type $M(24)$ or of type $\tilde{M}(22)$ or $\tilde{M}(23)$.

Proof: Let $H = \langle D_d \rangle$, $\Gamma = D_{a,d}$, and $L = \langle \Gamma \rangle$. Then $H/Z(H) \cong U_6(2)$ or $H/Z(H)$ is of type $M(22)$ or $M(23)$, while $L/Z(L) \cong PO_6^{+,\pi}(3)$, $PO_7^{\pi,\pi}(3)$, or $S_3/P\Omega_8^+(3)$ for G of type $M(22)$, $M(23)$, or $M(24)$, respectively. Thus we can appeal to 25.4 and conclude that L is transitive on $D_d - \Gamma$ and L_b is a D-subgroup of L with $L_b \cong S_6/E_{16}$, $\mathrm{Sp}_6(2)$, or $S_3/\Omega_8^+(2)$, respectively. In particular $L = L_b E(L)$, so (2) is established. Also $L_b = \langle C_D(S) \rangle$, so (3) holds.

Next each 3-string of G is conjugate to (d, a, u) for some $u \in D_d \cap A_a = D_d - \Gamma$, so (2) implies (1). Finally under the hypothesis of (4) and by 16.13.3 if G is of type $M(24)$, $H = C_G(d)$. Therefore $C_G(S) = C_H(a) \cap C_H(b) \ge L_b$. Further by 16.1, 16.3, and 16.11, $L\langle d \rangle$ is maximal in H, so $L = C_H(a)$ and hence $L_b = C_G(S)$, establishing (4).

(25.7) *Let $n = 19 + t$ for $t = 3, 4, 5$, G a group of type $M(n)$, with G of type $\tilde{M}(n)$ if $n = 22$ or 23, $T \in \mathrm{Syl}_2(G)$, $A = \langle T \cap D \rangle$, and $M = N_G(A)$. Then*

    (1) $|T \cap D| = n$.
    (2) $M/A \cong M_n$ and M is t-transitive on $T \cap D$.
    (3) A is the Todd module for M/A of dimension $7 + t = n - 12$.
    (4) $\prod_{d \in T \cap D} d = 1$.
    (5) The map $\{a, b\} \mapsto ab$ is an injection from the set of 2-subsets of $T \cap D$ into A.

Proof: We proved (1) and (4) in 16.7–16.12. Further we proved (2) for $M(24)$ in 16.12 and the proof in the remaining two cases is the same. (We need G of type $\tilde{M}$ to conclude $H = C_G(d)$ and hence the stabilizer X of $t - 2$ points of $T \cap D$ satisfies $X/A \cong L_3(4)$.) Notice this proof also shows $m(A) = 7 + t$.

Let $d \in T \cap D$. By (2), A is an image of the permutation module V for $M^* = M/A$ on $T \cap D$. From the discussion in Section 19 of [SG] and from

Section 22, if $n = 24$ then V is indecomposable and has a chief series

$$(*) \qquad\qquad 0 = V_0 < V_1 < V_2 < V_3 < V_4 = V$$

with $\dim(V_4/V_3) = 1$ and V_3/V_2 the 11-dimensional Todd module. As the bilinear form on V making $T \cap D$ into an orthonormal basis defines an isomorphism of V/V_2 with V_2^*, we have $\dim(V_1) = 1$ and $\dim(V_2/V_1) = 11$. In particular V/V_2 is the unique 12-dimensional image of V, so $A \cong V/V_2$ is the 12-dimensional Todd module for M^*.

Similarly restricting the M_{24}-module to M_{23}, if $n = 23$ we conclude

$$V = C_V(M^*) \oplus [V, M^*]$$

with $C_V(M^*)$ a point and $[V, M^*]$ an indecomposable with two 11-dimensional chief factors W and $[V, M^*]/W$, so $A \cong [V, M^*]/W$ is the unique 11-dimensional image of V, and hence is the Todd module for M_{23}. Finally when $n = 22$, restricting the M_{24}-module to M_{22}, we get a series $(*)$ with V_3/V_2 the 10-dimensional Todd module, and the same argument works.

So (3) is established. Now (3), 22.1.2, 22.3.2, and 22.4.2 imply (5) for $M(24)$, $M(22)$, and $M(23)$, respectively.

(25.8) *Assume $G = \langle D \rangle$ is primitive on D; let $a \in A_d$; $b \in D_{d,a}$; and assume $\langle D_{d,a} \rangle$ is transitive on $D_{d,a}$ and $C_G(b) \cap N_G(\langle ad \rangle) = \langle d, a, b^\perp \cap D_{d,a} \rangle$. Then $N_G(\langle da \rangle) = \langle d, a \rangle \times \langle D_{d,a} \rangle$.*

Proof: Let $L = \langle D_{d,a} \rangle$. By 8.4, $D_{d,a} = C_D(da)$, so $L \trianglelefteq N_G(\langle da \rangle)$. Then as L is transitive on $D_{d,a}$, by a Frattini argument,

$$N_G(\langle da \rangle) = L(N_G(\langle da \rangle) \cap C_G(b)) = L \langle d, a \rangle$$

by hypothesis.

(25.9) *Let G be of type $M(n)$ for $n = 22$, 23, or 24, with G of type $\tilde{M}$ if $n = 22$ or 23. Let $a \in A_d$. Then $N_G(\langle ad \rangle) = \langle a, d \rangle \times \langle D_{a,d} \rangle$.*

Proof: Let $b, c \in D_d$ with $c \in A_b$. By 25.8 it suffices to show $C_G(d) \cap N_G(\langle bc \rangle) = \langle b, c, d^\perp \cap D_{b,c} \rangle$. But from 16.12.3 and as G is of type $\tilde{M}$ when $n = 22$ or 23, $C_G(d)/\langle d \rangle \cong U_6(2)$, $M(22)$, $M(23)$ for G of type $M(22)$, $M(23)$, $M(24)$, respectively, so proceeding by induction on the order of G, it suffices to show that 25.9 holds for $G = U_6(2)$.

Let V be the natural projective module for G. Then from Section 11, $d = t(v)$, $a = t(w)$ are the transvections with centers v, $w \in V$ and by 11.7, $U = \langle v, w \rangle$ is a nondegenerate line of V. Then $U = [V, ad]$ and $U^\perp = C_V(ad)$, so $N_G(\langle ad \rangle) = (N_G(\langle ad \rangle) \cap SU(U)) \times SU(U^\perp) = \langle a, d \rangle \times \langle D_{a,d} \rangle$.

(25.10) *Let G be of type $M(24)$, $a \in A_d$, and $L = E(\langle D_{a,d} \rangle)$. Then $C_G(L) = \langle a, d \rangle$ and $N_G(L) = \langle a, d \rangle \times \langle D_{a,d} \rangle$.*

Proof: Let $M = N_G(L)$, $D_1 = C_D(L)$, and $D_2 = N_D(L) - D_1$. Then $N_D(L) = D_1 \cup D_2$ is an M-invariant partition, so by 7.3, $[D_1, D_2] = 1$. So as $a, d \in D_1$ and $\langle D_{a,d} \rangle$ is transitive on $D_{a,d}$ and faithful on L, we conclude $D_2 = D_{a,d}$. Next by 15.4, $\{a, d, a^d\} = C_D(D_2)$, so $D_1 = \{a, d, a^d\}$. Thus $\langle a, d \rangle \trianglelefteq M$, so 25.9 completes the proof.

(25.11) *Let $d^\perp \subseteq M \leq G$ with $\langle D_d \rangle = H$ transitive on D_d, $w_D(H) > 1$, $d^M \cap D_d \neq \emptyset$, and $H/O_\infty(H) \not\cong Sp_6(2)$ or $PO_7^{-\pi,\pi}(3)$. Assume*

$$O_3(C_G(d)) \leq M.$$

Then

(1) $\langle D \rangle O_3(G) \leq M$.
(2) If $C_G(d) \leq M$ then $G = M$.

Proof: By 25.2.2, $\langle D \rangle$ is transitive on D, so by a Frattini argument, $G = \langle D \rangle C_G(d)$ with $O_3(G) \leq \langle D \rangle C_{O_3(G)}(d) \leq \langle D \rangle O_3(C_G(d)) \leq \langle D \rangle M$ and with $C_G(d) \leq M$ in (2). Thus replacing G by $\langle D \rangle$, we may assume $G = \langle D \rangle$, and it remains to show $G = M$.

By 25.1, $d^{O_3(G)} = W_d = d^{O_3(M)}$, so $[O_3(M), d] \leq O_3(G)$ and by a Frattini argument, $O_3(G) = [O_3(M), d] C_{O_3(G)}(d) \leq M$. Thus passing to $G/O_3(G)$, we may assume $O_3(G) = 1$.

As H is transitive on D_d and $w_D(H) > 1$, $V_d = \emptyset$, so $O_2(G) \leq Z(G)$ by 9.2. Then by 9.4 and 17.2, G is primitive on D. Let $b \in d^M \cap D_d$ and $a \in A_b \cap D_d$. Then by 9.5.5, $G = \langle b^\perp, a \rangle \leq M$.

Exercises

1. Assume D is a set of 3-transpositions of G such that G is primitive on D and $O_p(G) \not\leq Z(G)$ for some prime p. Then $G \cong S_3$.
2. Assume D is a set of 3-transpositions of G, $d \in D$, $b \in D_d$, and

$$\langle D_d \rangle / Z(\langle D_d \rangle) \cong \langle D_b \rangle / Z(\langle D_b \rangle) \cong X.$$

Prove
(1) If $X \cong U_4(2)$ then $G/Z(G) \cong PO_6^{+,\pi}(3)$.
(2) If $X \cong PO_6^{\mu,\pi}(3)$ then $G/Z(G) \cong PO_7^{-\mu\pi,\pi}(3)$.
(3) If $X \cong PO_7^{-\pi,\pi}(3)$ then $G/Z(G) \cong PO_8^{+,\pi}(3)$ or $S_3/P\Omega_8^+(3)$.
(4) If $X \cong S_6/E_{16}$ then $G/Z(G) \cong Sp_6(2)$.

    (5) If $X \cong \mathrm{Sp}_6(2)$ then $G/Z(G) \cong O_8^\epsilon(2)$ or $S_3/\Omega_8^+(2)$.

    (6) If $X \cong S_6$ then $G/Z(G) \cong O_6^\epsilon(2)$.

    (7) If $X \cong S_5$ or S_7 then $G/Z(G) \cong S_7$ or S_9, respectively.

    (8) If $\langle D_d \rangle / \langle d \rangle \cong \mathbf{Z}_2/3^{1+2}/Q_8^2$ then $G/Z(G) \cong U_5(2)$.

       [*Hint:* Observe there is $g \in \langle D_b \rangle$ with $d^g \in D_d$. Then use 25.2.3 and appeal to Fischer's Theorem and 17.2.]

3. Prove that under the hypotheses of Lemma 25.4, $c^\perp \cap D \cap L \neq \emptyset$ for all $c \in D$.

4. Let G be a finite group, D a G-invariant set of involutions of G such that D is a set of 3-transpositions of $\langle D \rangle$, and t an involution in G with $C_D(t) = \emptyset$. Prove $d^t \in D_d$ for each $d \in D$.

5. Assume G is a finite group, D is a conjugacy class of G such that D is a set of 3-transpositions of $\langle D \rangle$, and $w_D(G) = 2$. Prove $\langle D \rangle / O_3(\langle D \rangle) \cong E_4$, S_4, or S_5.

    [*Hint:* Let $d \in D$, $H = \langle D_d \rangle$, and G be a minimal counterexample. Use 25.2.3 to conclude $\langle D \rangle$ is transitive on D. Then use minimality of G to conclude $G = \langle D \rangle$ and $O_3(G) = 1$. Use 9.6.5 to conclude $|D_d| > 1$ and hence $O_3(H) \not\leq Z(H)$. Then by Exercise 3.2, $O_2(G) \leq Z(G)$. Use 9.4 and 17.2 to conclude G is primitive on D and finally appeal to 13.4.]

9

Uniqueness systems and coverings
of graphs

A *covering* of a graph Δ is a surjective morphism $d : \tilde{\Delta} \to \Delta$ that is a local isomorphism. A precise definition appears in Section 26. The graph is *simply connected* if it admits no proper connected coverings. In Section 26 we recall basic results on coverings from Chapter 12 of [SG] including criteria for deciding when a graph is simply connected. Then in Section 27 we recall the notion of a uniqueness system from Chapter 13 of [SG], which supplies machinery for reducing the question of the uniqueness of a group G (subject to suitable hypotheses) to the question of whether certain graphs associated to the group are simply connected. In Chapter 11 we use this machinery to establish the uniqueness of the Fischer groups subject to hypotheses on centralizers of involutions. These uniqueness results also allow us to identify a certain section of the Monster as a group of type $M(24)$, hence establishing the existence of the Fischer groups.

The material on coverings and uniqueness systems appeared originally in the paper [AS] by Yoav Segev and the author.

26. Coverings and simple connectivity of graphs

In this section Δ is a graph with vertex set Δ. We direct the reader to Section 6 for our most basic notational conventions and terminology for graphs.

In addition recall that a *path of length n* in Δ from a vertex u to a vertex v is a sequence $u = u_0, \ldots, u_n = v$ of vertices such that $u_{i+1} \in u_i^\perp$ for each i. We write $u_0 \cdots u_n$ for this path.

Denote by $d(u, v)$ the minimal length of a path from u to v. If no such path exists set $d(u, v) = \infty$. Then $d(u, v)$ is the *distance* from u to v in Δ. The *diameter* of a graph Δ is the maximum distance between vertices in the graph.

Let $P = P(\Delta)$ be the set of paths in Δ. For each $p = x_0 \cdots x_r \in P$ we write $\text{org}(p)$, $\text{End}(p)$ for the *origin* x_0 and End x_r of p, respectively. Write $p \cdot q$ or simply pq for the concatenation of paths p and q such that $\text{End}(p) = \text{org}(q)$. Thus if $q = y_0 \cdots y_s$ with $y_0 = x_r$ then pq is the path $pq = z_0 \cdots z_{r+s}$ of length $r + s$ such that $z_i = x_i$ for $0 \le i \le r$ and $z_i = y_{i-r}$ for $r \le i \le r+s$.

Write p^{-1} for the path $x_r \cdots x_0$. A path $p = x_0 \cdots x_r$ is a *cycle* or *circuit* if $x_r = x_0$.

Define a subset S of P to be *closed* if S is a set of cycles satisfying the following six properties:

(C1) $rr^{-1} \in S$ for all $r \in P$.

(C2) If $p \in S$ then $p^{-1} \in S$.

(C3) If $p, q \in S$ and $\text{org}(p) = \text{org}(q)$, then $pq \in S$.

(C4) If $p \in S$ then $r^{-1}pr \in S$ for each $r \in P$ with $\text{org}(r) = \text{End}(p)$.

(C5) If p is a cycle and $r \in P$ with $\text{org}(p) = \text{End}(r)$ and $r^{-1}rp \in S$, then $p \in S$.

(C6) $xx \in S$ for all $x \in \Delta$.

The *closure* $\langle T \rangle$ of a set T of cycles is the intersection of all closed subsets containing T. It is easy to check that the intersection of closed sets is closed using the axioms (C1)–(C6). Thus

(26.1)   The closure of $\langle T \rangle$ of any set T of cycles is the smallest closed set containing T.

Write $\Delta^n(x)$ for the set of vertices at distance n from some vertex x in Δ. For $X \subset \Delta$, let $\Delta(X) = \bigcap_{x \in X} \Delta(x)$. Thus for example $\Delta(x, y) = \Delta(x) \cap \Delta(y)$.

In the remainder of this section S will be a closed subset of P. We say a cycle p is *trivial* if $p \in S$ and *nontrivial* otherwise. As a consequence of axioms (C1), (C2), and (C6):

(26.2) S contains all cycles of length at most 2.

For n a positive integer define $C_n(\Delta)$ to be the closure of the set of all cycles of Δ of length at most n. Define Δ to be *n-generated* if $C_n(\Delta)$ is the set of all cycles. We say Δ is *triangulable* if Δ is 3-generated. In Section 34 of [SG] we find various conditions that tend to force Δ to be triangulable.

(26.3)

(1) If pq, pr, and $r^{-1}q$ are cycles with pr, $r^{-1}q \in S$. Then $pq \in S$.

(2) Let $a_i, b_i, c_j \in P$, $1 \le i \le n$, $1 \le j < n$, such that $\text{org}(a_i) = x$, $\text{End}(b_i) = u$, $\text{End}(a_i) = \text{org}(b_i) = \text{org}(c_i) = \text{End}(c_{i-1})$ for $1 \le i \le n$. Assume $a_i c_i a_{i+1}^{-1}$ and $b_i^{-1} c_i b_{i+1}$ are in S for $1 \le i < n$. Then $a_n b_n b_1^{-1} a_1^{-1} \in S$.

Proof: See 34.3 in [SG].

Given integers n, m with $n \geq 2$, define $|m|_n = r$, where $0 \leq r \leq n/2$ and $m \equiv r$ or $-r$ mod n. Then define a cycle $p = x_0 \cdots x_n$ of length n to be an *n-gon* if $d(x_i, x_j) = |i - j|_n$, for all $i, j, 0 \leq i, j \leq n$.

(26.4) *Let p be a nontrivial cycle of minimal length r. Then p is an r-gon.*

Proof: See 34.4 in [SG].

(26.5) *Let d be the diameter of the graph Δ. Then S is the set of all cycles of Δ if and only if for all $r \leq 2d + 1$, each r-gon is in S.*

Proof: If p is an r-gon in Δ then $r \leq 2d + 1$, so the lemma follows from 26.4.

(26.6) *Let $p = x_0 \cdots x_4$ be a cycle such that $\Delta(x_0, x_2)$ is connected. Then $p \in C_3(\Delta)$.*

Proof: See 34.6 in [SG].

(26.7) *Let $p = x_0 \cdots x_5$ be a pentagon in Δ such that $x_0^{\perp} \cap x_2^{\perp} \cap x_3^{\perp} \neq \emptyset$. Then $p \in C_4(\Delta)$.*

Proof: See 34.8 in [SG].

(26.8) *Let $p = x_0 \cdots x_r$ be an r-gon in Δ, $m = r/2 + 1$ if r is even, and $m = (r - 1)/2 + 1$ if r is odd. Assume $x_0^{\perp} \cap x_2^{\perp} \cap \Delta^2(x_m) \neq \emptyset$. Then $p \in C_{m+1+\epsilon}(\Delta)$, where $\epsilon = 0, 1$ for r even, odd, respectively.*

Proof: This is a slight extension of 34.8 in [SG].

Define a morphism $d : \Lambda \to \Delta$ of graphs to be a *local isomorphism* if for all $\alpha \in \Lambda$,

$$d_\alpha = d_{|\alpha^{\perp}} : \alpha^{\perp} \to d(\alpha)^{\perp}$$

is an isomorphism of graphs. Define d to be a *covering* if d is a surjective local isomorphism. The covering is *connected* if its domain Λ is connected. We say Δ is *simply connected* if Δ is connected and Δ possesses no proper connected coverings.

Theorem 26.9. Δ *is simply connected if and only if Δ is triangulable.*

Proof: This is 35.14 in [SG].

The following example is due to Alperin and Glauberman.

Example 26.10: Let L be a quasisimple group, d an involution acting on L and inverting $Z(L)$ of odd order, $D = d^L$, $G = \langle d \rangle L$, and $G^* = G/Z(L)$. Assume $\mathcal{D}(D^*)$ is connected and let $f : G \to G^*$ be the natural map $f : g \mapsto g^*$. Then $\mathcal{D}(D)$ is connected and f induces a covering $f : \mathcal{D}(D) \to \mathcal{D}(D^*)$ of graphs with fibers $aZ(D), a \in D$.

The proof is as follows. Visibly $f : \mathcal{D}(D) \to \mathcal{D}(D^*)$ is a surjective morphism of graphs. As Z is of odd order, $C_{G^*}(d) = C_G(d)^*$ and as d inverts $Z(D)$, $f : C_G(d) \to C_G(d)^*$ is an isomorphism and $dZ(L) = f^{-1}(d^*)$ is the fiber of d^*. Therefore $f_d : \mathcal{D}(d^\perp) \to \mathcal{D}(d^{*\perp})$ is an isomorphism. Thus $f : \mathcal{D}(D) \to \mathcal{D}(D^*)$ is a covering.

It remains to show $\mathcal{D}(D)$ is connected. As $\mathcal{D}(D^*)$ is connected, $D^* = \Gamma(d)^*$, where $\Gamma(d)$ is the connected component of d in $\mathcal{D}(D)$ containing d. As L^* is simple, $G^* = \langle D^* \rangle = \langle \Gamma(d)^* \rangle$, so $G = \langle \Gamma(d) \rangle Z(G)$, and then as L is quasisimple, $G = \langle \Gamma(d) \rangle$. Therefore $D = d^G = d^{\langle \Gamma(d) \rangle} \subseteq \Gamma(d)$, so that $\mathcal{D}(D)$ is connected.

Example 26.11: From Corollary 35.2 there exist a quasisimple group L and an involution d acting on L such that d inverts $Z(L) \cong \mathbf{Z}_3$ and $G^* = G/Z(L) \cong M(24)$ with d^* a 3-transposition in G^*, where $G = L\langle d \rangle$. Such a group G will be said to be of *type* $\hat{M}(24)$.

Let $D = d^G$ and $f : G \to G^*$ the natural map $f : g \mapsto g^*$. Then by Example 26.10, f induces a connected covering $f : \mathcal{D}(D) \to \mathcal{D}(D^*)$ of degree 3 of the 3-transposition graph of $M(24)$. We see in the next lemma that $\mathcal{D}(D)$ is simply connected, so f is the universal covering of the $M(24)$ graph. In particular the $M(24)$ graph is not simply connected and indeed its fundamental group is of order 3, since the fundamental group is regular on the fibers of the universal covering map.

Theorem 26.12.   *Let G be of type $M(22)$, $M(23)$, or $\hat{M}(24)$; D the set of 3-transpositions of G in the first two cases; and D the class of involutions in G projecting on the 3-transpositions of $G/O_3(G)$ in the third case. Then $\mathcal{D}(D)$ is simply connected.*

Proof: Let $\mathcal{D} = \mathcal{D}(D)$. By 26.9, it suffices to show $\mathcal{D}$ is triangulable; that is each cycle of $\mathcal{D}$ is in the closure $C_3(\mathcal{D})$ of the triangles of $\mathcal{D}$ as defined previously.

Assume first G is of type $M(22)$ or $M(23)$. Then the graph $\mathcal{D}(D)$ has diameter 2, so by 26.5, it suffices to show each square and pentagon of $\mathcal{D}$ is in $C_3(\mathcal{D})$. Let $p = x_0 \cdots x_r$ be a square or pentagon. If p is a square then $r = 4$ and by 26.6, it suffices to show the subgraph D_{x_0,x_2} is connected. But this holds as $x_2 \in A_{x_0}$ so $\langle D_{x_0,x_2} \rangle$ is $PO_6^{+,\pi}(3)$ or $PO_7^{-\pi,\pi}(3)$. So let p be a pentagon. Then $r = 5$ and by 26.7, it suffices to show $D_{x_0,x_2,x_3} \neq \emptyset$. But (x_2, x_0, x_3) is a 3-string in D, so this follows from 25.6.3.

Thus we may take $G = \hat{M}(24)$ and let $Z = Z(E(G)) = \langle z \rangle$ and $G^* = G/Z$. Let $a \in D$. We write D_{a*}^*, A_{a*}^* for the set of $d^* \in D^*$ with $|a^* d^*| = 2, 3$, respectively, and D_{a*}, A_{a*} for the set of vertices $x \in D$ with $x^* \in D_{a*}^*$, A_{a*}^*, respectively. Also let $D_a^m = \{b \in D : d(a, b) = m\}$ be the set of vertices at distance m from a in $\mathcal{D}$. Thus $D_a^1 = D_a \cong D_{a*}^*$ is an orbit under $\langle D_a \rangle$.

Let abc be a path in $\mathcal{D}$ with $d(a, c) = 2$. Then $a, c \in D_b$, so ac is of order 3 and $c^* \in A_{a*}^*$. Also $\langle a^\perp \cap D_b \rangle$ is transitive on $D_b \cap A_{a*} = D_b \cap D_a^2$, so $\langle a^\perp \rangle = C_G(a)$ is transitive on D_a^2 and $D_a^2 \subseteq A_{a*}$.

Next $\langle D_{a*,c*}^* \rangle \cong S_3/P\Omega_8^+(3)$ has three connected components $\Delta_{a*,c*,i}^*$, $i = 1, 2, 3$, with $\langle \Delta_{a*,c*,i}^* \rangle \cong PO_8^{+,\pi}(3)$. Let $b^* \in \Delta_{a*,c*,1}^*$, K the preimage in G of $E(\langle D_{a*,c*}^* \rangle)$, and $L = C_K(a)$. As $C_G(a)$ is a complement to Z in $C_G(a^*)$, L is a complement to Z in K and hence $K = Z \times L$ and $L = E(K)$. In particular $L = C_K(c)$ too. Then $b^L \subseteq D_{a,c}$. Also by 25.6.2, L^* is transitive on $D_{c*}^* \cap A_{a*}^*$, so L is transitive on $D_c \cap A_{a*}$ and hence as there is $e \in D_{b,c} \cap A_{a*}$, we conclude $D_c \cap A_{a*} \subseteq D_a^2$.

Claim $b^L = D_{a,c}$. If not there is $b_2 \in D_{a,c}$ with $b_2^* \in \Delta_{a*,c*,2}^*$. But then

$$D_c \cap D_{a*} = D \cap L\langle b, b_2 \rangle \subseteq D_a,$$

so

$$D_c = (D_c \cap D_{a*}) \cup (D_c \cap A_{a*}) \subseteq D_a \cup D_a^2.$$

It follows that if $abce$ is a path in $\mathcal{D}$ of length 3 then $d(a, e) \leq 2$, and hence $\mathcal{D}$ is of diameter 2, so that $D = a^\perp \cup D_a^2$. This is impossible as $az \in D - (a^\perp \cup D_a^2)$ since $D_a^2 \subseteq A_{a*}$.

Therefore $b^L = D_{a,c}$ and then as $\langle D_{a*,c*}^* \rangle = C_{G*}(\langle a^*, c^* \rangle)$ by 25.9, $L\langle b \rangle = C_G(\langle a, c \rangle)$, so as $C_G(a)$ is transitive on D_a^2, we have

$$|D_a^2| = |C_G(a) : L\langle b \rangle| = 3|A_{a*}^*| = |A_{a*}|,$$

so $A_{a*} = D_a^2$. Then as $az \notin D_c$, $d(z, az) \geq 4$.

Next let $u \in D_a$ with $u^* \in \Delta_{a*,c*,2}^*$; then uz or uz^2 is in D_c, say $uz \in D_c$. Hence $uz \in D_a^3$. Notice auz is of order 6, $(uz)^a = uz^2$, and $(uz)^b \in D_c$ with $(uz)^{*b} \in \Delta_{a*,c*,3}^*$, so $C_G(a)$ is transitive on D_a^3 and $D_a^3 = D_{a*} - D_a$.

Finally $D_a^4 = \{az, az^2\}$ is of order 2. Thus $\mathcal{D}$ is of diameter 4, so to show $\mathcal{D}$ is simply connected it suffices by 26.5 to show that for each r, $4 \le r \le 9$, and each r-gon $p = x_0 \cdots x_r$ in $\mathcal{D}$, $p \in \mathcal{C}_3(\mathcal{D})$. Without loss, $x_0 = a$, $x_1 = b$, and $x_2 = c$. Now there exist no 9-gons, since the only pair of vertices az and az^2 at distance 4 from a are not adjacent in $\mathcal{D}$. If $r = 4$ then $p \in \mathcal{C}_3(\mathcal{D})$ by 26.6, since $D_{a,c} = b^D$ is connected. Similarly if $r = 5$ we have $x_3 \in D_c \cap D_a^2 = D_c \cap A_{a^*}$. Then as L is transitive on $D_c \cap A_{a^*}$ and $D_{b,c} \cap A_{a^*} \ne \emptyset$, we conclude $\emptyset \ne D_{a,c,x_3}$. Therefore 26.7 says $p \in \mathcal{C}_3(\mathcal{D})$ in this case.

So $r = 6, 7, 8$. By 26.8, it suffices to show $D_{x_{r/2+1}}^2 \cap D_{a,c} \ne \emptyset$ when $r = 6, 8$ and $D_{x_4}^2 \cap D_{a,b} \ne \emptyset$ if $r = 7$. If $r = 6$ and the condition fails then $D_{a,c}^* \cap A_{x_4^*} = \emptyset$ so $D_{a,c}^* \subseteq D_{x_4^*}$ and hence $x_4^* \in C_{G^*}(L^*) = \langle a^*, c^* \rangle$ by 25.10. Therefore $x_4^* = a^{*c}$. Thus if $e \in D_{x_3,x_5}$ with $e^* \notin \langle a^*, c^* \rangle$ then $q = x_0 \cdots x_3 e x_4 \cdots x_6$ and $x_3 x_4 x_5 e x_3$ are in $\mathcal{C}_3(\mathcal{D})$ so by 26.3, $p \in \mathcal{C}_3(\mathcal{D})$.

If $r = 7$ then as D_{a^*,b^*} is not contained in $D_{x_4^*}$, we have $D_{x_4}^2 \cap D_{a,b} \ne \emptyset$. Similarly if $r = 8$ then $x_5 \in D_{a,c}^3$ so $x_5^* \in D_{a^*,c^*}$ so $\emptyset \ne A_{x_5^*} \cap D_{a,c} = D_{x_5}^2 \cap D_{a,c}$. Hence the lemma is at last established.

27. Uniqueness systems

We begin by recalling the notion of a uniqueness system from Section 37 of [SG].

A *uniqueness system* is a 4-tuple $\mathcal{U} = (G, H, \Delta, \Delta_H)$ such that Δ is a graph, G is an edge and vertex transitive group of automorphisms of Δ, $H \le G$, Δ_H is a graph with vertex set xH and edge set $(x, y)H$ for some $x \in \Delta$ and $y \in \Delta(x) \cap xH$, and

(U) $G = \langle H, G_x \rangle$, $G_x = \langle G_{x,y}, H_x \rangle$, $H = \langle H(\{x, y\}), H_x \rangle$.

Associated to the uniqueness system is a family $\mathcal{F}(\mathcal{U}) = (G_i : 1 \le i \le 3)$ of subgroups of G defined by $G_1 = G_x$ the stabilizer in G of $x \in \Delta$, $G_2 = G(\{x, y\}) = G_{x,y}\langle t \rangle$ the stabilizer in G of the unordered edge $\{x, y\}$ of Δ (where $t \in H$ has cycle (x, y)) and $G_3 = H$. For $J \subset \{1, 2, 3\}$, let $G_J = \bigcap_{j \in J} G_j$.

Also associated to $\mathcal{U}$ is an *amalgam* $A((U)) = A(\mathcal{F}(\mathcal{U}))$ that encodes the data induced by the inclusions $G_J \to G_K$ for $K \subset J$ and the *universal completion* of the amalgam $A(\mathcal{U})$, which is the largest group generated by a family $\mathcal{F}$ of subgroups satisfying these data. See Section 36 of [SG] for a precise definition of these notions.

The first lemma in this section shows how to construct uniqueness systems from suitable data.

(27.1) *Let G be a group, G_1 and G_3 subgroups of G such that $G = \langle G_1, G_3 \rangle$, and $t \in G_3$ such that $t^2 \in G_1$, $G_1 = \langle G_1 \cap G_1^t, G_1 \cap G_3 \rangle$, and $G_3 = \langle G_1 \cap G_3, t \rangle$. Let $G_2 = \langle G_1 \cap G_1^t, t \rangle$. Then $(G, G_3, \Delta, \Delta_{G_3})$ is a uniqueness system, where Δ is the graph on G/G_1 defined by the orbital $(G_1, G_1 t)G$ and Δ_{G_3} is the graph on the coset space $G_1 G_3/G_1 \cong G_3/(G_1 \cap G_3)$ defined by the orbital $(G_1 \cap G_3, (G_1 \cap G_3)t)G_3$. Further $\mathcal{F}(\mathcal{U}) = (G_1, G_2, G_3)$.*

In the remainder of this section we assume groups of type $\hat{M}(24)$ exist. In particular groups of type $M(24)$ exist, so Fischer's Theorem says there exists a unique group of type $M(n)$ for $n = 22, 23$, and 24. Thus it makes sense to write $G = M(n)$ to indicate G is of type $M(n)$, and we do so in the remainder of this section.

Also by Remark 16.10, groups of type $M(22)$ are of type $\tilde{M}(22)$ and groups of type $M(23)$ are of type $\tilde{M}(23)$. Thus we can use our lemmas incorporating those hypotheses.

Finally by 26.12, each group of type $\hat{M}(24)$ is the group of automorphisms of the universal covering $f : \tilde{\mathcal{D}} \to \mathcal{D}(D)$ of the graph of $M(24)$, so there is a unique group of type $\hat{M}(24)$ and we write $G = \hat{M}(24)$ to indicate G is such a group.

Throughout the remainder of the section we work with the following example.

Example 27.2: Let $\hat{G}$ be $M(23)$, $M(24)$, or the group $\hat{M}(24)$ of Example 26.11. Let $G = E(\hat{G})$ and D the set of 3-transpositions of $\hat{G}$ when $\hat{G}$ is a Fischer group and the involutions of $\hat{G}$ projecting on the 3-transpositions of $\hat{G}/Z(G) \cong M(24)$ when $\hat{G}$ is $\hat{M}(24)$. Let $a \in D$, $G_1 = C_G(a)$, $a \in T \in Syl_2(\hat{G})$, $A = \langle T \cap D \rangle$, $G_3 = N_G(A)$, t an involution in $G_3 - G_1$ and $G_2 = (G_1 \cap G_1^t)\langle t \rangle$. Let Δ be the graph on G/G_1 defined by the orbital $(G_1, G_1 t)G$ and Δ_{G_3} the graph on the coset space $G_1 G_3/G_1 \cong G_3/(G_1 \cap G_3)$ defined by the orbital $(G_1 \cap G_3, (G_1 \cap G_3)t)G_3$. Let $\mathcal{U} = (G_1, G_3, \Delta, \Delta_{G_3})$ and $\mathcal{F} = (G_1, G_2, G_3)$. As usual for $J \subset \{1, 2, 3\}$ let $G_J = \bigcap_{j \in G} G_J$.

We assume the hypotheses and notation of Example 27.2 in the remainder of this section.

(27.3) (a, a^t) *is an edge in $\mathcal{D}(D)$ and $G_2 = O^{2'}(C_G(aa^t)) = G_{12}\langle t \rangle$, where $G_{12} = C_G(a) \cap C_G(a^t)$ is quasisimple with $G_{12}/\langle a, a^t \rangle \cong U_6(2)$ if $\hat{G} = M(23)$ and $G_{12}/\langle aa^t \rangle$ is isomorphic to $M(22)$ if $\hat{G} = M(24)$ or $\hat{M}(24)$.*

Proof: As A is abelian, $a^t \in D_a$, so (a, a^t) is an edge in $\mathcal{D}(D)$. Let $\hat{G}^* = \hat{G}/Z(G)$. Then as $a^{*t} \in D_{a^*}^*$, $C_{G^*}(a^* a^{*t}) = C_{G^*}(\langle a^*, a^{*t} \rangle)\langle t^* \rangle$ by 25.7.5.

That is $C_{G^*}(a^*a^{*t}) = G_2^*$. Further a inverts $Z(G)$ and $G_{1,2}^*$ is quasisimple and described in 27.3, so $Z(G) \cap G_2 = 1$ and $G_2^* \cong G_2 = G_{12}\langle t \rangle = O^{2'}(C_G(aa^t))$ with G_{12} quasisimple and described in the lemma.

(27.4) *The map*

$$\Delta \to \mathcal{D}(D)$$

$$G_1 g \mapsto a^g$$

is a $\hat{G}$-equivariant isomorphism of graphs that restricts to an isomorphism $\Delta_{G_3} \cong T \cap D$ of Δ_{G_3} with the complete graph $T \cap D$. Further up to conjugation under G, $T \cap D$ is the unique maximal complete subgraph of $\mathcal{D}(D)$ so each triangle of Δ is G-conjugate to a triangle of Δ_{G_3}.

Proof: The first statement follows as G is edge- and vertex-transitive on Δ and $\mathcal{D}$ with G_1 the stabilizer in G of $a \in D$ and (a, a') is an edge of $\mathcal{D}$. The second follows from the first and 8.5.

(27.5)

(1) $\mathcal{U}$ is a uniqueness system.
(2) $M(23)$, $E(\hat{M}(24))$ is the universal completion of the amalgam $\mathcal{A}(\mathcal{F})$ for $\hat{G} \cong M(23)$, $M(24)$, or $\hat{M}(24)$, respectively.

Proof: To prove (1) we must verify the hypotheses of 27.1. First G_3 is 2-transitive on $T \cap D$ so $G_3 = \langle G_{13}, t \rangle$. As $\mathcal{D}$ is connected with $G_1 = G_a$ and $t \in G_3$, $G = \langle G_1, G_3 \rangle$ and as G_1 is primitive on D_a with G_{12} the stabilizer in G_1 of a^t, $G_1 = \langle G_{12}, G_{13} \rangle$. Hence (1) is established.

Next if $\hat{G} = M(23)$ or $\hat{M}(24)$ then by 26.12, $\mathcal{D}(D)$ is simply connected, so by 27.4, Δ is simply connected and each triangle of Δ is conjugate to a triangle of Δ_{G_3}. Hence Δ_{G_3} is a base for Δ in the sense of Section 37 of [SG], so by 37.5.4 in [SG], G is the universal completion for $\mathcal{A}(\mathcal{F})$. As $M(24)$ and $\hat{M}(24)$ have isomorphic amalgams, (2) follows.

(27.6) $\text{Aut}(G_{13}) = G_{13}$ *if* $\hat{G} = M(24)$ *or* $\hat{M}(24)$, *whereas*

$$\text{Aut}(G_{13}) = \text{Aut}_{\text{Aut}(G_1)}(G_{13}) \cong \text{Aut}(M_{22})/E_{2^{10}}$$

if $\hat{G} \cong M(23)$.

Proof: First if $\hat{G}$ is $M(24)$ or $\hat{M}(24)$ then $G_{13} = N_{G_1}(A) \cong M_{23}/E_{2^{11}}$ by 25.7. Then G_{13} is absolutely irreducible on $A \cap G$ (cf. 22.5) and $M_{23} = \text{Aut}(M_{23})$ (cf. Exercise 7.5 in [SG]), so $\text{Aut}(G_{13}) = G_{13}C_{\text{Aut}(G_{13})}(A \cap G)$. Further by

22.8.1, $H^1(G_{13}/(A \cap G), A \cap G) = 0$ so $A \cap G$ is self-centralizing in $\text{Aut}(G_{13})$, and hence the lemma holds in this case.

So let $\hat{G} = M(23)$. Then $G_{13} \cong M_{22}/E_{2^{11}}$ with $\langle a \rangle = Z(G_{13})$ and G_{13} indecomposable on A and absolutely irreducible on $A/\langle a \rangle$. Therefore as in the proof of 21.13, $C_{GL(A)}(G_{13}/A) = 1$ and then as $H^1(G_{13}/A, A) = 0$ by 22.7 the lemma holds in this case too.

(27.7) $\text{Aut}(G_{12}) \cap C(G_{123}) = 1$.

Proof: By 27.3, $G_{12} = C_G(a) \cap C_G(a^t)$ is quasisimple. Further $G_{123} = N_{G_{12}}(A)$. In particular if $\hat{G} \cong M(23)$ then $\text{Aut}(G_{12}) = \text{Aut}(U_6(2))$ is the extension of $U_6(2)$ by S_3 and $N_{\text{Aut}(G_{12})}(A) \cong S_3/L_3(4)/E_{2^9}$ has a trivial center, so the lemma holds. Similarly if $\hat{G}$ is $M(24)$ or $\hat{M}(24)$ then $\text{Aut}(G_{12}) \cong \text{Aut}(M(22))$ and $N_{\text{Aut}(G_{12})}(A) \cong \text{Aut}(M_{22})/E_{2^{10}}$ has a trivial center, so the lemma holds here too.

Theorem 27.8.   *Let $\bar{G}$ be a group with subgroups $\bar{G}_i$, $i = 1, 3$, such that $\bar{G} = \langle \bar{G}_1, \bar{G}_3 \rangle$, $G_i \cong \bar{G}_i$, $G_1 \cap G_3 \cong \bar{G}_1 \cap \bar{G}_3$, and $G_1 \cap G_1^t \cong \bar{G}_1 \cap \bar{G}_1^{\bar{t}}$ for some involution $\bar{t} \in \bar{G}_3 - \bar{G}_1$. Then $\bar{G} \cong G$ or $\hat{G} \cong M(24)$ or $\hat{M}(24)$ and $\bar{G} \cong E(M(24))$ or $E(\hat{M}(24))$.*

Proof: Without loss $\hat{G} \cong \hat{M}(24)$ or $M(23)$. Let $\bar{G}_2 = (\bar{G}_1 \cap \bar{G}_1^{\bar{t}})\langle \bar{t} \rangle$ and as usual set $\bar{G}_J = \bigcap_{j \in J} \bar{G}_j$. By hypothesis there are isomorphisms $\alpha : G_1 \to \bar{G}_1$ and $\zeta : G_3 \to \bar{G}_3$. As $\bar{G}_{13} \cong G_{13}$, $\bar{G}_{13} = N_{\bar{G}_1}(\bar{A})$ for $\bar{A} = J(\bar{T})$ and $\bar{T} \in \text{Syl}_2(\bar{G}_1)$, so $\bar{G}_{13}$ is conjugate under $\bar{G}_1$ to $G_{13}\alpha$, and hence composing α with a suitable inner automorphism of $\bar{G}_1$, we may assume $G_{13}\alpha = \bar{G}_{13}$. Similarly we may assume $G_{13}\zeta = \bar{G}_{13}$.

Next as $\bar{G}_3$ is 2-transitive on the coset space $\bar{G}_3/\bar{G}_{13}$, we may assume $t\zeta \in \bar{G}_{13}\bar{t}$. Thus $\bar{G}_2 = (\bar{G}_1 \cap \bar{G}_1^{t\zeta})\langle t\zeta \rangle$, so without loss $t\zeta = \bar{t}$. Hence $G_{23}\zeta = (G_{13} \cap G_{13}^t)\zeta \langle t\zeta \rangle = (\bar{G}_{13} \cap \bar{G}_{13}^{\bar{t}})\langle \bar{t} \rangle = \bar{G}_{23}$ and similarly $G_{123}\zeta = \bar{G}_{123}$.

Next $G_{12} \cong \bar{G}_{12}$ by hypothesis and from 27.3, $G_{12} = E(C_{G_1}(B))$ for $B = Z(G_{12}) \leq A$, so $\bar{G}_{12} = E(C_{\bar{G}_1}(\bar{B}))$ for $\bar{B} = Z(\bar{G}_{12})$. As $B \leq A$, $B\alpha \leq A\alpha = \bar{A}$, while $\bar{B} = Z(\bar{G}_{123}) = Z(G_{123})\zeta = B\zeta \leq A\zeta = \bar{A}$. Then as $G_{123} \cong \bar{G}_{123}$, $\bar{B}$ is conjugate to $B\alpha$ in $\bar{G}_{13}$ and thus composing α with the inner automorphism of $\bar{G}_1$ induced by a suitable member of $\bar{G}_{13}$, we may take $B\alpha = \bar{B}$. Therefore $G_{12}\alpha = E(C(B))\alpha = E(C(B\alpha)) = E(C(\bar{B})) = \bar{G}_{12}$.

Now as G_1, G_3 satisfy the hypotheses of 27.1, applying the isomorphisms α and ζ, we conclude that $\bar{G}_1$ and $\bar{G}_3$ also satisfy the hypotheses of 27.1, so applying 27.1 we obtain a uniqueness system $\bar{\mathcal{U}}$ for $\bar{G}$. We will show $\mathcal{U}$ and $\bar{\mathcal{U}}$ are equivalent in the sense of Section 37 of [SG]. Then by 27.5.2 and 37.5.2 in

[SG], $\bar{G}$ is a nontrivial homomorphic image of G. Then as G is quasisimple, $G \cong G/Z$ for $Z \leq Z(G)$, and the lemma holds.

So it remains to show $\mathcal{U}$ and $\bar{\mathcal{U}}$ are equivalent. For that we use Theorems 37.9 and 37.11 in [SG]. In particular if Hypothesis V of Section 37 of [SG] is satisfied then those theorems and 27.7 complete the proof. Thus it suffices to establish Hypothesis V. First by 27.6, $\mathrm{Aut}(G_{13}) = G_{13}$ if $\hat{G} = \hat{M}(24)$, whereas $\mathrm{Aut}(G_{13}) = \mathrm{Aut}_{\mathrm{Aut}(G_1)}(G_{13})$ if $\hat{G} \cong M(23)$. In particular (V1) of Hypothesis V holds. Also $G_{13} = N_{\mathrm{Aut}(G_3)}(G_{13})$ so (V3) holds. Similarly $N_{\mathrm{Aut}(G_1)}(G_{13})$ is G_{13} if $\hat{G} = \hat{M}(24)$ and is G_{13} extended by an involutory outer automorphism s of G_1 acting on G_{12} if $\hat{G} = M(23)$. Therefore (V2) holds. Finally G_{123} is self-normalizing in G_{13}, so (V4) holds. Therefore the proof is complete.

REMARK 27.9: Theorem 27.8 depends upon the existence of the groups $M(23)$ and $\hat{M}(24)$, which is established in Corollaries 35.2 and 35.3. Namely we used the simple connectivity of the 3-transposition graph of $M(23)$ and of the 3-fold cover of the 3-transposition graph of $M(24)$ to prove that the universal completion for the amalgam of the family $\mathcal{F}$ is finite and isomorphic to our group G.

10

$U_4(3)$ as a subgroup of $U_6(2)$

The perfect central extension of $\mathbf{Z}_3$ by $U_4(3)$ is an absolutely irreducible subgroup of $SU_6(2)$, which is in turn naturally embedded in the orthogonal group $O_{12}^+(2)$. These embeddings are important to us. For one thing, the perfect central extension of $\mathbf{Z}_2$ by $U_6(2)$ is the group $\langle D_d \rangle$ for a 3-transposition d in $M(22)$ and for $a \in A_d$, $\langle D_{a,d} \rangle \cong PO_6^{+,\pi}(3)$ is $U_4(3)$ extended by a reflection. For another the centralizer H of a 2-central involution z in $G = M(24)'$ satisfies $F^*(H) = Q \cong 2^{1+12}$ with $H^* = H/Q \cong \mathbf{Z}_2/U_4(3)/\mathbf{Z}_3$. Then $\tilde{Q} = Q/\langle z \rangle$ is a 12-dimensional orthogonal space over $GF(2)$ of sign $+1$ and $H^* \leq O(\tilde{Q})$ and indeed $E(H^*) \leq C_{O(\tilde{Q})}(O_3(H^*)) = SU_6(2)$.

In this chapter we prove the existence and uniqueness of the embedding of $U_4(3)$ in $U_6(2)$ and study the properties of this embedding by constructing a 6-dimensional lattice Λ over $\mathbf{Z}[\omega]$, $\omega = e^{2\pi i/3}$, whose automorphism group is $U_4(3)/\mathbf{Z}_6$ extended by a reflection. This construction is carried out and explored in Section 28. The lattice is described in a somewhat different way in the *Atlas* [At]. Our treatment resembles that of Conway for the Leech lattice. In Section 29 we establish various properties of the embedding. For example we see in Theorem 29.9 that $U_6(2)$ has three classes of $U_4(3)$-subgroups and $\text{Aut}(U_6(2))$ is transitive on the $U_4(3)$-subgroups.

28. The $U_4(3)$ lattice

In this section we construct a lattice whose automorphism group G is a 6-fold cover of $U_4(3)$ extended by a reflection. The lattice is described in somewhat different terms on page 52 of the *Atlas* [At]. The treatment given here resembles that of Conway for the Leech lattice (cf. Section 22 in [SG]).

Throughout this section ω is a primitive cube root of 1 and $R = \mathbf{Z}[\omega]$. For $r = a + b\omega \in R$ let $\bar{r}$ be the complex conjugate of r, so that

$$\bar{r} = a + b\omega^2 = (a - b) - b\omega.$$

157

Then

(28.1) *For $r = a + b\omega \in R$, $|r|^2 = a^2 + b^2 - ab$.*

Let $\theta = 1 + 2\omega \in R$ and observe $\theta^2 = -3$ and $\bar{\theta} = -\theta$, so $|\theta|^2 = 3$. Let $R_0 = \{0, 1, \omega, \omega^2\}$ and observe R_0 is a set of coset representatives for $2R$ in R.

Let R^6 be the free R-module on a basis X of order 6. For $v = \sum_{x \in X} v_x x \in R^6$, let $m(v) = \sum_x v_x$. Thus

(28.2) *$m : R^6 \to R$ is an R-homomorphism.*

Let Λ be the subset of R^6 consisting of all $v = \sum_x v_x x$ such that the following two conditions are satisfied:

($\Lambda 1$) $m(v) \in 2R$.
($\Lambda 2$) $v_x \equiv \omega m(v)/2 \bmod 2R$ for each $x \in X$.

(28.3) *Λ is an R-submodule of R^6.*

Proof: This follows from 28.2.

We call Λ the $U_4(3)$-*lattice*.

Let Λ' denote the set of vectors $v = \sum_x v_x x \in R^6$ such that there exists $t(v) \in R_0$ with $v_x \equiv t(v) \bmod 2R$ for all $x \in X$. Observe that if $v \in \Lambda'$ then $t(v)$ is determined; namely $t(v) = R_0 \cap (v_x + 2R)$ for each $x \in X$. Define $B(v) = \sum_x (v_x - t(v))/2$.

Notice also that Λ' is an R-submodule of R^6 and $\Lambda \leq \Lambda'$ by ($\Lambda 2$).

(28.4) *Let $v = \sum_x v_x x \in R^6$. Then the following are equivalent:*

(1) $v \in \Lambda$.
(2) $m(v) \in 2R$, $v \in \Lambda'$, and $t(v) \equiv \omega m(v)/2 \bmod 2R$.
(3) $v \in \Lambda'$ and $B(v) \equiv \omega t(v) \bmod 2R$.

Proof: Notice that if v satisfies any one of conditions (1)–(3) then $v \in \Lambda'$, so we may assume $v \in \Lambda'$. Let $t = t(v)$ and $B = B(v)$. Then

$$\frac{m(v)}{2} = 3t + B \equiv t + B \bmod 2R,$$

so $m(v) \in 2R$. Further $v_x \equiv t \bmod 2R$ for each $x \in X$, so (1) and (2) are equivalent and $v_x \equiv \omega m(v)/2$ if and only if $t \equiv \omega(t + B)$ if and only if $B \equiv \omega t$. That is (1) and (3) are equivalent.

Let $\Lambda_0 = \{v \in \Lambda' : t(v) = 0 \text{ and } m(v) \in 4R\}$. For $Y \subset X$, let $e_Y = \sum_{y \in Y} y$ and for $x \in X$ let $\lambda_x = e_X + 2\omega e_x$.

(28.5) $\Lambda = \Lambda_0 + R\lambda_x$ *for each* $x \in X$.

Proof: Let $L = \Lambda_0 + R\lambda_x$. Then L is an R-submodule of R^6. Further if $v \in \Lambda_0$ then $m(v) \in 4R$, so

$$t(v) = 0 \equiv \frac{\omega m(v)}{2} \bmod 2R$$

and hence $\Lambda_0 \le \Lambda$ by 28.4. Similarly $\lambda_x \in \Lambda$ as $B(\lambda_x) = \omega$ and $t(\lambda_x) = 1$. Therefore $L \le \Lambda$. Conversely suppose $v \in \Lambda - L$. Then

$$t(v - t(v)\lambda_x) = t(v) - t(v)t(\lambda_x) = 0,$$

so replacing v by $v - t(v)\lambda_x$ if necessary, we may assume $t(v) = 0$. Thus $m(v)/2 = B(v)$ and by 28.4, $B(v) \in 2R$, so $v \in \Lambda_0$.

Let $(\ ,\ )$ be the hermitian symmetric sesquilinear form on R^6 defined by

$$(u, v) = \frac{\sum_{x \in X} u_x \bar{v}_x}{4}$$

and for $v \in R^6$ define $q(v) = (v, v)$.

(28.6) *For all* $u, v \in \Lambda$, $(u, v) \in R$ *and* $q(v) \in \mathbf{Z}$.

Proof: By 28.5 it suffices to assume $u, v \in \Lambda_0 \cup \{\lambda_y\}$ for some fixed $y \in X$. But $q(\lambda_y) = 2$, so we may take $u \in \Lambda_0$; that is $u_x = 2a_x$ and $\sum_x a_x = B(u) \in 2R$. Now if $v = \lambda_y$ then $(u, v) = B(u)/2 - \omega a_y \in R$, whereas if $v \in \Lambda_0$ then $v_x = 2b_x$ and $(u, v) = \sum_x a_x \bar{b}_x \in R$, with $(u, u) = \sum_x |a_x|^2 \in \mathbf{Z}$.

For $Y \subseteq X$ let $\epsilon_Y \in GL(R^6)$ be the map with $x\epsilon_Y = -x$ if $x \in Y$ and $x\epsilon_Y = x$ if $x \notin Y$. Identify $\mathrm{Sym}(X)$ with a subgroup of $GL(R^6)$ via

$$g : \sum_x v_x x \mapsto \sum_x v_x xg.$$

Let $\omega I \in GL(R^6)$ be scalar multiplication by ω and for $y \in X$ let $\tau_y = \epsilon_y \sigma$ be the composition of ϵ_y with the field automorphism σ of R^6 induced by X and complex conjugation. That is

$$\tau_y : \sum_x v_x x \mapsto \sum_x \bar{v}_x x - 2\bar{v}_y y.$$

Finally write $O(R^6)$ for the isometry group of the form $(\ ,\ )$ and let $G = \mathrm{Aut}(\Lambda)$ be the subgroup of $O(R^6)$ acting on Λ. Notice $\omega I \in Z(G)$.

Let $O(R^6)^+$ be $O(R^6)$ extended by σ; thus if $\alpha \in O(R^6)^+$ then α is semi-linear on R^6 and if $\alpha \notin O(R^6)$ then $(u\alpha, v\alpha) = \overline{(u, v)}$. Similarly let G^+ be the subgroup of $O(R^6)^+$ acting on Λ.

(28.7)

(1) Let $E_0 = \{\epsilon_Y : Y \subseteq X \text{ and } |Y| \text{ is even}\}$. Then the map $Y \mapsto \epsilon_Y$ is an isomorphism of the core of the permutation module for $\mathrm{Sym}(X)$ on X with E_0 and $E_0 \leq G$. Thus $E_0 \cong E_{32}$ and $S_6 \cong \mathrm{Sym}(X) \leq N_G(E_0)$ with $\epsilon_Y^g = \epsilon_{Yg}$ for $g \in \mathrm{Sym}(X)$.

(2) For $y \in X$, $\tau_y \in G^+$ and $E = E_0\langle \tau_y \rangle \cong E_{64}$ is independent of $y \in X$. Further $\tau_y^g = \tau_{yg}$ for $g \in \mathrm{Sym}(X)$, so $\mathrm{Sym}(X) \leq N_G(E)$.

Proof: The first remark in (1) is a straightforward calculation, as is the fact that E_0 and $\mathrm{Sym}(X)$ are contained in $O(R^6)$ and each g in E_0 or $\mathrm{Sym}(X)$ acts on Λ' with $t(vg) = t(v)$ for each $v \in \Lambda'$. Indeed $m(vg) \equiv m(v) \bmod 4R$, using the fact that $|Y|$ is even when $g = \epsilon_Y$. Thus g acts on Λ, so (1) is established.

Similarly $\tau = \tau_y \in O(R^6)^+$ and τ acts on Λ' with $t(v\tau) = \bar{t}(v)$. Also $m(v\tau) = \overline{m(v)} - 2\bar{v}_y$. Thus if $v \in \Lambda_0$, $t(v\tau) = 0$ and $m(v\tau) \in 4R$, so $v\tau \in \Lambda_0$. Finally $\lambda_y \tau = \lambda_y$, so τ acts on Λ and (2) holds.

Let $M^+ = \langle \omega I \rangle \mathrm{Sym}(X)E$. Then by 28.7, $M^+ \leq G^+$. Let $M = M^+ \cap G$. For $0 \neq w \in R^6$ define $r_w : R^6 \to R^6$ by

$$r_w : v \mapsto \frac{v - 2(v, w)w}{(w, w)}.$$

It is a standard fact (and easy to check) that $r_w \in O(R^6)$. Indeed r_w is a "reflection"; that is r_w inverts w and centralizes $w^\perp$.

(28.8)   For each $w \in \Lambda$ with $q(w) = 2$, $r_w \in G$.

Proof: We have already observed $r_w \in O(R^6)$. Further if $q(w) = 2$ then for $v \in \Lambda$, $r_w : v \mapsto v - (v, w)w \in \Lambda$, so r_w acts on Λ.

Example: Let $(x, y) \in \mathrm{Sym}(X)$ be a transposition. Then $(x, y) = r_{2e_x - 2e_y}$ and $2e_x - 2e_y \in \Lambda$ with $q(2e_x - 2e_y) = 2$.

(28.9)   Let $r \in R^\#$. Then

(1) If $r \in 2R$ then $|r|^2 \equiv 0 \bmod 4$.

(2) If $r \notin 2R$ then $|r|^2$ is odd.

(3) If $|r|^2 \leq 11$ then one of the following holds:
- *(a) $|r| = 1$ and $\pm r \in R_0$.*
- *(b) $|r|^2 = 3$ and $\pm r \in R_0 \theta$.*
- *(c) $|r|^2 = 4$ and $\pm r \in 2R_0$.*
- *(d) $|r|^2 = 7$ and $\pm r \in (1 - 2\omega) R_0 \cup (3 + 2\omega) R_0$.*
- *(e) $|r|^2 = 9$ and $\pm r \in 3R_0$.*

Proof: First if $r = 2s$ then $|r|^2 = 4|s|^2 \equiv 0 \bmod 4$. On the other hand if $r \notin 2R$ then $r = t + 2s$ with $0 \neq t \in R_0$, so $|r|^2 = 1 + 4|s|^2 + 2(t\bar{s} + \bar{t}s)$ is an odd integer. Thus (1) and (2) hold.

Let $r = a + b\omega$ and assume $|r|^2 \leq 11$. Define $[r] = \pm R_0 r \cup \pm R_0 \bar{r} - \{0\}$. Then $[r]$ consists of the elements $c + d\omega$ where

$$\pm(c, d) \in \{(a, b), (b, a), (b, b - a), (b - a, b), (a - b, a), (a, a - b)\}.$$

So in proving (3), replacing r by another member of $[r]$ if necessary, we may assume $b \geq a \geq 0$. Now by 28.1,

$$|r^2| = a^2 + b^2 - ab = a^2 + b(b - a),$$

and both terms in this sum are nonnegative by our reduction. If $a = 0$ or $b = a$ then (a), (c), or (e) holds, so we may assume otherwise. Hence $a^2 + b(b - a) \geq 3$ with equality if and only if $r = \theta$, so the lemma holds if $|r|^2 \leq 3$.

Similarly if $|r|^2 = p$ with $5 \leq p \leq 11$ prime then $1 \leq a \leq p^{1/2}$ and we check the possible choices for a and determine (d) holds. Finally if $|r|^2 = pq$ with p, q prime then as R is a unique factorization domain, either $p = q$ is prime in R and $\pm r \in pR_0$, or $r = st$ where $|s|^2 = p$ and $|t|^2 = q$. Thus we have reduced to a previous case.

For $v = \sum_x v_x x$ and $r \in R$, let

$$S_r(v) = \{x \in X : v_x = \pm r\}$$

and define the *shape* of v to be $(r_1^{n_1}, r_2^{n_2}, \ldots)$, where $n_i = |S_{r_i}(v)|$.

Let Λ_2^2 consist of the vectors in Λ of shape $(2^2, 0^4)$ and Λ_2^1 consist of the vectors of shape $(1^5, \theta)$.

(28.10) *Let $v \in \Lambda^{\#}$. Then $q(v) \geq 2$ with equality if and only if some R_0-multiple of v is in Λ_2^i for $i = 1$ or 2.*

Proof: Suppose $v \in \Lambda_0$. Then $v_x = 2c_x$ for each $x \in X$ and some $c_x \in R$, so $q(v) = \sum_x |c_x|^2$ and $B(v) = \sum_x c_x \in 2R$. Thus either at least two of the c_x

are nonzero or $c_x \in 2R$ for the unique nonzero c_x, so by 28.9, $q(v) \geq 2$ with equality if and only if some R_0-multiple of v is in Λ_2^2.

Thus we may assume $v \notin \Lambda_0$ and multiplying by $t(v)^{-1}$ we may even assume $t(v) = 1$. As $t(v) = 1$, $v_x \notin 2R$, so $|x_x|^2$ is odd by 28.9.2. Therefore $q(v) = (\sum_x |v_x|^2)/4 > 2$ by 28.9 unless either

(i) $|v_x|^2 = 1$ and $v_x \in \pm R_0$ for all $x \in X$, or
(ii) for some $y \in X$ and for all $y \neq x \in X$, $|v_y|^2 = 3$ and $v_y \in \pm R\theta$ and $|v_x|^2 = 1$ and $v_x \in \pm R_0$.

As $t(v) = 1$, $v_x = \pm 1$ if $v_x \in \pm R_0$ and $v_y = \theta$ or $\bar\theta$ if $v_y \in \pm R_0 \theta$. Therefore in case (ii), $v \in \Lambda_2^1$, and in case (i), $v \notin \Lambda$.

(28.11)

(1) M^+ has two orbits P_2^i, $i = 1, 2$, on the set P_2 of R-points Rv with $q(v) = 2$. Further M^∞ is transitive on each of these orbits.

(2) P_2^i consists of the points Rv with $v \in \Lambda_2^i$.

(3) $|P_2^1| = 96$, $|P_2^2| = 30$, and $|P_2| = 126$.

Proof: Evidently M^+ acts on P_2^i, and by 28.10, $P_2 = P_2^1 \cup P_2^2$. Each vector of shape $(2^2, 0^4)$ is in Λ by 28.4, and there are $4 \cdot \binom{6}{2} = 60$ such vectors, with each point of P_2^2 containing two members of Λ_2^2. Thus $|P_2^2| = 30$ and visibly M^∞ is transitive on P_2^2.

Similarly if v has shape $(1^5, \theta)$ then $v = e_{X+Y} - e_Y + 2\epsilon\omega y$ for some $Y \subseteq X$ and $y \in X$ with $\epsilon = -1$ if $y \in Y$ and $\epsilon = 1$ if $y \in Y + X$. Then $B(v) = -|Y| + \epsilon\omega$, so by 28.4, $v \in \Lambda$ if and only if $-|Y| + \epsilon\omega \equiv \omega \bmod 2R$ if and only if $|Y|$ is even. So there are $6 \cdot 2^5$ members of Λ_2^1 and each member of P_2^1 contains exactly two such vectors, so $|P_2^1| = 96$. Hence (3) is established. Again it is easy to see M^∞ is transitive on P_2^1, so (1) and (2) also hold.

(28.12)   G is transitive on P_2.

Proof: By 28.11 it suffices to show P_2^1 is fused into P_2^2 under G. But if x, y, z are distinct in X then the reflection r_{λ_y} maps $2(x + z) \in \Lambda_2^2$ into Λ_2^1, and by 28.8, $r_{\lambda_y} \in G$.

(28.13)   Let $\Lambda_0^0 = \{v \in \Lambda_0 : B(v) = 0\}$ and $f_x = 4x$. Then

(1) $\Lambda_0 = \Lambda_0^0 + Rf_y$ for each $y \in X$.

(2) $\Lambda_0^0 = \langle 2(x - y) : x, y \in X \rangle \leq \langle \lambda_x : x \in X \rangle$.

(3) $\Lambda = \langle \lambda_x : x \in X \rangle + Rf_y$.

Proof: If $v \in \Lambda_0$ then $m(v) \in 4R$ by definition of Λ_0. Thus $v - m(v)/4 \cdot f_y \in \Lambda_0^0$, so (1) holds. Next

$$\omega^2(\lambda_x - \lambda_y) = 2(x - y) \quad \text{and} \quad \Lambda_0^0 = \langle 2(x - y) : x, y \in X \rangle,$$

so (2) holds. Finally (1), (2), and 28.5 imply (3).

Let $\hat{R} = R/\theta R$ and $\hat{\Lambda} = \Lambda/\theta\Lambda$. Thus $\hat{R}$ is the field of order 3 and $\hat{\Lambda}$ a vector space over $\hat{R}$ via $\hat{r} \cdot \hat{v} = rv + \theta\Lambda$, where for $v \in \Lambda$, $r \in R$, $\hat{v}$ and $\hat{r}$ denote the image of v and r in $\hat{\Lambda}$ and $\hat{R}$, respectively.

Further q induces a quadratic form on $\hat{\Lambda}$ via $\hat{q}(\hat{v}) = q(v) + \theta R$. Of course the bilinear form for $\hat{q}$ is β, where

$$\beta(\hat{u}, \hat{v}) = (u, v) + \overline{(u, v)} + \theta R.$$

That is

$$\hat{q}(\hat{u} + \hat{v}) = \hat{q}(u) + \hat{q}(v) + \beta(\hat{u}, \hat{v}).$$

Notice we have an induced representation of G^+ on $\hat{\Lambda}$ and G^+ preserves $\hat{q}$ in this representation.

(28.14) *$\hat{\Lambda}$ is a 6-dimensional orthogonal space of sign -1 over the field $\hat{R}$ of order 3 with orthonormal basis $\{\hat{f}_x : x \in X\}$.*

Proof: First

$$\sum_x f_x + 8\omega y = 4\lambda_y = 3\lambda_y + \lambda_y \equiv \lambda_y \bmod \theta\Lambda,$$

so that $\hat{\lambda}_y$ is in the span of $B = \{\hat{f}_x : x \in X\}$. Hence by 28.13.3, B spans $\hat{\Lambda}$. Further if $a_x \in R$ and $\sum_x a_x f_x = \theta v$ for some $v \in \Lambda$ then as θ is a prime in R, θ divides a_x for each $x \in X$, so $\hat{a}_x = 0$. That is B is linearly independent. Therefore $\dim(\hat{\Lambda}) = 6$. Of course B is orthonormal so $\hat{q}$ is nondegenerate and $\hat{\Lambda}$ has sign -1.

(28.15) *The map $\zeta : Rv \mapsto \hat{R}\hat{v}$ is a bijection of P_2 with the set of points $\hat{R}\hat{v}$ of $\hat{\Lambda}$ such that $\hat{q}(\hat{v}) = 2$.*

Proof: Suppose $u, v \in \Lambda_2^1 \cup \Lambda_2^2$ with $\hat{R}\hat{v} = \hat{R}\hat{u}$ but $Ru \neq Rv$. Then $u - rv = w \in R\theta$ for some $r \in R$ and as $0, \pm 1$ are coset representatives for $R\theta$ in R and $-\Lambda_2^i = \Lambda_2^i$, we may take $r = 1$. By 28.11 and 28.12 we may assume $u = 2(x + y)$. Now θ divides w_z for each $z \in X$. But unless $v = \pm 2(x - y)$, there exists $z \in X$ with $w_z = \pm 1$ or ± 2, whereas if $v = \pm 2(x - y)$ then $w = 4x$ or $4y$.

So ζ is an injection. However by 28.14, $\hat{\Lambda}$ is a 6-dimensional orthogonal space of sign -1, and hence it has exactly 126 points $\hat{R}\hat{v}$ with $\hat{q}(\hat{v}) = 2$. Then 28.11.3 completes the proof.

(28.16) $\langle -\omega I \rangle \cong \mathbf{Z}_6$ *is the kernel of the representation of G^+ on P_2.*

Proof: As $-\omega I$ induces scalar action on R^6, it fixes each point of R^6. Conversely if $g \in G$ fixes each point in P_2 then for each $x, y \in X$, $v_{xy}g = a_{xy}v_{xy}$ for some $a_{xy} \in R$, where $v_{xy} = 2(x - y)$. Then

$$a_{xz}v_{xz} = v_{xz}g = (v_{xy} + v_{yz})g = a_{xy}v_{xy} + a_{yz}v_{yz},$$

so $a_{xy} = a_{xz} = a_{yz}$. Hence g induces scalar action on $\mathbf{C}^6$. Then as the scalar is a unit in R, $g \in \langle -\omega I \rangle$.

(28.17)

(1) $C_{G^+}(\hat{\Lambda}) = \langle \omega I \rangle$.

(2) G induces $\Omega_6^-(3)$ extended by the reflection r_{λ_y} on $\hat{\Lambda}$ and G^+ induces $O(\hat{\Lambda}, \hat{q})$ on $\hat{\Lambda}$.

(3) τ_y acts on $\hat{\Lambda}$ as the reflection with center $\hat{R}\hat{f}_y$.

Proof: Part (1) follows from 28.15 and 28.16. Further (3) is easy to check from the action of τ_y on the orthonormal basis of 28.14.

Next let $\hat{G}^+ = G^+/C_{G^+}(\hat{\Lambda})$ be the group induced by G^+ on $\hat{\Lambda}$. Then as we observed earlier, $\hat{G}^+ \leq O(\hat{\Lambda}, \hat{q})$. Now by 28.14, $\hat{\Lambda}$ is a 6-dimensional orthogonal space of sign -1. Further by 28.15 and 28.8, $\hat{G}$ contains the reflection $r_{\hat{v}} = \hat{r}_v$ for each point $\hat{R}\hat{v}$ with $\hat{q}(\hat{v}) = 2$, so the group $\hat{G}_0$ generated by those reflections is $\Omega_6^-(3)$ extended by the reflection r_{λ_y}. Thus $\hat{G}_0\langle \tau_y \rangle = O(\hat{\Lambda}, \hat{q})$, completing the proof.

Let $\Lambda_3^4, \Lambda_3^3, \Lambda_3^7$ be the set of vectors in Λ of shape $(2, 2\omega, 2\omega^2, 0^3)$, $(\theta^3, 1^3)$, $(\alpha, 1^7)$, respectively, where $|\alpha|^2 = 7$. Let P_3^i be the set of points Rv in Λ with $v \in \Lambda_3^i$.

Similarly let $\Lambda_4^3, \Lambda_4^7, \Lambda_4^9$ be the set of vectors in Λ of shape $(1, \theta^5)$, $(1^3, \theta^2, \alpha)$, $(1^4, \theta, 3)$, respectively, and let $\Lambda_4^{16}, \Lambda_4^6, \Lambda_4^4, \Lambda_4^{4\omega}$ be the set of vectors of shape $(4, 0^5)$, $(2\theta, 2, 0^4)$, $(2^4, 0^2)$, $(2^2, (2\omega)^2, 0^2)$, respectively. Let P_4^i be the set of points Rv with $v \in \Lambda_4^i$.

For $m = 3, 4$, let P_m be the set of R-points Rv with $q(v) = m$.

(28.18)

(1) M^+ has three orbits P_3^i, $i = 3, 4, 7$ on P_3.

(2) M^∞ is transitive on P_3^3 and M^∞ has two orbits on P_3^4 fused in M and M^∞ has two orbits on P_3^7 fused under τ_y.

(3) $|P_3^i| = 320, 160, 192$, for $i = 3, 4, 7$, respectively.

(4) $|P_3| = 672 = 2^5 \cdot 3 \cdot 7$.

(5) G is transitive on P_3.

(28.19)

(1) M^+ has seven orbits P_4^i, $i = 3, 4, 4\omega, 6, 7, 9, 16$, on P_4.

(2) M^∞ is transitive on P_4^i for $i \neq 7$, and P_4^7 is the union of two M^∞ orbits fused under τ_y.

(3) $|P_4^i| = 96, 120, 720, 60, 1920, 480, 6$, respectively.

(4) $|P_4| = 3402 = 2 \cdot 3^5 \cdot 7$.

(5) G is transitive on P_4.

We prove 28.18 and 28.19 together; more precisely we sketch proofs. First let $v \in \Lambda_m$, $m = 3$ or 4; that is $v \in \Lambda$ with $q(v) = m$. Suppose first that $v \in \Lambda_0$. Then $v = 2u$ with $\sum_x |u_x|^2 = m$ and $\sum_x u_x \in 2R$. We conclude that up to multiplication via R_0, $v \in \Lambda_3^4$ if $m = 3$ and $v \in \Lambda_4^i$ for $i = 16, 6, 4$, or 4ω if $m = 4$. Moreover the signs of the entries are arbitrary, so we can count the number of vectors of each shape. Then when $m = 4$, Rv contains exactly two vectors from Λ_4, so $|P_4^i|$ is as claimed. Similarly Rv contains exactly six vectors from Λ_3^4, allowing us to calculate $|P_3^4|$. The transitivity statements about M^+ and M^∞ are also easy.

Therefore we may assume $v \notin \Lambda_0$; thus we may take $t(v) = 1$. Now by 28.9.2, $|v_x|^2$ is an odd integer for each $x \in X$. Of course $\sum_x |v_x|^2 = 4m$. Thus

$$|v_x|^2 \leq 4m - 5 \leq 11,$$

so the possibilities for v_x are described in 28.9.3. Then it is straightforward to verify that v has one of the shapes described in Lemmas 28.18 and 28.19. As $B(v) \equiv \omega \mod 2R$ by 28.4, we determine that only half of the possible signs are possible for each shape, and then calculate the order of Λ_m^i. Finally as each point Rv contains exactly two members of Λ_m^i, we obtain $|P_m^i|$. Again the transitivity statements for M^+ and M^∞ are easy. Finally to show G is transitive on P_m, simply use 28.8 and check the M-orbits P_m^i are fused under the reflections r_{λ_y}.

Next let $\tilde{R} = R/2R$ and $\tilde{\Lambda} = \Lambda/2\Lambda$. Thus as R_0 is a set of coset representatives for $2R$ in R, $\tilde{R} = \tilde{R}_0$ is the field of order 4, and $\tilde{\Lambda}$ is a vector space over $\tilde{R}$ via $\tilde{r} \cdot \tilde{v} = rv + 2\Lambda$. Also the hermitian symmetric form $(\ ,\ )$ on Λ induces a hermitian symmetric form on $\tilde{\Lambda}$, which we also write as $(\ ,\ )$ via $(\tilde{v}, \tilde{u}) = (u, v) + 2R$. Observe that G preserves this form and τ_y preserves it up to a twist by a field automorphism.

(28.20)

(1) $\tilde{\Lambda}$ *is a 6-dimensional unitary space over* $\tilde{R} = GF(4)$.

(2) *G has three orbits* $\tilde{P}_m$, $m = 2, 3, 4$, *on the points of* $\tilde{\Lambda}$ *of length* 126, 672, 567, *respectively*.

(3) *For* $m = 2, 3$ *the map* $Rv \mapsto \tilde{R}\tilde{v}$ *is a bijection of* P_m *with* $\tilde{P}_m$, *and* $P_4 \mapsto \tilde{P}_4$ *has fibers of order* 6. *The fiber of* $\tilde{f}_x$ *is* $\{Rf_x : x \in X\}$.

(4) *G is irreducible on* $\tilde{\Lambda}$.

(5) *The orbits of* M^+ *on* $\tilde{P}_4$ *are* $\{\langle \tilde{f}_x \rangle\} = \tilde{P}_4^{16}$, $\tilde{P}_4^4 = \tilde{P}_4^6$, $\tilde{P}_4^{4\omega}$, $\tilde{P}_4^3 = \tilde{P}_4^9$, $\tilde{P}_4^7$ *of length* 1, 30, 120, 96, 320, *respectively*.

(6) $(\tilde{v}, \tilde{v}) = 0$ *if* $\tilde{R}\tilde{v} \in \tilde{P}_2$ *or* $\tilde{P}_4$ *and* $(\tilde{v}, \tilde{v}) = 1$ *if* $\tilde{R}\tilde{v} \in \tilde{P}_3$.

Proof: We have already observed that $(,)$ is a hermitian symmetric form $\tilde{\Lambda}$. Suppose $u, v \in \Lambda$ with $q(u), q(v) \leq 4$, $u \neq \pm v$, and $u - v \in 2\Lambda$. Then $u - v = 2w$ for some $w \in \Lambda^\#$. Then

$$4q(w) = q(u - v) = q(u) + q(v) - \gamma(u, v),$$

where $\gamma(u, v) = (u, v) + \overline{(u, v)} \in \mathbf{Z}$. Replacing v by $-v$ if necessary, we may assume $\gamma(u, v) \geq 0$. Then as $q(u), q(v) \leq 4$ and $q(w) \geq 2$, we conclude $q(u) = q(v) = 4$, $\gamma(u, v) = 0$, and $q(w) = 2$. Further $\gamma(u, v) = 0$ if and only if $(u, v) \in \mathbf{Z}\theta$. Thus to complete the proof of (3) it remains to show that in the last case for fixed u there are exactly five choices for v. By 28.19.5, we may take $u = f_x$. Then we use the description in 28.19 of P_4 to complete the proof of (3). Notice $(f_x, f_y) = 0$ for $x \neq y$, so for $u \in \Gamma_4$, the fiber of Rv consists of Rv together with the points Ru such that $q(u) = 4$, $(u, v) = 0$, and $q(u - v) = 2$.

By 28.11, 28.18, 28.19, and (3), $|\tilde{P}_m| = 126, 672$, and 567, for $m = 2, 3$, and 4, respectively, and G is transitive on $\tilde{P}_m$ for each m. Now

$$126 + 672 + 567 = 1,365 = \frac{4^6 - 1}{3}$$

so to complete the proof of (2) it suffices to show $\dim_{\tilde{R}}(\tilde{\Lambda}) \leq 6$, or equivalently $\dim_{GF(2)}(\tilde{\Lambda}) \leq 12$. But $\dim_{\mathbf{Z}}(\Lambda) \leq \dim_{\mathbf{Z}}(R^6) = 12$, so $\dim_{GF(2)}(\tilde{\Lambda}) \leq 12$.

Thus (2) is established. Further the sum of no proper set of orbits of G on points of $\tilde{\Lambda}$ is of the form $(4^k - 1)/3$, so G is irreducible on $\tilde{\Lambda}$. Thus as the form $(,)$ on $\tilde{\Lambda}$ is nonzero and G-invariant, it is even nondegenerate. This proves (1). Part (5) follows from 28.19 once we check how the fiber of a point $\tilde{R}\tilde{v}$ under the map $\varphi : Rv \mapsto \tilde{R}\tilde{v}$ intersects the various orbits P_4^i, and hence how by (3) they are identified under φ. Recall from an earlier paragraph that the fiber of Rv consists of Rv together with the points Ru such that $q(u) = 4$, $(u, v) = 0$, and $q(u - v) = 2$.

Theorem 28.21. *$F^*(G^+)$ is quasisimple with center $Z = \langle -\omega I \rangle \cong \mathbf{Z}_6$ and G^+/Z is $P\Omega_6^-(3)$ extended by the reflections τ_y and r_{λ_y}.*

Proof: By 28.17, $G/\langle \omega I \rangle$ is $\Omega_6^-(3)$ extended by the two reflections. Thus it remains to show $\omega I \in E(G)$. But this follows as $|P\Omega_6^-(3)|_3 = 3^6 = |U_6(2)|_3$, so, as $SU_6(2)$ does not split over its center, neither does its subgroup $E(G)\langle \omega I \rangle$.

29. The embedding of $U_4(3)/\mathbf{Z}_3$ in $SU_6(2)$

In this section we continue the hypotheses and notation of Section 28. From Section 28, $\tilde{R}$ is the field of order 4, $\tilde{\Lambda}$ is a 6-dimensional vector space over $\tilde{R}$, and $(\ ,\ )$ defines a unitary form on $\tilde{\Lambda}$ making $\tilde{\Lambda}$ into a unitary space over $\tilde{R}$. Write $SU(\tilde{\Lambda})$ for the special unitary group of this space. Thus $\tilde{G} \leq SU(\tilde{\Lambda})$ and $\tilde{\tau}_y$ induces a graph-field automorphism on $SU(\tilde{\Lambda})$.

We investigate the representation of $\tilde{G}^+$ on $\tilde{\Lambda}$. Let $L = G^\infty$. By Theorem 28.21, $\tilde{L} = F^*(\tilde{G}^+)$ is quasisimple with $Z(\tilde{L}) = Z(SU(\tilde{\Lambda})) \cong \mathbf{Z}_3$ and $\tilde{L}/Z(\tilde{L}) \cong U_4(3)$. In particular $\tilde{L} \leq SU(\tilde{\Lambda})$. Let $\tilde{G}^- = \tilde{L}\langle \tilde{\tau}_x \rangle$. By Theorem 28.21, $\tilde{G}/Z(\tilde{G}) \cong \tilde{G}^-/Z(\tilde{G}) \cong PO_6^{+,\pi}(3)$. We take our index set X of order 6 to be $X = \{1, \ldots, 6\}$.

(29.1)

    (1) $\tilde{f}_x = \tilde{f}_y$ for all $x, y \in X$.

    (2) $\tilde{M}^+$ is the stabilizer of $\tilde{R}\tilde{f}_x \in \tilde{P}_4$ in $\tilde{G}^+$.

    (3) $\tilde{M}^+$ is the split extension of $\tilde{E} \cong E_{32}$ by $\langle \omega \tilde{I} \rangle \times \widetilde{\mathrm{Sym}}(X) \cong \mathbf{Z}_3 \times S_6$.

Proof: Part (1) holds by 28.20.3. As M^+ permutes $\{Rf_x : x \in X\}$, $\tilde{M}^+$ stabilizes $\tilde{R}\tilde{f}_x$ and then as $\tilde{G}^+$ is transitive on $\tilde{P}_4$ with $|\tilde{G}^+ : \tilde{M}^+| = 567 = |\tilde{P}_4|$, (2) holds. Part (3) follows from 28.7.

(29.2)

    (1) $\tilde{\Lambda}_0 = \tilde{f}_x^\perp$ and $\tilde{\Lambda} = \tilde{\Lambda}_0 \oplus \tilde{R}\tilde{\lambda}_x$.

    (2) $(\tilde{f}_{6,x} : 1 \leq x < 6)$ is a basis for $\tilde{\Lambda}_0$, where $f_{y,x} = 2(y - x)$.

Proof: First for $v = \sum_x v_x x \in \Lambda$, $(f_x, v) = v_x$, so $\tilde{v} \in \tilde{f}_x^\perp$ if and only if $v_x \in 2R$ if and only if $t(v) = 0$, so $\tilde{\Lambda}_0 \leq \tilde{f}_x^\perp$. On the other hand by 28.5, $\tilde{\Lambda} = \tilde{\Lambda}_0 \oplus \tilde{R}\tilde{\lambda}_x$ so the proof of (1) is complete.

Further $\dim(\tilde{\Lambda}_0) = \dim(\tilde{f}_x^\perp) = 5$, so to prove (2) it suffices to show

$$(\tilde{f}_{6,x} : 1 \leq x < 6) \quad \text{is linearly independent.}$$

But for $a_x \in R$, $\sum_{x<6} a_x f_{6x} = 2v$ where $v = \sum_x v_x x$, $v_y = -a_y$ if $y \neq 6$, and $v_6 = \sum_{x<6} a_x$. Then $m(v) = 0$ so by 28.4, if $v \in \Lambda$ then $t(v) = 0$ and $a_y \in 2R$ for all y. That is $\tilde{a}_y = 0$ for all y, completing the proof.

(29.3) *The transposition* $(x, y) \in \mathrm{Sym}(X)$ *induces an* $\tilde{R}$-*transvection on* $\tilde{\Lambda}$ *with center* $\tilde{f}_{x,y}$.

Proof: Observe (x, y) fixes $\tilde{\lambda}_z$ and $\tilde{f}_{6,z}$ for $z \neq x, y, 6$. Further

$$(x, y) : \tilde{f}_{6,x} \mapsto \tilde{f}_{6,y} = \tilde{f}_{6,x} + \tilde{f}_{x,y}.$$

Now use 29.2.

(29.4)

(1) $\tilde{L}$ has one class of involutions with representative $\tilde{e}_{12}$.
(2) $C_{\tilde{\Lambda}}(\tilde{e}_{12}) = \langle \tilde{f}_{6,x}, \tilde{f}_{1,2} : x = 3, 4, 5 \rangle$ is of $\tilde{R}$-dimension 4 with

$$\tilde{e}_{12} : \tilde{\lambda}_x \mapsto \tilde{\lambda}_x - \tilde{f}_{1,2} + \tilde{f}_1$$

and $\tilde{f}_{6,1} \mapsto \tilde{f}_{6,1} + \tilde{f}_1$ for $x \neq 1, 2$.

Proof: Part (1) holds as $U_4(3)$ has one class of involutions (cf. 45.3.4 and 45.5.1 in [SG]). Part (2) is an easy calculation using 29.2.

(29.5) *If* $\tilde{R}\tilde{v} \in \tilde{P}_2$ *then* $\tilde{G}_{\tilde{v}}^+ \cong \mathbf{Z}_2 \times O_6^-(2)$ *with* $O_2(\tilde{G}_{\tilde{v}}^+) = \tilde{r}_v$.

Proof: By 28.17, $G_{R v}^+/\langle -\omega I \rangle \cong \mathbf{Z}_2 \times PO_5(3) \cong \mathbf{Z}_2 \times O_6^-(2)$ with $O_2(G_{Rv}^+) = \langle r_v, -I \rangle$.

(29.6) *If* $0 \neq \tilde{v}$ *is a singular vector in* $\tilde{\Lambda}$ *and* $\tilde{L} \leq H \leq \tilde{G}^+$ *with* $m_2(O_2(H_{\tilde{v}})) \geq 5$ *then* $\tilde{R}\tilde{v} \in \tilde{P}_4$, $O_2(H_{\tilde{v}}) \cong E_{32}$, $O_2(H_{\tilde{f}_x}) = \tilde{E}$, *and* $H = \tilde{G}^+$ *or* $\tilde{G}^-$.

Proof: By 28.20.6, $\tilde{R}\tilde{v} \in \tilde{P}_m$, $m = 2$ or 4, and by 29.5 it must be the latter. Thus we may take $v = f_x$. Now 29.1 completes the proof.

(29.7) *Let* $\tilde{T} = \tilde{E} \langle (1, 2), (1, 3)(2, 4), (5, 6) \rangle \leq \tilde{M}^+$. *Then*

(1) $\tilde{T} \in \mathrm{Syl}_2(\tilde{G}^+)$.
(2) $\mathcal{A}(\tilde{T} \cap \tilde{L}) = \{U_1, U_2\}$ where $U_1 = \tilde{E}_0$ and

$$U_2 = \langle (1, 2)(3, 4), (3, 4)(5, 6), \tilde{e}_{1,2}, \tilde{e}_{3,4} \rangle.$$

(3) $C_{\tilde{\Lambda}}(U_2) = \tilde{R}\tilde{f}_{1,2} + \tilde{R}\tilde{f}_{3,4} + \tilde{R}\tilde{f}_{5,6}$ is a totally singular 3-dimensional $\tilde{R}$-subspace of $\tilde{\Lambda}$ with $C_{\tilde{G}^+}(C_{\tilde{\Lambda}}(U_2)) = U_2\langle \tilde{r}_{\lambda_x}\rangle$ and $N_{\tilde{G}^+}(C_{\tilde{\Lambda}}(U_2))/C_{\tilde{G}^+}(C_{\tilde{\Lambda}}(U_2)) \cong \hat{S}_6$.

(4) $\langle \tilde{f}_x \rangle = C_{\tilde{\Lambda}}(U_1)$.

Proof: First $\tilde{T} \in \mathrm{Syl}_2(\tilde{M}^+)$ and $|\tilde{G}^+ : \tilde{M}^+|$ is odd so (1) holds. Let $T_L = T \cap L$. As $\tilde{L}/Z(\tilde{L}) \cong U_4(3)$, $|A(\tilde{T}_L)| = 2$ and $m(\tilde{T}_L) = 4$ (cf. 45.5.3 in [SG]). So as $U_1 \cong U_2 \cong E_{16}$, $A(\tilde{T}_L) = \{U_1, U_2\}$. By 29.4,

$$W = C_{\tilde{\Lambda}}(\langle \tilde{\epsilon}_{1,2}, \tilde{\epsilon}_{3,4}\rangle) = \tilde{R}\tilde{f}_{1,2} + \tilde{R}\tilde{f}_{3,4} + \tilde{R}\tilde{f}_{5,6}$$

and $\tilde{R}f_x = \tilde{R}(\sum_{i<6} \tilde{f}_{i,6}) = C_{\tilde{\Lambda}}(U_1)$. An easy calculation shows $[W, U_2] = 0$, so $W = C_{\tilde{\Lambda}}(U_2)$. Then $N_{\tilde{G}^+}(U_2)/Z(\tilde{L}) \cong S_6/E_{32}$ by 29.1 as U_1 and U_2 are conjugate in $\mathrm{Aut}(\tilde{L}/Z(\tilde{L}))$. Then $N_{\tilde{G}^+}(U_2)$ is maximal in $\tilde{G}^+$ (as it is the normalizer of a maximal parabolic of $\tilde{L}$) so $N_{\tilde{G}^+}(U_2) = N_{\tilde{G}^+}(W)$. Also W is a maximal totally singular subspace of the unitary space $\tilde{\Lambda}$, so the actions of $N_{\tilde{G}^+}(W)$ on W and $\tilde{\Lambda}/W$ are dual. In particular $N_{\tilde{G}^+}(W)/O_2(N_{\tilde{G}^+}(W))$ is faithful on W so $N_{\tilde{G}}(W)/O_2(N_{\tilde{G}}(W)) \le SL_3(4)$ and hence $N_{\tilde{G}}(W)/O_2(N_{\tilde{G}}(W)) \cong \hat{A}_6$ is irreducible on W, so $E_{32} \cong O_2(N_{\tilde{G}}(W)) = C_{\tilde{G}}(W) = C_{\tilde{G}^+}(W)$ and the lemma holds.

(29.8) Let $\mathcal{U}$ be the set of $E_{2^k} \cong U \le \tilde{G}^-$ with $m(C_{\tilde{\Lambda}}(U)) + m(U) \ge 10$. Then $\mathcal{U} = \mathcal{U}_2^{\tilde{L}}$.

Proof: If $U \not\le \tilde{L}$ then $U = (U \cap \tilde{L})\langle s \rangle$ with $s \in U - \tilde{L}$ inverting $\tilde{Z} = Z(\tilde{L})$. So $C_{\tilde{\Lambda}}(U) = C_{\tilde{\Lambda}}(U \cap \tilde{L}) \cap C(s)$ and as s inverts $\tilde{Z}$ with $[C_{\tilde{\Lambda}}(U \cap \tilde{L}), \tilde{Z}] = C_{\tilde{\Lambda}}(U \cap \tilde{L})$, we have $m(C_{\tilde{\Lambda}}(U)) = m(C_{\tilde{\Lambda}}(U \cap \tilde{L}))/2 \le 4$ by 29.4. Then as $m_2(\tilde{G}^-) = 5$,

$$m(C_{\tilde{\Lambda}}(U)) + m(U) \le 9 < 10,$$

a contradiction.

So $U \le \tilde{L}$ and then $m(U) \le m_2(\tilde{L}) = 4$, so $m(C_{\tilde{\Lambda}}(U)) \ge 6$. By 29.4, $m(C_{\tilde{\Lambda}}(t)) = 8$ for $t \in U^{\#}$ and as $U \le \tilde{L}$, $C_{\tilde{\Lambda}}(U)$ is an $\tilde{R}$-subspace of $\tilde{\Lambda}$ so $m(C_{\tilde{\Lambda}}(U))$ is even. Hence either $m(C_{\tilde{\Lambda}}(U)) = 6$ and $m(U) = 4$ or $m(C_{\tilde{\Lambda}}(U)) = 8$, $m(U) \ge 2$, and $C_{\tilde{\Lambda}}(U) = C_{\tilde{\Lambda}}(t)$ for each $t \in U^{\#}$. As $C_{\tilde{L}}(t)$ is maximal in $\tilde{L}$ the second case is impossible, so $m(U) = 4$. Then by 29.7.2, $U \in \mathcal{U}_i^{\tilde{L}}$ for $i = 1$ or 2. Finally $C_{\tilde{\Lambda}}(U_1) = \tilde{R}\tilde{f}_x$, so $U \in \mathcal{U}_2^{\tilde{L}}$.

Theorem 29.9. *Let (V, q) be a 12-dimensional orthogonal space over $GF(2)$ and $\mathcal{L}$ the set of subgroups K of $O(V, q)$ with $\mathbf{Z}_3 \cong Z = Z(K)$ and*

$K/Z \cong U_4(3)$. *Let $K \in \mathcal{L}$. Then*

(1) K is quasisimple and irreducible on V with $Z(K) = F^{\#}$, where

$$F = \text{End}_{GF(2)K}(V) \quad \text{is of order } 4.$$

(2) K preserves a unitary form f on V regarded as an F-space V_F such that $GU_6(2) \cong O(V_F, f) = C_{O(V,q)}(Z)$ is determined up to conjugacy in $O(V, q)$.

(3) $O(V_F, f)$ is transitive on $\mathcal{L} \cap O(V_F, f)$, $S = O(V_F, f) \cap SL(V_F) \cong SU_6(2)$ has three orbits on $\mathcal{L} \cap O(V_F, f)$, and $\text{Aut}(S)$ permutes the three orbits as S_3.

(4) $O(V, q)$ is transitive on $\mathcal{L}$.

Proof: Let $Z \leq Y \leq K$ with $Y/Z \cong A_6/E_{81}$ (cf. 45.9 in [SG]) and let $A = O_3(Y)$. First observe that $A \cong E_{3^5}$. For if not, then as Y is irreducible on A/Z, A is extraspecial. But then the minimal degree of a faithful $GF(2)A$-module is $2 \cdot 3^2 = 18$ (cf. 34.9 in [FGT]), a contradiction.

Next $V = \bigoplus_{B \in \mathcal{B}} C_V(B)$, where $\mathcal{B}$ is the set of hyperplanes B of A with $C_V(B) \neq 0$ (cf. Exercise 4.1.3 in [FGT]). As $V = [V, Z]$, $m(C_V(B))$ is even, so $12 = \dim(V) \geq 2|\mathcal{B}|$ and hence $|\mathcal{B}| \leq 6$. So as $\mathcal{B}$ is Y-invariant and 6 is the minimal degree of a nontrivial permutation representation of $Y/A \cong A_6$, we conclude $\mathcal{B}$ is an orbit of Y of length 6 and $\dim(C_V(B)) = 2$ for each $B \in \mathcal{B}$. It follows that K is irreducible on V and $F = \text{End}_{GF(2)K}(V)$ is of order 4. Also $|K|_3 = |SU_6(2)|_3$, so as $SU_6(2)$ does not split over its center, neither does K by Gaschütz's Theorem 2.6, once we establish (2). Thus K is quasisimple, so that (1) will follow once we prove (2).

Next part (2) is a special case of a general fact about embeddings in classical groups; see for example [A3]. For completeness we sketch a proof here. As $F = \text{End}_{GF(2)K}(V)$, V has a unique F-space structure V_F preserved by K and a generator z of Z induces scalar action ωI on V_F via some generator ω of $F^{\#}$.

Let $\sigma : a \mapsto a^2$ be the generator of $Gal(F/GF(2))$ and $V^F = F \otimes_{GF(2)} V$. Then $V^F = W \oplus W^{\sigma}$ with $W \cong V_F$ as an FK-module (cf. 25.10 in [FGT]). Let b be the bilinear form associated with q, and q^F and b^F the quadratic and bilinear forms on V^F agreeing with q, b on $1 \otimes V$. Then (V^F, q^F) is an orthogonal space over F and $K \leq O(V^F, q^F)$. Also σ induces a semilinear map on V^F commuting with $D = C_{O(V^F, q^F)}(Z)$ with fixed space $1 \otimes V$ and with $b^F(x^{\sigma}, y^{\sigma}) = b^F(x, y)^{\sigma}$ for all $x, y \in V^F$.

As K is irreducible on V and hence on W, (W, q^F) is either nondegenerate or totally singular. In the former case $K \leq O(W, q^F) \cong O_6^{\epsilon}(4)$, impossible as

$O_6(4)$ does not contain the scalar map $z = \omega I$. So W is totally singular. Notice W and W^σ are the eigenspaces for Z on V^F so D acts on W and W^σ.

Define $f : W \times W \to F$ by $f(u, v) = b^F(u, v^\sigma)$. As $\sigma : W \to W^\sigma$ is an FD-isomorphism,

$$f(ug, vg) = b^F(ug, (vg)^\sigma) = b^F(ug, v^\sigma g) = b^F(u, v^\sigma) = f(u, v)$$

for each $g \in D$. That is $D \leq O(W, f)$. As b^F is linear in its first variable, so is f, while

$$f(v, u) = b^F(v, u^\sigma) = b^F(u^\sigma, v) = b^F(u, v^\sigma)^\sigma = f(u, v)^\sigma,$$

so f is hermitian symmetric and sesquilinear. As K is irreducible on W and $f \neq 0$, f is nondegenerate, so (W, f) is a 6-dimensional unitary space over F. Conversely $O(W, f)$ acts on W^σ via $v^\sigma g = (vg)^\sigma$, and then $O(W, f) \leq D$. So $O(W, f) = D$.

Next let $T = Tr^F_{GF(2)} = 1 + \sigma$ be the trace of F to $GF(2)$. Now $1 \otimes V = W(1 + \sigma)$ with the map $w \mapsto w(1 + \sigma)$ a $GF(2)D$-isomorphism of W with $1 \otimes V \cong V$. Further for $w, v \in W$,

$$b(w(1 + \sigma), v(1 + \sigma)) = b^F(w(1 + \sigma), v(1 + \sigma))$$

$$= b^F(w, v^\sigma) + b^F(w^\sigma, v) = f(w, v)(1 + \sigma)$$

$$= (Tr \circ f)(w, v).$$

That is $b = Tr \circ f$, so as D preserves f it preserves b. Similarly $Tr(f(w, w)) = q(w)$, so $D \leq O(V, q)$. Then as $C_{O(V, q)}(Z) \leq C_{O(V^F, q^F)}(Z) = D$, we have $D = O(V, q)$. Finally as the representation of D on V is determined up to quasiequivalence, D is determined up to conjugacy in $GL(V)$ by 1.1, and then as q is the unique quadratic form on V preserved by D, D is also determined up to conjugacy in $O(V, q)$. Therefore (2) is established.

Notice (2) and (3) imply (4), so it remains to prove (3). First

$$Y = Y^\infty \leq D^\infty = S \cong SU_6(2),$$

and indeed each subgroup A of S isomorphic to E_{3^5} is the pointwise stabilizer of a decomposition $V = \bigoplus_{B \in \mathcal{B}} C_V(B)$ of the unitary space V as an orthogonal sum of points, so, as Y induces S_6 on $\mathcal{B}$, $Y = N_S(A)$ and A are determined up to conjugacy in S and $N_D(A) = YA_0$ with $A \leq A_0 \cong E_{3^6}$ the pointwise stabilizer in D of the decomposition.

Next if $P \in \mathrm{Syl}_3(Y)$ and $I = Z_2(P)$ then $I = I_1 \times I_2 \cong E_9$ with $V = V_1 \oplus V_2$ the orthogonal direct sum of nondegenerate 3-subspaces $V_i = C_V(I_i)$ and

$$D_I = O^2(N_D(I)) = D_1 \times D_2$$

with $D_i \cong GU_3(2)$ the centralizer of V_{3-i}. Then $S_I = S \cap D_I$ is of index 3 in D_I and $O^2(N_K(I)) = K_I$ is of index 8 in S_I with $K_I \cong SL_2(3)/(3^{1+2} \times 3^{1+2})$. Further $O_{3,2}(K_I)$ is one of the three P-invariant subgroups of $O_{3,2}(S_I)$ distinct from $O_{3,2}(D_i)O_3(S_I)$ and hence determined up to conjugation under A_0. So $K = \langle Y, K_I \rangle$ is determined up to conjugation in D.

Thus the first part of (3) is established. Then as $\text{Out}(S) \cong S_3$ and $\text{Aut}(S)$ is transitive on the three orbits of S on $\mathcal{L} \cap S$, $\text{Aut}(S)$ permutes these orbits as S_3, completing the proof.

(29.10)   *Regard $\tilde{\Lambda}$ as a 12-dimensional space over $GF(2)$ and define*

$$q : \tilde{\Lambda} \to GF(2) \quad and \quad f : \tilde{\Lambda} \times \tilde{\Lambda} \to GF(2)$$

by $q(\tilde{v}) = (\tilde{v}, \tilde{v})$ and $f(\tilde{u}, \tilde{v}) = Tr^{\tilde{R}}_{GF(2)}((\tilde{u}, \tilde{v}))$. Then

(1) q is a quadratic form on $\tilde{\Lambda}$ with bilinear form f.
(2) $\tilde{G}^+ = N_{O(\tilde{V},q)}(\tilde{L})$.

Proof: As $(\ ,\ )$ is a unitary form on the $\tilde{R}$-space $\tilde{\Lambda}$, $q(\tilde{v}) \in GF(2)$ and f is a symmetric bilinear form. Also

$$q(\tilde{u} + \tilde{v}) = q(\tilde{u}) + q(\tilde{v}) + f(\tilde{u}, \tilde{v}),$$

so q is a quadratic form with bilinear form f. As $(\tilde{u}\tilde{\tau}_x, \tilde{v}\tau_x) = (\tilde{u}, \tilde{v})^2$, $\tilde{\tau}_x \in O(\tilde{\Lambda}, q)$, while $\tilde{G} \leq SU(\tilde{V}) \leq O(\tilde{\Lambda} q)$, so $\tilde{G}^+ = \tilde{G}\langle \tilde{\tau}_x \rangle \leq O(\tilde{\Lambda}, q)$. By 29.9.1, $C_{O(\tilde{\Lambda},q)}(\tilde{L}) = Z(\tilde{L})$, so $N_{O(\tilde{\Lambda},q)}(\tilde{L})/Z(\tilde{L}) \leq \text{Aut}(\tilde{L}/Z(\tilde{L})) \cong P\Gamma_6^-(3)$. Thus $\tilde{G}^+/Z(\tilde{L})$ is of index 2 in $\text{Aut}(\tilde{L}/Z(\tilde{L}))$ and

$$\alpha \in \text{Aut}(\tilde{L}/Z(\tilde{L})) - \tilde{G}^+/Z(\tilde{L})$$

does not act on $\tilde{L}^l_{\tilde{v}}$ for $\tilde{R}\tilde{v} \in \tilde{P}_2$, so $\alpha \notin O(\tilde{\Lambda}, q)$. Hence (2) is established.

11

The existence and uniqueness of the Fischer groups

In this chapter we show that there do indeed exist 3-transposition groups of type $M(22)$, $M(23)$, and $M(24)$. In addition we prove the uniqueness of $M(22)$, $M(23)$, and $M(24)'$ as groups with a suitable involution centralizer. Indeed we use the uniqueness results to establish the existence of the Fischer groups.

More precisely in 32.4 in [SG] it is shown that there exists a subgroup X of the Monster such that $E(X)$ is quasisimple with center Z of order 3, X is the split extension of $E(X)$ by an involution inverting Z, and $E(X)/Z$ and X/Z are groups of type F_{24} and Aut(F_{24}), respectively, in the sense of Sections 34 and 35 or the Introduction to Part II. In Theorem 35.1 we show each group of type Aut(F_{24}) is generated by 3-transposition and is of type $M(24)$. Hence Fischer's three 3-transposition groups exist, and, by Fischer's Theorem, groups of type Aut(F_{24}) are unique up to isomorphism.

We characterize the Fischer groups (and $M(24)' = F_{24}$) via the centralizer of an involution by showing each group of type $\tilde{M}$ is of type M, for $M = M(22)$ and $M(23)$, and by showing all groups of type F_{24} are isomorphic. See the Introduction to Part II for an outline of the proofs of these results.

30. A characterization of $U_6(2)$

In this section we establish the following characterization of $U_6(2)$, which will be used later in the chapter to prove the existence of the Fischer groups as 3-transposition groups and the uniqueness of the Fischer groups as groups with a suitable involution centralizer.

Theorem 30.1. Let G be a finite group, z an involution in G, $H = C_G(z)$, $Q = F^*(H)$, and $\tilde{H} = H/\langle z \rangle$. Assume

(1) $Q \cong 2^{1+8}$ and $\tilde{Q}$ is the natural module for $H/Q \cong U_4(2)$.

(2) There is $E_{2^9} \cong A \leq G$ with $z \in A$ and A is the 9-dimensional Todd module for $N_G(A)/A \cong L_3(4)$.

Then $G \cong U_6(2)$.

We establish Theorem 30.1 via a series of reductions. Thus throughout this section we assume the hypotheses of Theorem 30.1. Let $M = N_G(A)$ and $A \leq T \in \mathrm{Syl}_2(H)$. We conclude from 22.2 that

(30.2) *M has three orbits I_i, $1 \leq i \leq 3$, on $A^\#$ of length 21, 210, 280, respectively, as described in Lemma 22.2.*

As $Q \cong 2^{1+8}$ is extraspecial the commutator map and power map on Q induce an orthogonal space structure on $\tilde{Q}$ over $GF(2)$ preserved by H/Q in which elementary abelian subgroups U of Q correspond to totally singular subspaces $\tilde{U}$ in the orthogonal space $\tilde{Q}$ (cf. Section 8 in [SG]). As $\tilde{Q}$ is the natural module for $H/Q \cong U_4(2)$, $\tilde{Q}$ has the structure of a unitary space over the field of order 4 preserved by H/Q with the singular vectors in this unitary space the singular vectors of the orthogonal space $\tilde{Q}$.

(30.3)

(1) $z \in I_1$.

(2) $A \cap Q \cong E_{32}$, $\widetilde{A \cap Q}$ is a totally singular subspace of the unitary space $\tilde{Q}$, and $(H \cap M)/Q \cong A_5/E_{16}$ is the stabilizer of $\widetilde{A \cap Q}$ in H/Q.

(3) $A = J(T)$.

Proof: Let $H^* = H/Q$ and $E_{2^k} \cong B \leq T$ with $k \geq 9$. Then

$$m(Q \cap B) \leq m(Q) \leq 5 \quad \text{and} \quad m(B^*) \leq m(H^*) = 4,$$

so all inequalities are equalities. In particular $m(T) = 9$ and $A \in \mathcal{A}(T)$. Indeed $J(T^*) \cong E_{16}$ with $N_{H^*}(J(T^*))$ the maximal parabolic of H^* isomorphic to A_5/E_{16} and stabilizing the totally singular line $C_{\tilde{Q}}(J(T^*))$ of the unitary space $\tilde{Q}$. Therefore $B^* = J(T^*)$ and $\widetilde{B \cap Q} \leq C_{\tilde{Q}}(B^*) = C_{\tilde{Q}}(J(T^*))$ so $B \cap Q$ is the full preimage of $C_{\tilde{Q}}(J(T^*))$. Finally $B = C_H(B \cap Q)$ is determined so $\{B\} = \mathcal{A}(T)$. Thus as $A \in \mathcal{A}(T)$, (3) is established. Notice we have also proved (2).

Next the preimage M_0 of $N_{H^*}(J(T^*))$ is contained in M as $A = J(T)$ is normal in that preimage. Notice $M_0 = H \cap M = C_M(z)$ by maximality of M_0 in H. Therefore $|z^M| = |M : M_0| = 21$ and hence (1) follows from 30.2.

(30.4) *M controls G-fusion in A and indeed* $K \cap M$ *controls K-fusion in A for each subgroup K of G containing A, so* $I_1 = z^G \cap A$ *and* $d^K \cap A = d^{K \cap M}$ *for* $d \in A$.

Proof: This follows from 30.3.3 and 2.8.1.

(30.5) *z is weakly closed in Q with respect to G.*

Proof: As M is irreducible on A (cf. 22.5), $A = \langle z^M \rangle = \langle I_1 \rangle$. But by 30.2 and 22.2, M is 2-transitive on I_1, so $A = \langle z, u^{H \cap M} \rangle$, where $u \in I_1 - \{z\}$. So, as $Q \cap A \neq A$, $u \notin Q \cap A$ and hence z is weakly closed in $Q \cap A$ with respect to G. Finally as the involutions in Q correspond to the singular vectors of $\tilde{Q}$ and each is contained in an H-conjugate of $\widetilde{Q \cap A}$, the lemma holds.

Let $D = z^G$, $\mathcal{D}$ the commuting graph on D, $D_z = C_D(z) - \{z\}$, and $A_z = D - z^\perp$.

(30.6)

    (1) H *is transitive on* D_z.
    (2) D_z^* *is the set of transvections in* $H^* = H/Q$.
    (3) For $d \in D_z$, $C_{H^*}(d^*) = C_H(d)^*$ *and* $D \cap dQ = d^Q$ *is of order 4.*

Proof: First by 23.2, $H^* \cong U_4(2)$ has two classes $\mathcal{J}_m$, $m = 1, 2$, of involutions where $j_m \in \mathcal{J}_m$ has m Jordon blocks of size 2 on the $GF(4)$-space $\tilde{Q}$. Thus $|\tilde{Q} : C_{\tilde{Q}}(j_m)| = 4^m$. Further $\mathcal{J}_m \cap A^* \neq \emptyset$ for $m = 1, 2$.

Let $\tilde{d} \in I_1 - \{z\}$ so that $1 \neq d^* \in D_z^* \cap A^*$ by 30.5. Then

$$|H \cap M : C_{H \cap M}(d)| = |d^{H \cap M}| = 20$$

by 2-transitivity of M on I_1. Therefore $|Q : C_Q(d)|$ divides 4, so as

$$|\tilde{Q} : C_{\tilde{Q}}(j_2)| = 16,$$

we conclude $d^* \in \mathcal{J}_1$ and

$$|Q : C_Q(d)| = |\tilde{Q} : C_{\tilde{Q}}(d^*)| = |C_{(H \cap M)^*}(d^*) : C_{H \cap M}(d)^*| = 4.$$

In particular $C_Q(d)/\langle z \rangle = C_{\tilde{Q}}(d^*)$ and $C_{(H \cap M)^*}(d^*) = C_{H \cap M}(d)^*$, so that by 30.4

$$d^{C_H(d^*)} \cap A = d^{C_{H \cap M}(d^*)} = d^Q.$$

Let t be an involution in dQ. Then $t = dx$, $x \in Q$ inverted by d. Now if $y \in C_Q(t)$ then $\tilde{y} \in C_{\tilde{Q}}(t) = C_{\tilde{Q}}(d) = C_Q(d)/\langle z \rangle$, so $y \in C_Q(d) \cap C_Q(x)$. Thus $|Q : C_Q(t)| = 4$ if and only if $C_Q(d) \leq C_Q(x)$ if and only if $x \in [Q,d] \leq A$. So as $d^{C_H(d^*)} \cap A = d^Q$, we have $d^{t_H(d^*)} = d^Q$ and hence $C_H(d)^* = C_{H^*}(d^*)$. However $C_{H^*}(d^*)$ is transitive on singular vectors in $C_{\tilde{Q}}(d) - [\tilde{Q},d]$, so $C_H(d)Q$ has three orbits on involutions in dQ with representatives dx_i, $1 \leq i \leq 3$, where $x_1 = 1$, $x_2 = z$, and x_3 is an involution in $A - [Q,d]$. Therefore all involutions in dQ are fused into A under H and hence

$$D \cap dQ = D \cap dQ \cap A = d^M \cap dQ$$

by 30.4. But as M is 2-transitive on $\mathcal{I}_1 = d^M$, we have $d^M = \{z\} \cup d^{H \cap M}$, so

$$D \cap dQ = d^M \cap dQ = d^{H \cap M} \cap dQ = d^{C_{H \cap M}(d^*)} = d^Q.$$

Let $a \in A$ with $a^* \in \mathcal{J}_2$. Then $m(\tilde{Q}/C_{\tilde{Q}}(a)) = m(\tilde{Q})/2$, so by Exercise 2.8 in [SG], each involution in aQ is conjugate under Q to a or az. In particular each involution t with $t^* \in \mathcal{J}_2$ is fused into A under H, so as $\mathcal{I}_1 = A \cap D$ and we showed $e \in d^H$ for each $e \in \mathcal{I}_1 - \{z\}$, it follows that $D_z = D \cap H - \{z\} = d^H$, completing the proof of the lemma.

(30.7) D_z *is a set of 3-transpositions of* H.

Proof: Let $a,d \in D_z$. As D_z^* is a set of 3-transpositions of H^* by 30.6.2, $|a^*d^*| \leq 2$. If $a^* = d^*$ then by 30.6.3, $a \in d^Q \subseteq A$, so $|ad| \leq 2$. Similarly if $|a^*d^*| = 2$ then as $C_{H^*}(d^*)$ is transitive on $C_{D_z^*}(d^*)$, we may take $a^* \in A^*$ and then by 30.6.3, $a,d \in A$ so that $|ad| = 2$.

Thus we may assume a^*d^* has order 3. Then d inverts x of order 3 with $x^* = a^*d^*$ and $a \in d^xQ \cap D = d^{xQ}$. Thus there are $16 = |d^Q||a^Q| = |(a,d)^Q|$ pairs (b,c) in $(d^Q \cap D) \times (a^Q \cap D)$, so Q is transitive on such pairs. In particular as (d,d^x) is such a pair, $|ad| = |dd^x| = 3$.

(30.8) D *is a set of 3-transpositions of* G.

Proof: We appeal to 24.4.2. To do so we first prove that if $d \in D_z$ and $a \in D_d \cap A_z$ then $b^{\perp} \cap D_{a,z} \neq \emptyset$ for all $b \in D_z$. Let $G_d = C_G(d)$, $Q_d = O_2(G_d)$, $\hat{G}_d = G_d/Q_d$, and $H^* = H/Q$. By 30.7, D_d is a set of 3-transpositions of G_d, so as $a \in D_d \cap A_z$, az is of order 3. Then as $\hat{D}_d$ is the set of transvections in $\hat{G}_d \cong U_4(2)$, $\langle \hat{D}_{d,z,a} \rangle = \langle \hat{b}, \hat{c} \rangle \cong S_3$ for $b,c \in D_{d,z,a}$. Then by 30.6.3, $\langle D_{d,z,a} \rangle = \langle b,c \rangle C_{Q_d}(\langle z,a \rangle)$ with $C_{Q_d}(\langle z,a \rangle) \cong Q_8^2$. In particular $\langle d \rangle$ is the unique minimal normal subgroup of $\langle D_{d,a,z} \rangle$, so as $d \notin Q$, $\langle D_{d,z,a} \rangle \cong \langle D_{d,z,a} \rangle^*$. Then $\langle D_{d,z,a} \rangle^*$ contains a subgroup $B^* \cong E_{16}$ generated by the five transvections in a Sylow 2-subgroup of H^*. Also for $e \in D_{d,z,a}$, $\langle D_{e,z,a} \rangle^* \cong \langle D_{d,z,a} \rangle^*$ by symmetry, so $\langle D_{e,z,a} \rangle^* \not\leq C_{H^*}(d^*)$. Therefore by Exercise 7.2,

$\langle D_{a,z} \rangle^* = H^*$. Now $z \notin G_a$ whereas as H is irreducible on $\tilde{Q}$, we would have $H \leq \langle D_{z,a} \rangle \leq G_a$ if $Q \cap \langle D_{z,a} \rangle \not\leq \langle z \rangle$. Therefore $\langle D_{a,z} \rangle$ is a complement to Q in H, so in particular for $b \in D_z$ there is a unique member c of b^Q in $D_{a,z}$ and $c \in b^\perp \cap D_{z,a}$, establishing the claim.

It remains to verify the remaining hypotheses of 24.4.2. First 30.7 establishes hypothesis (i) of 24.4. Next by 30.6, D_z^* is the set of transvections in $H^* \cong U_4(2)$, so $A^* = \langle D_z^* \cap T^* \rangle$, and then by 30.6.3, $A = \langle T \cap D \rangle$, and of course A is abelian. By 22.5, M is irreducible on A, so $A = \langle ab : a, b \in T \cap D \rangle$. Further by 30.1–30.4 and 22.2, the map $\{a, b\} \mapsto ab$ is injective on 2-subsets of $T \cap D$, $C_M(t)$ is intransitive on $T \cap D$ for each involution $t \in A$, and $T \cap D$ is of order 21. In particular $T \cap D \neq A^\#$ and each involution in T fixes a member of $T \cap D$ as 21 is odd. Thus we have verified all hypotheses of 24.4.2, completing the proof of the lemma.

We are now in a position to complete the proof of Theorem 30.1. Namely by 30.8, D is a set of 3-transpositions of G and by 8.12 in [SG], G is simple. Therefore Theorem Q in Section 14 completes the proof.

31. The uniqueness of groups of type $\tilde{M}(22)$

Define a finite group G to be of type $\tilde{M}(22)$ if G possesses an involution d such that $H = C_G(d)$ is quasisimple with $H/\langle d \rangle \cong U_6(2)$ and d is not weakly closed in H with respect to G. In this section we establish the uniqueness of groups of type $\tilde{M}(22)$ by proving

Theorem 31.1. *Let G be a finite group of type $\tilde{M}(22)$ and d the involution supplied by the defining hypotheses. Then $D = d^G$ is a set of 3-transpositions of G and G is of type $M(22)$, so all groups of type $\tilde{M}(22)$ are isomorphic.*

We establish Theorem 31.1 via a series of reductions. Throughout this section we assume the hypotheses of Theorem 31.1. Thus d is an involution of G, $H = C_G(d)$, H is quasisimple, $\tilde{H} = H/\langle d \rangle \cong U_6(2)$, and d is not weakly closed in H with respect to G.

We recall from Section 23 that $\tilde{H} = H_0/Z(H_0)$ where

$$H_0 = SU(V) \cong SU_6(2)$$

is the special unitary group of a 6-dimensional unitary space V over the field F of order 4. Thus V is a projective space for $H_0/Z(H_0) = \tilde{H}$.

Let U be a maximal totally singular subspace of V, so that $\dim_F(U) = 3$ and from the discussion preceding 23.3, $\tilde{K} = N_{\tilde{H}}(U)$ is a maximal parabolic subgroup of $\tilde{H}$ with $\tilde{K}$ the split extension of $\tilde{A} = O_2(\tilde{K}) \cong E_{2^9}$ by $L_3(4)$.

From 23.2, $\tilde{H}$ has three classes $\mathcal{J}_m$, $1 \le m \le 3$, of involutions, where $j_m \in \mathcal{J}_m$ has m Jordon blocks of size 2 on V. In particular $\mathcal{J}_1$ is the set of transvections of $\tilde{H}$ and a set of 3-transpositions of $\tilde{H}$. Also

$$\mathcal{I}_m = \mathcal{J}_m \cap \tilde{A} \ne \emptyset$$

and by 23.3

(31.2)   *The sets $\mathcal{I}_m$, $1 \le m \le 3$, are the orbits of K on $\tilde{A}^\#$ of length* 21, 210, 280, *respectively, and described in Lemma* 22.2.

(31.3)

    (1) $\tilde{K}$ is irreducible on $\tilde{A}$.
    (2) $A \cong E_{2^{10}}$.
    (3) $A = J(S)$ for $A \le S \in \mathrm{Syl}_2(H)$.
    (4) $[K, A] = A$.

Proof: Parts (1) and (3) are 23.4.1 and 23.4.2; parts (2) and (4) are 23.5.1 and 23.5.5.

    Let $M = N_G(A)$ and $D = d^G$.

(31.4) $A \cap D = d^M \ne \{d\}$.

Proof: By 31.3.3 and 2.8.2, $A \cap D = d^M$. As d is not weakly closed in H with respect to G and $\tilde{A} \cap \mathcal{J} \ne \emptyset$ for each class $\mathcal{J}$ of involutions of $\tilde{H}$, d is not weakly closed in A with respect to G.

(31.5)

    (1) Each involution in $\tilde{H}$ lifts to an involution in H.
    (2) d is not a square in G.

Proof: See 23.5.2 and 23.5.3.

(31.6)   *If $a \in A$ with $\tilde{a} \in \mathcal{J}_1$ then*

    (1) $ad \notin a^H$, and
    (2) a^H is a set of 3-transpositions of H.

Proof: See 23.5.4.

(31.7) $A \cap D = \{d\} \cup a^K$ *is of order 22 for some* $a \in A$ *with* $\tilde{a} \in \mathcal{J}_1$.

Proof: Let $I_0 = \{d\}$ and I_m the set of $a \in A$ with $\tilde{a} \in \mathcal{I}_m$. By 31.2 and 31.6, $I_1 = I_{1,1} \cup I_{1,2}$ is the union of two orbits of length 21 under K.

Next if $j \in I_2$ then from 31.2 and 22.2, $j = ab$ or abd for some $a, b \in I_{1,1}$. As K acts 2-transitively as $L_3(4)$ on $I_{1,1}$ there is $t \in K$ with cycle (a, b) and $C_K(\tilde{j}) = C_K(\langle \tilde{a}, \tilde{b} \rangle)\langle t \rangle$. By 31.6,

$$C_K(\langle \tilde{a}, \tilde{b} \rangle) = C_K(\langle a, b \rangle) \le C_K(ab)$$

and of course $t \in C_K(ab)$, so $C_K(\tilde{j}) = C_K(ab) = C_K(j)$. Thus $I_2 = I_{2,1} \cup I_{2,2}$ is the union of two K-orbits of length 210 by 31.2.

Finally by 31.2 either K is transitive on I_3 of length 560 or $I_3 = I_{3,1} \cup I_{3,2}$ is the union of orbits of length 280.

Next $j_m \in \mathcal{J}_m$ is a square in $\tilde{H}$ for $m = 1, 2$, so some member of I_m is a square in H. Hence by 31.5.2, at most one of $I_{m,1}$ and $I_{m,2}$ is contained in D.

Now $A \cap D = \bigcup_{J \in \mathcal{O}} J$ for some collection $\mathcal{O}$ of orbits of K on $A^\#$ with $\{d\} \in \mathcal{O}$ and $|\mathcal{O}| > 1$ by 31.4. Further by 31.4, M is transitive on $A \cap D$, so $n = |A \cap D|$ divides the order of $GL_{10}(2)$. As the prime divisors of $|GL_{10}(2)|$ are $2, 3, 5, 7, 11, 17, 31, 73$, and 127, n is a product of powers of these primes. Further $n = \sum_{J \in \mathcal{O}} |J|$ with the possible lengths $|J|$ of orbits discussed previously and with at most one of $I_{m,1}$ and $I_{m,2}$ in $\mathcal{O}$ for $m = 1, 2$. We conclude that one of the following holds:

(a) $A \cap D = \{d\} \cup I_{1,1}$ is of order 22.
(b) $A \cap D = \{d\} \cup I_3$ is of order $561 = 3 \cdot 11 \cdot 17$.
(c) $A \cap D = \{d\} \cup I_{1,1} \cup I_{2,1} \cup I_{3,1}$ is of order $512 = 2^9$.

In case (a) the lemma holds, so we must eliminate (b) and (c).

If (c) holds then $T \in \mathrm{Syl}_2(M)$ is transitive on $A \cap D$. But as $\tilde{K}$ is irreducible on $\tilde{A}$ by 31.3, $\tilde{A} = \langle \mathcal{I}_3 \rangle$, so $A = \langle A \cap D \rangle = \langle d^T \rangle$. Thus if B is a T-invariant hyperplane of A, $d^T \subseteq A - B$, so as $|d^T| = 2^9 = |A - B|$, $B = A - d^T \trianglelefteq M$. This is impossible by 31.3.4.

Finally assume case (b) holds and let $X \in \mathrm{Syl}_{11}(M)$ and $M^* = M/A$. As $K = C_M(d)$,

$$|M| = |A \cap D||K| = 2^{16} \cdot 3^3 \cdot 5 \cdot 7 \cdot 1 \cdot 17.$$

Then $|M : X| \equiv -2 \bmod 11$, so by Sylow's Theorem,

$$|N_M(X) : X| \equiv -2 \bmod 11.$$

On the other hand X is irreducible on A so

$$N_M(X) \cong N_{M^*}(X^*) \le N_{GL(A)}(X^*)$$

and $N_{GL(A)}(X^*)$ is isomorphic to $GF(2^{10})^{\#} \cong \mathbf{Z}_{2^{10}-1}$ extended by

$$\text{Aut}(GF(2^{10})) \cong \mathbf{Z}_{10},$$

so $|N_M(X) : X|$ divides 30. This contradicts $|N_M(X) : X| \equiv -2 \bmod 11$.

Let $\mathcal{D}$ be the commuting graph on D and as usual let $D_d = C_D(d) - \{d\}$ and $A_d = D - C_D(d)$. By 31.6 and 31.7

(31.8) *H is transitive on D_d and D_d is a set of 3-transpositions of H.*

(31.9) *D is a set of 3-transpositions of G and $\mathcal{D}$ is connected.*

Proof: We appeal to 24.4.2. To do so we first prove that if $b \in D_d$ and $a \in D_b \cap A_d$ then $c^{\perp} \cap D_{a,d} \neq \emptyset$ for all $c \in D_d$. Let $G_b = C_G(b)$ and $\hat{G}_b = G_b/\langle b \rangle$. Then as $\hat{D}_b$ is a set of transvections in $\hat{G}_b \cong U_6(2)$, $\langle \hat{D}_{b,d,a} \rangle \cong U_4(2)$. Thus by Exercise 8.2,

$$\langle D_{d,a} \rangle / Z(\langle D_{d,a} \rangle) \cong PO_6^{+,\pi}(3)$$

and then by Exercise 8.3, $c^{\perp} \cap D_{a,d} \neq \emptyset$ for each $c \in D_d$, as desired.

It remains to verify the remaining hypotheses of 24.4.2. First 31.8 establishes hypothesis (i) of 24.4. Indeed if $\mathcal{D}$ is connected then 24.4.1 now completes the proof so we may assume the connected component Δ of $\mathcal{D}$ containing d is proper in D. Then by induction on the order of G, $Y = N_G(\Delta)$ is of type $M(22)$ and type $\tilde{M}(22)$, and Δ is the set of 3-transpositions of Y. In particular by 25.7, $M/A \cong M_{22}$ with A the Todd module for M/A, so M is irreducible on A by 22.5 and M/A has one class of involutions, each of which fixes a point of $T \cap D$. Thus $Y \cap D = \Delta$, so for $A \leq T \in \text{Syl}_2(Y)$, we have $A = \langle T \cap D \rangle$, and of course A is abelian.

As M is irreducible on A,

$$A = \langle ab : a, b \in T \cap D \rangle.$$

By 25.7.5 applied to Y, the map $\{a, b\} \mapsto ab$ is injective on 2-subsets of $T \cap D$. As a subgroup of M of order 11 has no fixed points on A, $C_M(t)$ is intransitive on $T \cap D$ for each involution $t \in A$. Finally we have already observed each involution in M fixes a member of $T \cap D$. Thus we have verified all hypotheses of 24.4.2, completing the proof of the lemma.

We are now in a position to complete the proof of Theorem 31.1. Namely by 31.9, D is a set of 3-transpositions of G and $\mathcal{D}$ is connected. Therefore G is primitive on D by 9.4 and then 15.11 shows G is determined up to isomorphism and completes the proof.

32. The uniqueness of groups of type $\tilde{M}(23)$

Define a finite group G to be of *type* $\tilde{M}(23)$ if G possesses an involution d such that $H = C_G(d)$ is quasisimple with $H/\langle d \rangle$ of type $M(22)$ and $\tilde{M}(22)$, and d is not weakly closed in H with respect to G. In this section we establish the uniqueness of groups of type $\tilde{M}(23)$ by proving

Theorem 32.1. *Let G be a finite group of type $\tilde{M}(23)$ and d the involution supplied by the defining hypotheses. Then $D = d^G$ is a set of 3-transpositions of G and G is of type $M(23)$, so all groups of type $\tilde{M}(23)$ are isomorphic.*

The proof of Theorem 32.1 is much like that of Theorem 31.1; in particular we sometimes only sketch arguments that are essentially the same as analogous arguments in Section 31.

Throughout this section we assume the hypotheses of Theorem 32.1. Thus d is an involution of G, $H = C_G(d)$, H is quasisimple, $\tilde{H} = H/\langle d \rangle$ is of type $M(22)$ and $\tilde{M}(22)$, and d is not weakly closed in H with respect to G.

Let $\mathcal{J}_1$ be the set of 3-transpositions of $\tilde{H}$. We find in 37.4 that $\tilde{H}$ has three classes $\mathcal{J}_m$, $1 \leq m \leq 3$, of involutions, where $j_m \in \mathcal{J}_m$ is the product of m commuting members of $\mathcal{J}_1$. Let $S \in \mathrm{Syl}_2(H)$, $\tilde{A} = \langle \mathcal{J}_1 \cap S \rangle$, and $\tilde{K} = N_{\tilde{H}}(\tilde{A})$. By 25.7, $\tilde{A} \cong E_{2^{10}}$ and $\tilde{A}$ is the Todd module for $K/A \cong M_{22}$. Let $\mathcal{I}_m = \mathcal{J}_m \cap \tilde{A}$. By 2.8, $\mathcal{I}_m$, $1 \leq m \leq 3$, are the orbits of $\tilde{K}$ on $\tilde{A}^{\#}$. Indeed collecting these remarks and appealing to 22.3

(32.2) $\tilde{A}$ *is the Todd module for* $\tilde{K}/\tilde{A} \cong M_{22}$. *The sets* $\mathcal{I}_m$, $1 \leq m \leq 3$, *are the orbits of K on $\tilde{A}^{\#}$ of length 22, 231, 770, respectively, and described in Lemma 22.3.*

(32.3)

(1) $\tilde{K}$ is irreducible on $\tilde{A}$.

(2) $A \cong E_{2^{11}}$.

(3) $A = J(S)$ for $A \leq S \in \mathrm{Syl}_2(H)$.

Proof: Part (1) is 22.5 and (2) is 23.8.2. By Exercise 11.1, $\tilde{A} = J(\tilde{S})$, so (2) implies (3).

Let $M = N_G(A)$ and $D = d^G$.

(32.4)

(1) $A \cap D = d^M \neq \{d\}$.

(2) Each involution in $\tilde{H}$ lifts to an involution in H.

(3) d is not a square in G.

(4) If $a \in A$ with $\tilde{a} \in \mathcal{J}_1$ then $ad \notin a^H$, and a^H is a set of 3-transpositions of H.

Proof: The proof of (1) is the same of that of 31.4, while 23.8 implies the remaining parts.

(32.5) $A \cap D = \{d\} \cup a^K$ *is of order 23 for some $a \in A$ with $\tilde{a} \in \mathcal{J}_1$.*

Proof: The proof is very much like that of 31.7. Let $I_0 = \{d\}$ and I_m the set of $a \in A$ with $\tilde{a} \in \mathcal{I}_m$. By 32.2 and 32.4, $I_1 = I_{1,1} \cup I_{1,2}$ is the union of two orbits of length 22 under K, and arguing as in 31.7, $I_2 = I_{2,1} \cup I_{2,2}$ is the union of two K-orbits of length 231. Similarly either K is transitive on I_3 of length 1,540 or $I_3 = I_{3,1} \cup I_{3,2}$ is the union of orbits of length 770.

Next from 31.7, $j_2 \in \mathcal{J}_2$ is a square in $\tilde{H}$, so some member of I_2 is a square in H and hence by 32.4.3, at most one of $I_{2,1}$ and $I_{2,2}$ is contained in D.

We can now argue as in 31.7 to conclude that one of the following holds:

(a) $A \cap D = \{d\} \cup I_{1,1}$ is of order 23.

(b) $A \cap D = \{d\} \cup I_1$ is of order 45.

(c) $A \cap D = \{d\} \cup I_{1,1} \cup I_{2,1} \cup I_{3,1}$ is of order $1,024 = 2^{10}$.

(d) $A \cap D = \{d\} \cup I_{1,1} \cup I_{2,1}$ is of order $254 = 2 \cdot 127$.

(e) $A \cap D = \{d\} \cup I_1 \cup I_{2,1}$ is of order $276 = 2^2 \cdot 3 \cdot 23$.

In case (a) the lemma holds, so we must eliminate the remaining cases.

If (c) holds then arguing as in 31.7, there is an M-invariant hyperplane B of A with $A \cap D = A - B$. But then 23.8.6 supplies a contradiction, so case (c) is eliminated.

Next assume case (b) or (d) holds. Then M is rank 3 on $A \cap D$ and we use the theory of rank 3 groups such as in 6.3 or in Section 16 of [FGT] to obtain a contradiction. Namely we have M of degree 45 or 254 with M_d 2-transitive on a suborbit Δ of degree $k = 22$ and $k + 1$ and $l + 1$ do not divide the degree, so M is primitive on $A \cap D$ and $\lambda = 0$. Now 6.3 supplies a contradiction.

Finally assume case (e) holds and let $X \in \mathrm{Syl}_{23}(M)$ and $M^* = M/A$. Then

$$|M^* : X^*| \equiv 17 \bmod 23,$$

so by Sylow's Theorem, $|N_{M^*}(X^*) : X^*| \equiv 17 \bmod 23$. On the other hand X is irreducible on A so $N_{M^*}(X^*) \leq N_{GL(A)}(X^*)$ and $N_{GL(A)}(X^*)$ is isomorphic to $GF(2^{11})^\# \cong \mathbf{Z}_{2^{11}-1}$ extended by $\mathrm{Aut}(GF(2^{11})) \cong \mathbf{Z}_{11}$, so $|N_M(X) : X|$ divides 11. This contradicts $|N_{M^*}(X^*) : X^*| \equiv 17 \bmod 23$.

Let $\mathcal{D}$ be the commuting graph on D and as usual let $D_d = C_D(d) - \{d\}$ and $A_d = D - C_D(d)$. By 32.4.4 and 32.5

(32.6) *H is transitive on D_d and D_d is a set of 3-transpositions of H.*

(32.7) *D is a set of 3-transpositions of G and $\mathcal{D}$ is connected.*

Proof: The proof is much like that of 31.9; again we appeal to 24.4.2. We first prove that if $b \in D_d$ and $a \in D_b \cap A_d$ then $c^\perp \cap D_{a,d} \neq \emptyset$ for all $c \in D_d$. Let $G_b = C_G(b)$ and $\hat{G}_b = G_b/\langle b \rangle$. Then as $\hat{D}_b$ is a set of transvections in $\hat{G}_b$ with $G_b/\langle b \rangle$ of type $M(22)$, $\langle D_{b,d,a}, b \rangle/\langle b \rangle \cong PO_6^{+,\pi}(3)$. Then by Exercise 8.2, $\langle D_{d,a} \rangle/Z(\langle D_{d,a} \rangle) \cong PO_7^{-\pi,\pi}(3)$. Therefore by Exercise 8.3, $c^\perp \cap D_{a,d} \neq \emptyset$ for each $c \in D_d$, as desired.

To verify the remaining hypotheses of 24.4.2 we argue as in 31.9. Notice that as $T \cap D = A \cap D$ is of odd order 23 for $A \leq T \in \mathrm{Syl}_2(G)$, each involution in T fixes a member of $T \cap D$, establishing hypothesis (3) of 24.3.

We are now in a position to complete the proof of Theorem 32.1. Namely by 32.7, D is a set of 3-transpositions of G and $\mathcal{D}$ is connected. Therefore G is primitive on D by 9.4 and then 15.14 says G is of type $M(23)$ and determined up to isomorphism, completing the proof.

33. The uniqueness of groups of type $\tilde{M}(24)$

In this section we provide a characterization of $M(24)$ by proving

Theorem 33.1. *Let G be a finite group, d an involution in G, $H = C_G(d)$, and assume*

(1) $C_G(d) \cong \mathbf{Z}_2 \times E(C_G(d))$ *with* $E(C_G(d))$ *of type* $M(23)$ *and* $\tilde{M}(23)$.
(2) *There is* $E_{2^{12}} \cong A \leq G$ *with* $d \in A$ *and* A *is the 12-dimensional Todd module for* $N_G(A)/A \cong M_{24}$.
(3) $|C_G(t) : C_H(t)| = 2$ *for* t *a 3-transposition in* $E(H)$.

Then $D = d^G$ is a set of 3-transpositions of G and G is of type $M(24)$.

We say G is of *type* $\tilde{M}(24)$ if G satisfies the hypotheses of Theorem 33.1. Thus Theorem 33.1 and Fischer's Theorem say all groups of type $\tilde{M}(24)$ are isomorphic. We could eliminate hypotheses (2) and (3) in the definition of groups of type $\tilde{M}(24)$ at the expense of requiring that d not be weakly closed in H with respect to G, adding as conclusions the cases $G/O(G)$ isomorphic to the wreath product $M(23)$ *wr* $\mathbf{Z}_2$ or $M(24)$ with d inverting $O(G)$, and working much harder. Since the weak characterization of $M(24)$ of this section will be sufficient to establish the stronger characterization of $M(24)$ and F_{24} of Theorems 35.1 and 34.1, it seems best to include hypotheses (2) and (3).

The proof of Theorem 33.1 is much like that of Theorems 30.1, 31.1, and 32.1; in particular we sometimes only sketch arguments that are essentially the same as analogous arguments in those proofs.

Throughout this section we assume the hypotheses of Theorem 33.1 and we let $M = N_G(A)$, $D = d^G$, $\mathcal{D}$ the commuting graph on D, and $D_d = C_D(d) - \{d\}$.

We find in 37.4 that $E(H)$ of type $M(23)$ has three classes $\mathcal{J}_m$, $1 \le m \le 3$, of involutions, where $\mathcal{J}_1$ is the set of 3-transpositions of $E(H)$ and $j_m \in \mathcal{J}_m$ is the product of m commuting members of $\mathcal{J}_1$. Thus

(33.2)   H has seven classes of involutions: $\{d\}$, $\mathcal{J}_m$, *and* $d\mathcal{J}_m$, $1 \le m \le 3$.

Let $A \le S \in \mathrm{Syl}_2(H)$. From 32.3.3 and as $|\mathcal{J}_1 \cap S|$ is odd, $\mathcal{A}(S) = \{A_1\}$ with $A_1 \cap E(H) = \langle \mathcal{J}_1 \cap S \rangle$, $A_1 \cong E_{2^{12}}$, and $N_H(A_1)/A_1 \cong M_{23}$. So as $A \cong A_1$ we conclude $A = A_1 = J(S)$. Therefore our usual argument (cf. 31.4) shows $A \cap D = d^M$.

Now by hypothesis, A is the Todd module for $M/A \cong M_{24}$, so from 22.1, M has four orbits $\mathcal{I}_m$, $1 \le m \le 4$, with M 5-transitive on $\mathcal{I}_1$ of order 24, the map $\{a_1, \ldots, a_m\} \mapsto a_1 \cdots a_m$ is injective on m-subsets of $\mathcal{I}_1$ for $m \le 3$, and the stabilizer of a member of $\mathcal{I}_4$ in M/A is not M_{23}. Therefore as $C_M(d)/A \cong N_H(A)/A \cong M_{23}$, $A \cap M = d^M = \mathcal{I}_1$ and M is 5-transitive on d^M of order 24. Therefore $D_d \cap A = (d\mathcal{J}_1) \cap A$. Further by 33.2 and as each member of $\mathcal{J}_m$ is the product of m commuting members of $\mathcal{J}_1$, each involution in H is fused into A under H. So as $A \cap D = \mathcal{I}_1 = \{d\} \cup (d\mathcal{J}_1) \cap A$ and as $\mathcal{J}_1 \cap S \subseteq A$, we conclude $D_d = d\mathcal{J}_1$ is a set of 3-transpositions of H. We record all this as

(33.3)

(1) $A = J(S)$ *for* $A \le S \in \mathrm{Syl}_2(H)$.

(2) $D_d = d\mathcal{J}_1$ *is a conjugacy class of 3-transpositions of H.*

(3) *The map* $\{a_1, \ldots, a_m\} \mapsto a_1 \cdots a_m$ *is an injection of the set of m-subsets of* $A \cap D$ *into A for* $m \le 3$.

(33.4)   D is a set of 3-transpositions of G and $\mathcal{D}$ is connected.

Proof: The proof is similar to that of 31.9 and appeals to 24.4.1 and 24.3. We first observe that D is a set of odd transpositions of G. For let $b \in D_d \cap A$ and $t = db$. By 24.1 it suffices to prove $C_D(t) \subseteq C_H(t)$. By Hypothesis (3) of Theorem 33.1, $|C_G(t) : C_H(t)| = 2$, so $C_H(t) = \langle d \rangle \times I \trianglelefteq C_G(t)$, where $I = C_{E(H)}(t)$ with $I/\langle t \rangle$ of type $M(22)$. Let $s \in C_D(t)$. By 37.3, s fixes a 3-transposition $u\langle t \rangle$ in $I/\langle t \rangle$, so s acts on $\langle u, d, b \rangle$. Now $u = dc$ for some $c \in D_{d,b}$ and conjugating in $C_H(t)$ we may take $c \in A$. Then by 33.3.3, $\{d, b, c\} = \langle u, d, b \rangle \cap D$, so s centralizes $e \in \{d, b, c\}$. Then as D_e is a set of 3-transpositions of $C_G(e)$, s centralizes d, as desired.

We next prove that if $b \in D_d$ and $a \in D_b \cap A_d$ then $c^{\perp} \cap D_{a,d} \neq \emptyset$ for all $c \in D_d$. Let $G_b = C_G(b)$ and $\hat{G}_b = C_G(b)/\langle b \rangle$. Then by 33.3 and the definition of $M(23)$, $\langle D_{b,d,a} \rangle \cong \mathbf{Z}_2 \times PO_7^{\pi,\pi}(3)$. Then by Exercise 8.2, $K = \langle D_{d,a} \rangle \cong PO_8^{+,\pi}(3)$ or $S_3/P\Omega_8^+(3)$. Now by 16.6, $E(K)$ is determined up to conjugacy in H and then by 25.5, $N_H(E(K))$ is a D-subgroup of H with $N_H(E(K))/E(K)\langle d \rangle \cong S_3$. As $C_D(E(K)) \cup (N_D(E(K)) - C_D(E(K)))$ is an $N_G(E(K))$-invariant partition of $N_D(E(K))$ and D is a set of odd transpositions of G, it follows from 24.2.4 and 7.3 that $a \in C_D(E(K))$ centralizes $N_{D_d}(E(K)) \subseteq N_D(E(K)) - C_D(E(K))$, so $K \cong S_3/P\Omega_8^+(3)$. Now by Exercise 8.3, $c^{\perp} \cap D_{a,d} \neq \emptyset$ for each $c \in D_d$, as desired.

Next let Δ be the connected component of $\mathcal{D}$ containing d. If $\mathcal{D}$ is connected we are done by 24.4.1, so assume otherwise. Then by induction on the order of G, $Y = N_G(\Delta)$ is of type $M(24)$ and Δ is the set of 3-transpositions of Y. We now verify the remaining hypotheses of 24.3 to complete our proof.

First by 37.4 each involution in Y fixes a point of Δ. Thus $\Delta = Y \cap D$ and $A = \langle T \cap D \rangle$ for $A \leq T \in \mathrm{Syl}_2(G)$. We have seen $24 = |A \cap D|$, so hypothesis (1) of 24.3 is satisfied. By 33.3.3, the map $\{abc\} \mapsto abc$ is an injection on 3-subsets of $T \cap D$, so $C_M(abc)$ is intransitive on $T \cap D$. Hence we have established hypothesis (2) of 24.3 and completed the proof of the lemma.

We are now in a position to complete the proof of Theorem 33.1. Namely by 33.4, D is a set of 3-transpositions of G and $\mathcal{D}$ is connected. Therefore G is primitive on D by 9.4 and then 15.14 says G is of type $M(24)$ and completes the proof.

34. Groups of type F_{24}

In this section we begin the proof of a theorem that will establish the uniqueness of groups of type F_{24}:

Theorem 34.1. *Let G be a finite group, z an involution in G, $H = C_G(z)$, and assume $Q = F^*(H) \cong 2^{1+12}$ with $H/Q \cong \mathbf{Z}_2/U_4(3)/\mathbf{Z}_3$, and z is not weakly closed in Q with respect to G. Then G is of type $M(24)'$, so in particular G is determined up to isomorphism.*

A group satisfying the hypotheses of Theorem 34.1 will be said to be of *type* F_{24}. Notice G is of type F_{24} precisely when G satisfies Hypothesis

$$\mathcal{H}(6, \mathbf{Z}_2 / U_4(3) / \mathbf{Z}_3)$$

of [SG]. Moreover by 32.4 in [SG], if F is the Monster then there is a subgroup Y of F of order 3 such that $C_F(Y)/Y$ is of type F_{24}.

G is of *type* $M(24)'$ if G is isomorphic to the commutator group of a group of type $M(24)$.

We establish Theorem 34.1 via a series of reductions. Throughout this section we assume the hypotheses of Theorem 34.1. In addition let $\tilde{H} = H/\langle z \rangle$ and $H^* = H/Q$. By hypothesis

$$X^* = O_3(H^*) \cong \mathbf{Z}_3.$$

Let X be a Sylow 3-subgroup of the preimage of X^* in H.

As $Q \cong 2^{1+12}$ is extraspecial, the commutator map and power map on Q induce an orthogonal space structure on $\tilde{Q}$ over $GF(2)$ preserved by H^* in which elementary abelian subgroups U of Q correspond to totally singular subspaces $\tilde{U}$ of the orthogonal space $\tilde{Q}$ (cf. Section 8 in [SG]). Write $O(\tilde{Q})$ for the isometry group of this orthogonal space structure. Hence by 29.9,

(34.2) $C_{H^*}(X^*)$ *preserves a unitary space structure on* $\tilde{Q}$ *over* $GF(4)$ *and is determined up to conjugation in* $O(\tilde{Q})$.

Because $C_{H^*}(X^*)$ is determined up to conjugation in $O(\tilde{Q})$, it follows that we may identify the unitary space $\tilde{Q}$ with the unitary space $\tilde{\Lambda}$ of Section 29 and $H^* \leq H^+ = N_{O(\tilde{Q})}(X^*)$. By 29.10, there is a quasiequivalence of the action of H^+ on $\tilde{Q}$ with the action of the group $\tilde{G}^+$ of Section 29 on the space $\tilde{\Lambda}$ of Section 29. Therefore as $|\tilde{G}^+| = 2|H^*|$, we conclude

(34.3) H^* *is of index 2 in* H^+ *and there is a quasiequivalence of the action of* H^+ *on* $\tilde{Q}$ *with the action of* $\tilde{G}^+$ *on* $\tilde{\Lambda}$.

Next by 34.3 and 28.20.2 and 28.20.6, H^* has two orbits $\tilde{P}_2$ and $\tilde{P}_4$ on the singular points of the unitary space $\tilde{Q}$ and hence two orbits $\tilde{\Lambda}_2$ and $\tilde{\Lambda}_4$ on the singular vectors of the orthogonal space $\tilde{Q}$. As these vectors correspond to the involutions in $Q - \langle z \rangle$, H has two orbits on such involutions.

By hypothesis, there is $s = z^g \in Q - \langle z \rangle$. Let t be a representative for the second orbit of H on involutions in $Q - \langle z \rangle$. Let H^- be the subgroup of H^+ corresponding to the group $\tilde{G}^-$ of Section 29 under the quasiequivalence $\tilde{G}^+ \cong H^+$ of 34.3.

(34.4)

   (1) $H^* = H^-$.

   (2) $s^H = z^G \cap Q - \{z\}$.

   (3) $\tilde{s} \in \tilde{\Lambda}_4$, $C_{H^*}(\tilde{s}) = C_H(s)^* \cong A_6/E_{32}$, $(Q^g \cap H)^* = O_2(C_H(s))^*$, and $Q \cap Q^g \cong E_{2^7}$.

   (4) $C_{H^*}(\tilde{t}) \cong O_6^-(2)$.

Proof: Let $m+1 = m(Q \cap Q^g)$. By 8.15.8 in [SG], $m_2((Q^g \cap H)^*) = 12-m-1$, $(Q^g \cap H)^* \trianglelefteq C_{H^*}(\tilde{s})$, and $m \le 6$. Therefore $m_2(O_2(C_{H^*}(\tilde{s}))) \ge 5$, so by 29.6 and 34.3, (1) holds, $\tilde{s} \in \tilde{\Lambda}_4$ so (2) holds, and

$$(Q^g \cap H)^* = O_2(C_{H^*}(\tilde{s}) \cong E_{32}.$$

Therefore $m = 6$ and 29.1 completes the proof of (3). Then (4) follows from 29.5.

Let $T \in \mathrm{Syl}_2(H)$. Then $\langle z \rangle = Z(T)$, so $T \in \mathrm{Syl}_2(G)$. By 29.7.4, $C_{\tilde{Q}}(T)$ is generated by a member of $\tilde{\Lambda}_4$, which by 34.4 we may take to be $\tilde{s}$.

(34.5) $\mathcal{A}(T) = \{A\}$ with $A \cong E_{2^{11}}$, $m(A \cap Q) = 7$, $A^* \cong E_{16}$, $N_{H^*}(A^*)/A^* \cong \hat{S}_6$, and $N_H(A)/A \cong \hat{S}_6/E_{64}$.

Proof: Let U_2^* be the subgroup of T^* corresponding to the group U_2 of 29.7 under the quasiequivalence $\tilde{G}^+ \cong H^+$, and let $\tilde{A}_2 = C_{\tilde{Q}}(U_2^*)$ and $K^* = N_{H^*}(U_2^*)$. Then by 29.7.3, $A_2 \cong E_{2^7}$ and $K^* \cong \hat{S}_6/E_{2^4}$ with $U_2^* = C_{H^*}(\tilde{A}_2) = O_2(K^*)$. Next if U_2 is the preimage of U_2^* in H then $U_2 = C_{U_2}(X)Q$ and $K = N_K(X)Q$ with $\langle z \rangle = C_Q(X)$. As K is transitive on $U_2^{*\#}$, $C_{U_2}(X) \cong E_{32}$.

Next as $[U_2, \tilde{A}_2] = 1$, U_2 induces a group of transvections with center z on A_2, and then as Q induces the full group of such transvections, $U_2 = AQ$ where $A = C_{U_2}(A_2) = C_H(A_2)$ and $A \cap Q = C_Q(A_2) = A_2$. As $A = C_A(X)A_2$, $C_A(X) = C_{U_2}(X)$, so $A = C_{U_2}(X)A_2 \cong E_{2^{11}}$.

Now suppose $z \in B \le T$ with $B \cong E_{2^k}$ and $k \ge 11$. Then

$$m(B^*) + m(\widetilde{B \cap Q}) \ge k-1 \ge 10,$$

so by 29.8, $B^* = U_2^*$ and $B \cap Q = A_2$. Then $B \le C_H(A_2) = A$. Therefore $\{A\} = \mathcal{A}(T)$ and the proof of the lemma is complete.

In the remainder of this section let $A = J(T)$ and $M = N_G(A)$, so that $A \cong E_{2^{11}}$ is described in 34.5.

(34.6)   *Let $V_3 = \langle s^{Q^X} \rangle$. Then*

(1) *$\tilde{V}_3$ is the point of the unitary space $\tilde{Q}$ containing $\tilde{s}$ and $E_8 \cong V_3 = \langle z, z^g, z^k \rangle = Q \cap Q^g \cap Q^k$ for some $k \in G$.*

(2) *$N_G(V_3) = L_1 L_2$ where $L_1 = \langle Q, Q^g, Q^k \rangle$, $L_2 = C_G(V_3)$, $R = O_2(N_G(V_3)) = L_1 \cap L_2$ is special with $Z(R) = V_3$ and R/V_3 the tensor product of natural modules for $L_1/R \cong L_3(2)$ and $L_2/R \cong A_6$.*

Proof: Of course $V_3 \cong E_8$ and $\tilde{V}_3$ is the point of the unitary space $\tilde{Q}$ containing $\tilde{s}$ as X induces scalar multiplication on that space. Let $P = Q^g \cap H$. By 34.4.3, $Q \cap Q^g \cong E_{2^7}$ and $P^* = O_2(C_{H^*}(\tilde{s}))$. Then $[Q \cap Q^g, P] \le \langle s \rangle$, so from 29.2 and 29.4, $Q \cap Q^g \le \tilde{W} = \tilde{V}_3^{\perp}$, the subspace of the unitary space $\tilde{Q}$ orthogonal to $\tilde{V}_3$. Further there is $u \in P$ inverting X so if $v \in W - V_3$ and $\tilde{v} \notin C_{\tilde{Q}}(u)$ then $[\tilde{v}, u] \notin \tilde{V}_3$ and hence $[v, u] \ne s$. Therefore $Q \widetilde{\cap} Q^g = \tilde{V}_3 C_{\tilde{W}}(u)$. In particular $V_3 \le Q \cap Q^g = V_3 W_3$, where $\tilde{W}_3 = C_{\tilde{W}}(u)$.

Next $V_3 = \langle z, s, s^x \rangle$ for $X = \langle x \rangle$. Notice $s^x = z^k$ for $k = gx$ and $V_3 = V_3^x \le Q^{gx} = Q^k$. Further $C_{\tilde{W}}(u) \cap C_{\tilde{W}}(u)^x = 1$, so $V_3 = Q \cap Q^g \cap (Q \cap Q^g)^x = Q \cap Q^g \cap Q^k$ and (1) holds.

Now by (1) and 8.16 in [SG], $R_1 = C_{L_1}(V_3) = O_2(L_1)$ with $L_1/R_1 \cong L_3(2)$. Also by 8.16 in [SG], $L_1 \trianglelefteq N_G(V_3) = L_1 L_2$ with $N_G(V_3)/R_1 = L_1/R_1 \times L_2/R_1$. On the other hand $L_2/O_2(L_2) \cong A_6$ is irreducible on $(Q^g \cap Q)/V_3$ and $(Q \cap Q^k)/V_3$ of rank 4, so by 8.16.4 in [SG], $L_1 L_2$ has a chief section on R_1/V_3 of the form $V_3^* \otimes N$, where N is a natural 4-dimensional module for $L_2/O_2(L_2) \cong A_6$ and V_3^* is the dual of V_3 as an L_1/R_1-module. Therefore as $R_1 \le R$ and

$$|R| \le \frac{|T|}{|L_3(2) \times A_6|_2} = 2^{15} = |V_3||V_3^* \otimes N|,$$

we conclude $R = R_1$ with $R/V_3 \cong V_3^* \otimes N$ as an $L_1 L_2/R$-module. In particular $L_1 L_2$ is irreducible on R/V_3, so R_3 is special with center V_3.

(34.7)   *Let $K = N_M(V_3)$. Then $K = L_1 K_2$, where L_1 is defined in Lemma 34.6, $K_2 = C_M(V_3)$, $K/O_2(K) = L_1/O_2(K) \times K_2/O_2(K) \cong L_3(2) \times S_3$, $O_2(K)/A \cong E_{64}$ is the tensor product of natural modules for the factors of $K/O_2(K)$, and K has chief series $V_3 < A_3 < A$ on A with $A_3/V_3 \cong O_2(K)/A$ as a $K/O_2(K)$-module and $L_1 O_2(K) = C_K(A/A_3)$.*

Proof: Adopt the notation of the previous lemma. Then AR/R is a 4-subgroup of L_2/R, so as $A = J(T)$, a Frattini argument says

$$K/R = N_{L_1 L_2/R}(AR/R) = L_1/R \times N_{L_2/R}(AR/R) \cong L_3(2) \times S_4.$$

Then K_2 is the preimage in L_2 of $N_{L_2/R}(AR/R)$, $K/O_2(K) \cong L_3(2) \times S_3$, and $A/A_3 \cong E_4$, where $A_3 = A \cap R$. Finally $R/V_3 \cong V_3^* \otimes N$ as an $L_1 L_2/R$-module and has chief factors A_3/V_3 and $(O_2(K) \cap R)/A_3 \cong O_2(K)/A$ as a K-module, so $O_2(K)/A \cong A_3/V_3 \cong V_3^* \otimes N_2$ as a $K/O_2(K)$-module, where N_2 is the 2-dimensional natural module for $K_2/O_2(K) \cong S_3$.

(34.8) $M/A \cong M_{24}$.

Proof: Let $M^* = M/A$ and $K = N_M(V_3)$. By 34.7, K^* is the split extension of $O_2(K^*) = A_1^* \cong E_{64}$ by $K_1^* = K_{1,1}^* \times K_{1,2}^*$ with $K_{1,1}^* \cong L_3(2)$, $K_{1,2}^* \cong S_3$, and A_1^* the tensor product of natural modules for the factors. Let $\langle u_2^* \rangle = T^* \cap K_{1,2}^*$ and $U^* = C_{A_1^*}(u_2^*)\langle u_2^* \rangle$. Then $E_{16} \cong U^* \trianglelefteq K_{1,1}^* A_1^* T^* = L_1^* T^*$ with $L_1^* T^*/U^*$ the stabilizer of the hyperplane $U^* \cap A_1^*$ in $GL(U^*)$.

Next let $H_M = H \cap M$ and $A_2^* = O_2(H_M^*)$. By 34.5, H_M^* is the split extension of A_2^* by $H_1^* \cong \hat{S}_6$. Now the isomorphism types of H_M^* and K^* are determined and H_M^* and K^* are isomorphic to maximal parabolics H_M' and K' in M_{24} described in Section 39 of [SG]. Therefore T^* is isomorphic to a Sylow 2-subgroup of M_{24}. In particular by 39.1 in [SG], $\mathcal{A}(T^*) = \{A_1^*, A_2^*\}$. Then

$$H_K^* = H_M^* \cap K^* = N_{H_M^*}(A_1^*) = N_{K^*}(A_2^*).$$

By 39.1.8 in [SG], $C_{H_K^*}(A_1^* \cap U^*) = A_1^* U^*$ with A_1^* and U^* the maximal elementary abelian subgroups of $A_1^* U^*$. Thus the image U' of U^* under our isomorphism of H_M^* with H_M' is uniquely determined, so as $N_{H_M'}(U') \not\leq H_K'$, $N_{H_M^*}(U^*) \not\leq H_K^*$ and then as $\mathrm{Aut}_{H_K^*}(U^*)$ is a maximal parabolic of $GL(U^*)$, it follows that $\mathrm{Aut}_{M^*}(U^*) = GL(U^*)$.

Next u_2^* inverts $Y^* = O_3(K_{1,2}^*)$ and $V_3 = C_A(Y)$, so as $[u_2, V_3] = 0$, $[A, u_2] = B = [A, Y, u_2]$ is of rank 4. Also $U^* \cap A_1^*$ centralizes the series $0 < V_3 < A_3 < A$ of 34.7, and hence centralizes the hyperplane $B_3 = A_3 \cap B$ of B. If for some $1 \neq u^* \in U^* \cap A_1^*$, $[B, u] = 0$ then as u^* is fused to u_2^* under $N_{M^*}(U^*)$ and $m(C_A(u_2)) = 7$, we have $V_3 + B = C_A(u_2) = C_A(u)$. Then as $N_{M^*}(U^*)$ is 2-transitive on $U^{*\#}$, $V_3 + B = C_A(v^*)$ for all $v^* \in U^{*\#}$. But $U^* \cap A_2^* \cong E_4$ and A_2^* induces the full group of transvections with center z on $Q \cap A$, so $C_{Q \cap A}(U^* \cap A_2^*) \neq C_{Q \cap A}(v^*)$ for $1 \neq v^* \in U^* \cap A_2^*$.

Therefore $U^* \cap A_1^*$ induces the group of transvections on B with axis B_3. Now for $1 \neq v^* \in U^* \cap A_2^*$, $z \in [A, v] \leq A \cap Q$, so $C_{M^*}([A, v]) = C_{H_M^*}([A, v])$ is a 2-group since no nontrivial element of odd order in H_M centralizes a 4-dimensional subspace of $A \cap Q$. So as u_2^* is fused to v^* in M^*, $C_{M^*}(B)$ is a 2-group and thus

$$V^* = C_{M^*}(u_2^*) \cap C_{M^*}(B) \leq O_2(C_{M^*}(u_2^*)).$$

In particular V^* is contained in each Sylow 2-subgroup of $C_{M^*}(u_2^*)$. But as

$$|C_{GL(U^*)}(u_2^*)|_2|U^*| = 2^{10} = |T^*|$$

and $T^* \in \mathrm{Syl}_2(M^*)$, it follows that $I^* = N_{M^*}(U^*) \cap C(u_2^*)$ contains a Sylow 2-subgroup of $C_{M^*}(u_2^*)$, so $V^* \leq I^*$. Now I^* stabilizes the hyperplane B_3 of B, so I^*/V^* is contained in the stabilizer in $GL(B)$ of B_3. Then since $O_2(I^*) = C_{I^*}(B_3)$ and U^* induces the full group of transvections of B with axis B_3, we conclude $O_2(I^*) = V^*U^*$ and $|V^*| = 16$ with I^* inducing $GL(V^*/\langle u_2^* \rangle)$ on $V^*/\langle u_2^* \rangle \cong U^*V^*/U^*$. Therefore $V^* \cong E_{16}$ and $Q_8^3 \cong U^*V^*$ with I^*/V^* the stabilizer of B_3 in $GL(B)$. In particular I^*/V^* is maximal in $GL(B)$ so either $I^* = C_{M^*}(u_2^*)$ or $C_{M^*}(u_2^*)/V^* \cong GL(B)$. The latter is impossible as $V^* = C_{I^*}(V^*)$ so, as $GL(B)$ is simple, $C_{M^*}(u_2^*)/V^*$ is faithful on V^* and hence $GL(B) \cong C_{M^*}(u_2^*)/V^* = C_{GL(V^*)}(u_2^*)$, a contradiction.

So $C_{M^*}(u_2^*) = I^*$ is the split extension of U^* by the stabilizer of u_2^* in $GL(U^*)$. Therefore by Theorem 44.4 in [SG], $M^* \cong M_{24}$, as desired. Notice the other possible conclusions in that theorem are eliminated because of the structure of the 2-locals $N_{M^*}(A_i^*)$, $i = 1, 2$ (cf. the discussion at the end of Section 40 of [SG]).

(34.9) *A is the 11-dimensional Todd module for $M/A \cong M_{24}$.*

Proof: See Section 22 for a discussion of the Todd module; more details appear in Section 19 of [SG].

Let $M^* = M/A \cong M_{24}$ and write A^* for the dual of A as an M^*-module. As the dual of the 11-dimensional Todd module is the 11-dimensional Golay code module (discussed in Section 19 of [SG]), it suffices to show A^* is isomorphic to the Golay code module.

Let $N^* = N_{M^*}(U^*)$, where $U^* \leq M^*$ was defined during the proof of the previous lemma. Then N^* is the stabilizer of an octad of the Steiner system of M^*. We will show there exist $0 \neq a^* \in A^*$ and a 5-dimensional subspace A_0^* of A^* containing a^* such that N^* fixes a^*, N^* is faithful on A_0^*, and each nontrivial vector in A_0^* is conjugate to a^* under M^*. Then Exercise 7.2.2 in [SG] completes the proof.

Let $A_0 = C_A(U^*)$. We saw during the proof of 34.8 that $A_0 = V_3 + C_{[A,u_2^*]}(U^*)$ is of rank 6. So the annihilator A_0^* of A_0 in A^* is of dimension $\dim(A) - 6 = 5$.

Also we saw during the proof of 34.8 that $A_0 \cap [A, u_2^*]$ is a hyperplane of $[A, u_2^*]$, so u_2^* induces a transvection on A/A_0. Therefore the map

$$\pi : U^* \to A/A_0$$
$$u^* \mapsto [A/A_0, u^*]$$

is a nontrivial $GF(2)N^*/U^*$-homomorphism, so as U^* is an irreducible N^*/U^*-module, $\pi : U^* \to U^*\pi$ is an isomorphism. Let $a/A_0 = U^*\pi$ and a^* the annihilator of a in A^*. Then a is an N^*-invariant hyperplane of A, so $0 \neq a^* \in A^*$ is an N^*-invariant vector. Further A_0^*/a^* is the dual of $a/A_0 \cong U^*$ as an N^*/U^*-module, so N^* is transitive on $(A_0^*/a^*)^{\#}$. Also U^* induces the group of transvections on A/A_0 with axis a/A_0, so U^* induces the group of transvections on A_0^* with center a^*. Therefore N^* is transitive on $A_0^* - a^*$, so it remains to show a^* is fused into $A_0^* - a^*$ under M^*.

But K^* has a chief series $0 < V_3 < A_3 < A$ with $[U^* \cap A_1^*, A] \le A_3$, so

$$A_0 = V_3 + [A, u_2^*, U^* \cap A_1^*] \le A_3$$

and hence $A_3^* \le A_0^*$. Also $A_3 = [A, O^2(K^* \cap N^*)] \le a$, so $a^* \in A_3^*$. Then as K^* induce $GL(A/A_3)$ on A/A_3, K^* induce $GL(A_3^*)$ on A_3^*, so the proof is complete.

We choose notation so that our representative t of the second class of involutions under H in Q is contained in A and let $\tilde{E}$ be the $GF(4)$-point of the unitary space $\tilde{Q}$ containing $\tilde{t}$. Choose E to be an X-invariant complement to $\langle z \rangle$ in the preimage of $\tilde{E}$ in Q. Then as $X \le M$ and $t \in A$, $E = [t, X] \le A$.

(34.10)

(1) M controls fusion in A.

(2) z^M and t^M are the orbits of vectors of weight 4 and 2 in the Todd module A, respectively.

(3) $C_M(t)/A \cong \mathrm{Aut}(M_{22})$.

(4) $C_M(E)/A \cong L_3(4)$ with A/E the 9-dimensional Todd module for $C_M(E)/A$.

Proof: Part (1) follows from 34.5 and 2.8.1. By 22.1, M^* has two orbits $\mathcal{V}_m$, $m = 4, 2$, on $A^{\#}$ where $\mathcal{V}_m$ consists of the vectors of weight m in the Todd module A and with stabilizer in $M^* = M/A$ isomorphic to $\hat{S}_6/E_{64}$ and $\mathrm{Aut}(M_{22})$, respectively. Thus as $H \cap M$ is the stabilizer of z and $(H \cap M)^* \cong \hat{S}_6/E_{64}$, (1) implies (2) and (3). Further as X is transitive on $E^{\#}$ and $t \in E$, (4) follows from the discussion in Section 22.

(34.11) $C_G(E)$ *is quasisimple with* $C_G(E)/E \cong U_6(2)$.

Proof: Let $Y = C_G(E)$ and $\bar{Y} = Y/E$. By 34.4.4, $C_H(E)^* \cong U_4(2)$ with $C_Q(E) \cong E_4 \times 2^{1+8}$ and $C_Q(E)/E\langle z \rangle$ the natural module for $C_H(E)^*$. Now for $e \in E^{\#}$, $ez \in e^{Q}$ and $E^{\#} \subseteq t^{G} \neq z^{G}$, so $C_{\tilde{Y}}(\bar{z}) = (H \cap Y)/E \cong U_4(2)/2^{1+8}$ with $O_2(C_{\tilde{Y}}(\bar{z}))/\langle \bar{z} \rangle$ the natural module for $C_{\tilde{Y}}(\bar{z})/O_2(C_{\tilde{Y}}(\bar{z}))$.

Further by 34.10, $\bar{A}$ is the 9-dimensional Todd module for

$$N_{\tilde{Y}}(\bar{A})/\bar{A} \cong N_Y(A)/A \cong L_3(4).$$

Therefore by Theorem 30.1, $\bar{Y} \cong U_6(2)$. Finally $E \leq [C_M(E), A]$, so Y is quasisimple.

REMARK 34.12: As we observed at the start of this section, 32.4 in [SG] shows that the Monster contains a section of type F_{24}. In the next lemma we show each group of type F_{24} contains a section of type $\tilde{M}(22)$, whereas by Theorem 31.1, groups of type $\tilde{M}(22)$ are of type $M(22)$. By Fischer's Theorem, up to isomorphism there is a unique group of type $M(22)$. Thus a group X is of type $M(22)$ if and only if X is of type $\tilde{M}(22)$ and such groups exist. So from now on we denote the unique such group by $M(22)$ and write $X \cong M(22)$ to indicate X is of type $M(22)$. By 37.2, $|\text{Aut}(M(22)) : M(22)| = 2$. Lemma 34.13 shows $\text{Aut}(M(22))$ is a section of each group of type F_{24}.

(34.13)   *$F^*(C_G(t))$ is quasisimple with $C_G(t)/\langle t \rangle \cong \text{Aut}(M(22))$.*

Proof: Let $I = C_G(t)$, $H_I = H \cap I$, and $Q_I = Q \cap I$. Then $H_I/Q_I \cong O_6^-(2)$ and $Q_I \cong \mathbb{Z}_2 \times 2^{1+10}$ by 34.4.4. Let $r \in Q_I - C_Q(E)$ be an involution. Claim $r \notin O^2(I)$. For if $r \in O^2(I)$ then as $r \notin O^2(H_I)$, Exercise 13.1 in [FGT] says there exists $y \in I - H_I$ with $r^y \in H_I$ and $C_{H_I}(r^y)$ contains a Sylow 2-subgroup R of $C_I(r^y)$. Then by Sylow's Theorem we may choose y so that $C_{Q_I}(r)^y \leq R$. Now H_I/Q_I contains no 2^{1+8}-subgroups so $z^y \in Q_I$. Thus by 34.4, $z \in Q^y$ and $C_{Q^y}(\langle t, r^y \rangle) = C_{Q_I}(r)^y \leq R \leq H$, so $z \in Z(C_{Q^y}(\langle t, r^y \rangle)) = \langle z^y, t, r^y \rangle$. Then as $t \notin z^G$ and $tz^y \in t^{Q^y}$, $z \in \langle z^y, t \rangle r^y$, so replacing r by z^j, for $j = y^{-1}$ if necessary, we may take $z = r^y$. This is impossible since $t \in Q \cap Q^j$, whereas from the proof of 34.6, $Q \cap Q^j$ is contained in the subspace $\tilde{W}$ of the unitary space $\tilde{Q}$ orthogonal to $\tilde{r} = \tilde{z}^j$, whereas $\tilde{t} \notin \tilde{W}$ as $r \notin C_Q(E)$.

Thus the claim is established so there is I_0 of index 2 in I with $r \notin I$. Now $N_I(A)/A = C_M(t)/A \cong \text{Aut}(M_{22})$ by 34.10, and as A is the Todd module, $A = [N_I(A)^\infty, A]$, so $N_{I_0}(A) = N_I(A)^\infty$ and $I_0 = O^2(I)$. Further E is in the center of a Sylow 2-subgroup of $C_H(t)^\infty$, so $I_0 \cap H = C_H(E)$. Hence if we set $\bar{I}_0 = I_0/\langle t \rangle$, then $C_{\bar{I}_0}(\bar{E}) = C_G(E)/\langle t \rangle$ is quasisimple with $C_G(E)/E \cong U_6(2)$ by 34.11. Therefore by Theorem 31.1, $\bar{I}_0 \cong M(22)$. As r induces an outer automorphism on $N_{\bar{I}_0}(\bar{A})$, r induces an outer automorphism on $\bar{I}_0$, so $\bar{I} \cong \text{Aut}(M(22))$ by 37.2.1. Finally as $t \in C_G(E)^\infty$, $t \in I_0^\infty$, so $I_0 = F^*(I)$ is quasisimple.

35. The uniqueness of groups of type Aut (F_{24})

In this section we interrupt our proof of the uniqueness of groups of type F_{24} to prove the uniqueness of groups of type Aut(F_{24}). This uniqueness result has as a corollary the existence of 3-transposition groups of type $M(24)$. The existence of such groups will be used in turn to complete our proof of the uniqueness of groups of type F_{24} in the next section.

Theorem 35.1. *Let $\hat{G}$ be a finite group possessing a subgroup G of index 2 of type F_{24}. Let z be a 2-central involution in G, $\hat{H} = C_{\hat{G}}(z)$, and assume $F^*(\hat{H}) = F^*(C_G(z))$. Then $\hat{G}$ is a 3-transposition group of type $M(24)$, so in particular $\hat{G}$ and G are determined up to isomorphism.*

A group $\hat{G}$ satisfying the hypotheses of Theorem 35.1 will be said to be of *type* Aut(F_{24}). Before proving Theorem 35.1 we derive two important corollaries to the theorem.

Corollary 35.2. *Let F be the Monster, z a 2-central involution in F, $z \neq s \in z^F \cap O_2(C_F(z))$, and Y a 3-central subgroup of $C_F(\langle z, s \rangle)$ of order 3. Then $C_F(Y)$ is quasisimple with $C_F(Y)/Y$ of type $M(24)'$ and $N_F(Y)/Y$ of type $M(24)$ is generated by 3-transpositions.*

Corollary 35.3. *There exist 3-transposition groups of type $M(22)$, $M(23)$, and $M(24)$.*

Notice Corollary 35.2 implies Corollary 35.3 since it implies the existence of a 3-transposition group of type $M(24)$, and such a group has as sections 3-transposition groups of type $M(22)$ and $M(23)$. Further by 32.4 in [SG], under the hypotheses of Corollary 35.2, $C_G(Y)/Y$ is of type F_{24}, while by 26.6 in [SG], $C_F(Y)$ is of index 2 in $N_F(Y)$ and $N_F(Y)/Y$ is of type Aut(F_{24}). Therefore Theorem 35.1 implies Corollary 35.2.

After the proof of Theorem 35.1, we know that up to isomorphism there exist unique groups of type $M(22)$, $M(23)$, and $M(24)$, and we denote the unique such group by $M(22)$, $M(23)$, and $M(24)$, respectively. Further, by Remark 16.10, we know $M(22)$ is of type $\tilde{M}(22)$ and $M(23)$ is of type $\tilde{M}(23)$.

In the remainder of this section we prove Theorem 35.1. Thus we assume the hypotheses of Theorem 35.1. In addition as G is of type F_{24} we continue all the notation of the previous section, and we are free to appeal to the results in that section.

As in the statement of Theorem 35.1, we let $\hat{H} = C_{\hat{G}}(z)$. As $\langle z \rangle = Z(T)$ for $T \in \text{Syl}_2(H)$, by a Frattini argument we have $\hat{G} = G\hat{H}$ and H is of index 2 in

$\hat{H}$ as G is of index 2 in $\hat{G}$. Let $\hat{H}^* = \hat{H}/Q$. As $\hat{H}^* \le N_{O(\tilde{Q})}(X^*) = H^+$ we conclude from 34.3 and $|\hat{H}^* : H^*| = 2$ that

(35.4) $\hat{H}^* = H^+$.

By 35.4 and 29.3 there exists $d \in \hat{H} - H$ with d^* inducing a $GF(4)$-transvection in the unitary group $SU(\tilde{Q})$ with center the 1-dimensional $GF(4)$-subspace $\tilde{E}$ generated by $\tilde{i}$. Then d^* centralizes $N_{H^*}(\tilde{E})$. In particular d^* centralizes X^* and $C_Q(X) = \langle z \rangle$, so by a Frattini argument we may choose d so that $[N_H(XE), d] \le \langle z \rangle$. Then as $[C_Q(E), d] \le \langle z \rangle$, $[d, N_H(E)] \le \langle z \rangle$ and hence d centralizes a subgroup of $N_H(E)$ of index at most 2. In particular as $C_H(E) \le O^2(N_H(E))$ (cf. 34.4.4) d centralizes $C_H(E)$.

Next d acts on $C_G(E)$ and by 34.11, $C_G(E)$ is quasisimple with $C_G(E)/E \cong U_6(2)$. Further d centralizes $C_H(E)$, while $C_{\text{Aut}(C_G(E))}(C_H(E)) = \langle z \rangle$, so, replacing d by dz if necessary, d centralized $C_G(E)$. Therefore $[d, N_H(E)] \le \langle z \rangle \cap C_G(C_G(E)) = 1$.

Now d acts on $C_G(t)$ and by 34.13, $F^*(C_G(t)) = K$ is quasisimple and $C_G(t)/\langle t \rangle \cong \text{Aut}(M(22))$. Then as d centralizes $N_G(E)$ while

$$C_{\text{Aut}(K)}(C_K(E)) = E/\langle t \rangle,$$

replacing d by de for suitable $d \in E$ if necessary, we may assume $[d, K] = 1$. Notice that as d induces a transvection on $\tilde{Q}$ with center $\tilde{E}$, $[d, v] = t$ for $v \in C_Q(t) - C_Q(E)$. Therefore we have shown

(35.5) *There exists an involution $d \in \hat{H} - H$ such that d centralizes $E(C_G(t))$ and $[v, d] = t$ for $v \in C_G(t) - E(C_G(t))$.*

(35.6) $C_G(d)$ *is of type* $M(23)$ *and* $\tilde{M}(23)$ *with* $t^{C_G(d)}$ *a set of 3-transpositions of* $C_G(d)$.

Proof: Let $Y = C_G(d)$. By 35.5, $C_Y(t)$ is quasisimple with $C_Y(t)/\langle t \rangle \cong M(22)$. As d centralizes a conjugate of X in H transitive on $E^\#$, t is not weakly closed in $C_Y(t)$ with respect to Y. Therefore Theorem 32.1 completes the proof.

Next $A \le E(C_G(t))$, so d centralizes A. Thus $\hat{A} = A\langle d \rangle \cong E_{2^{12}}$. Then as $A = C_G(A)$, $\hat{A} = C_{\hat{G}}(A) \trianglelefteq N_{\hat{G}}(A) = M\hat{A} = \hat{M}$. Further as $C_G(d)$ is of type $M(23)$ and $\tilde{M}(23)$, $C_M(d)/A \cong M_{23}$ by 25.7. In particular $\hat{A}$ does not split over A as an M-module since $C_M(d)$ fixes no member of $A^\#$. Hence by 22.8.2, $\hat{A}$ is the 12-dimensional Todd module for M/A. We summarize these remarks:

(35.7)

(1) $\hat{A} = A\langle d \rangle \cong E_{2^{12}}$.

(2) $N_{\hat{G}}(A) = N_{\hat{G}}(\hat{A}) = M\hat{A}$.

(3) $\hat{A}$ is the 12-dimensional Todd module for M/A.

We are now in a position to complete the proof of Theorem 35.1. Namely 35.5–35.7 provide the hypotheses of Theorem 33.1, so by that theorem, d^G is a set of 3-transpositions of $\hat{G}$ and $\hat{G}$ is of type $M(24)$.

36. The uniqueness of groups of type F_{24}

In this section we complete our proof of the uniqueness of groups of type F_{24}. We continue the hypotheses and notation of Section 34. In addition, from Theorem 35.1 we know that there is a group $\hat{G}$ of type $M(24)$ generated by a set D of 3-transpositions and that the commutator group of $\hat{G}$ is of type F_{24}. Write G' for the commutator group of $\hat{G}$ and let $T' \in \mathrm{Syl}_2(G')$, $\langle z' \rangle = Z(T')$, $H' = C_{G'}(z')$, and so on.

(36.1)

(1) $C_H(X) \cong U_4(3)/\mathbf{Z}_6$ is quasisimple.

(2) $H \cong H'$.

Proof: First $H^* = H^-$ is determined up to conjugation in $O(\tilde{Q}) = \mathrm{Out}(Q)$ by 29.9.4, so $\tilde{H} = Inn(H)$ is determined up to conjugation in $\mathrm{Aut}(Q)$ and hence determined up to isomorphism.

Next let $K = H^{\infty}$ and $K_0 = Cov_2(K)$. Then by 23.1, $O_2(K_0) = Q_0 \times Z_0$, where $Q_0 \cong Q$ and $Z_0 \cong \mathrm{Schur}_2(U_4(3))$. By Exercise 7.3, $\mathrm{Schur}_2(U_4(3)) \cong \mathbf{Z}_2$, so by 23.1, $K \cong K_0/V$ for some complement V to $Z(Q_0)$ in $Z(Q_0) \times Z_0 \cong E_4$. In particular up to isomorphism there are two choices for V and hence two choices for K. Namely if $Z_0 = \langle z_0 \rangle$ and $Z(Q_0) = \langle z_1 \rangle$ then $V = Z_0$ or $\langle z_0 z_1 \rangle$. Further if $Y \in \mathrm{Syl}_3(O_{2,3}(K_0))$ then $C_{K_0}(Y)/V \cong C_K(X)$ and $C_{K_0}(Y)/V$ is quasisimple if and only if $V \neq Z_0$. Thus (1) implies K is determined up to isomorphism. Moreover $H = K\langle a \rangle$ where $a \in A - K$ and $\tilde{H}$ is determined up to isomorphism, so the automorphism induced on K by a is determined. Hence as $\langle a, z \rangle$ splits over $\langle z \rangle$, H is determined up to isomorphism by the hypothesis that G is of type F_{24}. Therefore as G' is also of type F_{24}, $H \cong H'$.

So it remains to prove (1). But as A is the Todd module for M/A, $C_A(X) = C_{QA}(X) = O_2(C_{H \cap M}(X))$ is a 5-dimensional indecomposable for

$$C_{H \cap M}(X)/XC_A(X) \cong A_6$$

by Exercise 7.4. On the other hand if $C_H(X)$ is not quasisimple then

$$C_H(X)/X \cong \mathbf{Z}_2 \times U_4(2)$$

and hence $O_2(C_{H \cap M}(X))$ splits over $\langle z \rangle$ as a $C_{H \cap M}(X)$-module. This establishes (1) and completes the proof of the lemma.

(36.2) $M \cong M'$.

Proof: We appeal to 21.10 with $G_1 = M$, $G_2 = M'$, $A = V_1$, $A' = V_2$, $H_1 = H \cap M$, and $H_2 = H' \cap M'$. By 36.1 there is an isomorphism $\varphi :$ $H \to H'$ and without loss $T\varphi = T'$, so $A\varphi = J(T)\varphi = J(T') = A'$. Then $H_1\varphi = N_H(A)\varphi = N_{H'}(A') = H_2$, so hypothesis (1) of 21.10 is satisfied. Next by 34.9 there is a quasiequivalence $\xi : G_1/V_1 \to G_2/V_2$, $\zeta : V_1 \to V_2$ and as $\langle z \rangle = C_A(T)$ and $\langle z' \rangle = C_{A'}(T')$, we may choose ζ so that $z\zeta = z'$. Hence

$$(H_1/V_1)\xi = C_{G_1/V_1}(z)\xi = C_{G_2/V_2}(z\zeta) = C_{G_2/V_2}(z') = H_2/V_2,$$

so hypothesis (2) of 21.10 is satisfied.

To verify hypothesis (3) of 21.10, we use Remark 21.9. As H_2/V_2 is up to conjugation the unique subgroup of G_2/V_2 of its isomorphism type, hypothesis (2) of 21.9 is satisfied for $H_2/V_2 \le G_2/V_2 \le GL(V_2)$. Finally Exercise 7.5 says hypothesis (1) of 21.9 is satisfied. Hence our lemma follows from 21.10.

Now $M^* = M/A \cong M_{24}$ is 5-transitive on a set $\Omega = \{1, \ldots, 24\}$ of order 24 and the automorphism group of a Steiner system (Ω, C). We regard A as the 11-dimensional Todd module for M^* defined by this Steiner system and use 34.10 to identify t with the vector $\tilde{e}_{12}$ of weight 2 in A, using the notation established in Section 22. Then from Section 22 and 34.10, we may take $E = \langle \tilde{e}_{12}, \tilde{e}_{1,3} \rangle$. Let L_3 be the preimage in M of the stabilizer in M^* of $1 \in \Omega$. Thus $L_3^* \cong M_{23}$ and A is the Todd module for L_3^*. Notice the isomorphism of 36.2 restricts to an isomorphism $L_3 \cong L_3'$.

Next let $L_1 = E(C_G(t))$. By 34.12, L_1 is quasisimple with $L_1/\langle t \rangle \cong M(22)$. Then by 23.8, L_1 is determined up to isomorphism, so we have an isomorphism $\psi : L_1 \cong L_1'$ and as $E/\langle t \rangle$ is a 3-transposition in $L_1/\langle t \rangle$ we can choose $E\psi = E'$.

Finally let $L_{1,2} = C_G(E)$. Then by 34.11, $L_{1,2}$ is quasisimple with $L_{1,2}/E \cong U_6(2)$. Let $j \in L_3$ have cycle $(2, 3)$ on Ω; then $\tilde{e}_{1,2}^j = \tilde{e}_{1,3}$, so j acts on E and $L_{12} = L_1 \cap L_1^j$. Let $j' = j\psi$. As $E\psi = E'$, $L_{1,2}\psi = L_{1,2}' = L_1' \cap (L_1')^{j'}$ and ψ induces an isomorphism $L_{1,2} \cong L_{1,2}'$.

Let $G_1 = \langle L_1, L_3 \rangle$ and $G_1' = \langle L_1', L_2' \rangle$. We have shown $L_i \cong L_i'$ for $i = 1, 3$, $L_1 \cap L_1^j \cong L_1' \cap (L_1')^{j'}$. Of course ψ restricts to an isomorphism of $L_1 \cap L_3 = N_{L_1}(A)$ with $L_1' \cap L_2' = N_{L_1'}(A')$. Thus we have achieved the hypotheses of Lemma 27.8. Further by Corollary 35.3, the Fischer group of type $M(23)$ exists so, apropos of Remark 27.9, we can legitimately appeal to Lemma 27.8 at this point, and conclude

(36.3) $\langle L_1, L_3 \rangle = G_1 \cong \langle L_1', L_3' \rangle = G_1' \cong M(23)$.

Next let $G_3 = M$. By 36.2, $G_3 \cong G_3'$ and by 36.3, $G_1 \cong G_1'$. By construction, $G_1 \cap G_3 = L_3 \cong L_3' = G_1' \cap G_3'$. Let s be an involution in G_3 with cycle $(1, 2)$ and s' the image of s under our isomorphism of G_3 with G_3'. Then s centralizes $t = \tilde{e}_{12}$, so $C_{G_1}(t) \le G_1 \cap G_1^s$. As the stabilizer of t in L_3 is $M_{22}/E_{2^{11}}$, t is a 3-transposition in $G_1 \cong M(23)$, so $C_{G_1}(t) \cong M(22)/\mathbf{Z}_2$. Therefore $C_{G_1}(t) = E(C_G(t)) = L_1$. Similarly $G_1' \cap (G_1')^{s'} = L_1' \cong L_1$. Thus we have achieved the hypotheses of Lemma 27.8 again, so appealing to that lemma we conclude

(36.4) $G_0 = \langle G_1, G_3 \rangle \cong \langle G_1', G_3' \rangle = G' \cong M(24)'$.

Notice $Z(G_0) = 1$ as $C_G(L_1) = \langle t \rangle$.

Now to complete the proof of Theorem 34.1 given 36.4, it remains to show $G_0 = G$. But by 37.4, G_0 has two classes of involutions with representatives z and t. Also $C_G(t) \cong C_{G_0}(t)$, so $C_G(t) \le G_0$ and $C_G(z) = \langle C_{G_1}(z), C_{G_3}(z) \rangle \le G_0$. So either $G = G_0$ or G_0 is strongly embedded in G. The latter is impossible by 7.6 in [SG] as G has two classes of involutions.

Thus the proof of Theorem 34.1 is at last complete.

Exercises

1. Let G be of type M and $\tilde{M}$ for $M = M(22)$ or $M(23)$, and let D be a set of 3-transpositions of G that make G of type M. Let $T \in \mathrm{Syl}_2(G)$ and $A = \langle T \cap D \rangle$. Prove
   (1) If $M = M(22)$ then $A = J(T)$.
   (2) D is $\mathrm{Aut}(G)$-invariant.
   (3) D is the unique class of 3-transpositions of G.
   [*Hint:* Use Exercise 7.9.3 to prove (1). Notice when $M = M(23)$ that $|T \cap D|$ is odd, so $A = J(T)$ by 32.3. Now show $T \cap D$ is $N_{\mathrm{Aut}(G)}(A)$-invariant and then use (1) and a Frattini argument to prove (2). Finally show (2), 15.11, and 15.14 imply (3).]

PART III

THE LOCAL STRUCTURE OF THE FISCHER GROUPS

Introduction

The author has a program to put in order the foundations of the theory of the sporadic groups. That program has three goals: Prove the existence and uniqueness of each of the sporadics as a finite group with a suitable involution centralizer, and determine the conjugacy classes and normalizers of subgroups of prime order in each of the sporadics.

In Part II we showed that up to isomorphism there exists a unique group of type $\tilde{M}$ for $M = M(22)$ and $M(23)$, and a unique group of type F_{24}. This provides the necessary existence and uniqueness proofs for the three Fischer groups.

In Part III we turn to the third phase of the program and determine the conjugacy classes of subgroups of prime order and their normalizers in each of the three Fischer groups. To do so we make heavy appeals to the theory of 3-transposition groups. The determination of the conjugacy classes and normalizers of subgroups of order p is fairly straightforward when $p > 3$. Thus almost all our time is spent with the 2-local and 3-local structure of the Fischer groups.

In Chapter 12 we determine the conjugacy classes of involutions and their centralizers in Aut($M(22)$), $M(23)$, and $M(24)$. We find that the classes of involutions in the Fischer groups are $\mathcal{J}_m$, $1 \leq m \leq k$, where $k = 3$ for the two smaller Fischer groups, $k = 4$ for $M(24)$, and $\mathcal{J}_m$ denotes the involutions that are the product of m distinct commuting 3-transpositions. Thus the classes of involutions and their centralizers are nicely described in terms of the 3-transposition theory.

Chapter 13 contains a discussion of conjugacy classes of elements g of order 3 in orthogonal groups $O(V)$ over the field of order 3. The classes are described in terms of the Jordon block structure of g on V together with a geometric invariant associated to that structure. This information is important, since $G = M(22)$ and $G = M(23)$ possess D-subgroups $L \cong PO_7^{-\pi,\pi}(3)$ and $L \cong S_3/P\Omega_8^+(3)$, respectively, containing a Sylow 3-subgroup of G. Thus each element of order 3 in G is fused into L, and indeed for some subgroups X of L of order 3, $N_G(X) = N_L(X)$.

In Chapter 14 we determine the conjugacy classes and normalizers of subgroups of the Fischer groups of odd prime order. Again the subgroups of order 3 are the only troublesome subgroups, and the group L of the previous paragraph provides us with a point of reference for studying the 3-local structure of the smaller Fischer groups.

In Theorem 40.13 we produce the 3-local subgroup of $M(23)$ that contains the universal 4-generator 3-transposition group of width 1 appearing in Marshall Hall's Theorem 19.14.

The basic structure of the Fischer groups seems to be part of the folklore. Presumably it was first derived by Fischer in unpublished work.

12

The 2-local structure of the Fischer groups

In this chapter we determine the automorphism group of each Fischer group, the conjugacy classes of involutions in each automorphism group, and the structure of the centralizer of each involution in the automorphism group. We also obtain a few extra pieces of information about the 2-locals in the Fischer groups.

Lemmas 37.2, 37.3, and 37.4 were used in Chapter 11 to prove the existence and uniqueness of the Fischer groups. Notice their proofs are valid under the hypothesis that the Fischer group of type M discussed is of type $\tilde{M}$, but do not use the existence of larger Fischer groups. Hence the appeals in Chapter 11 are legitimate.

37. Involutions and their centralizers in the Fischer groups

(37.1) $\text{Aut}(F_{24}) = M(24)$.

Proof: Let $L = F_{24}$ and $G = \text{Aut}(L)$. Let $T \in \text{Syl}_2(L)$, $A = J(T)$, $M = N_L(A)$, and $N = N_G(A)$. By a Frattini argument, $G = LN$.

By 34.5, $A \cong E_{2^{11}}$, by 34.8, $M/A \cong M_{24}$, and by 34.9, A is the 11-dimensional Todd module for M/A. By Exercise 7.5 in [SG], $M_{24} = \text{Aut}(M_{24})$, so $N = MC_N(M/A)$. As M is absolutely irreducible on A, $C_N(M/A)$ centralizes A. Therefore by 22.8.2, $|C_N(M/A) : AC_N(M)| \leq 2$.

Next $L \leq G_0 \leq G$ with $G_0 \cong M(24)$ and $N_{G_0}(A) = MA_0$ with $E_{2^{12}} \cong A_0$ and $C_{A_0}(M) = 1$. Therefore $N = MA_0 C_N(M)$ and to show $G = G_0$ it remains to show $C_N(M) = 1$.

Let $\langle z \rangle = Z(T)$, $H = C_L(z)$, and $K = C_G(z)$. Then $F^*(H) = Q \cong 2^{1+12}$ with $C_H(Q) = \langle z \rangle$, so, as $Q \leq T \leq M$, $C_N(M)\langle z \rangle = C_K(Q)$ and $[H, C_N(M)] \leq C_H(Q) = \langle z \rangle$. Then as $H = H^{\infty}$, $[H, C_N(M)] = 1$. Therefore $C_N(M)$ centralizes $\langle H, M \rangle$, while by 36.4, $L = \langle H, M \rangle$, so $C_N(M) = 1$, completing the proof.

REMARK: Let G be a Fischer group of type M, $M = M(22)$, $M(23)$, or $M(24)$. By Exercise 11.1, G has a unique class of 3-transpositions except possibly in the last case. But if G is of type $M(24)$ then by 37.1, $G = \text{Aut}(G)$, so by Theorem Z in Section 15, D *is* uniquely determined.

(37.2)

   (1) $|\text{Aut}(M(22)) : M(22)| = 2$.
   (2) $M(23) = \text{Aut}(M(23))$.

Proof: Let $L = M(22)$ or $M(23)$ and $G = \text{Aut}(L)$. By Exercise 11.1, the class D of 3-transpositions of L making L of type M is G-invariant. Let $d \in D$; then $G = LC_G(d)$ by a Frattini argument.

Let $H = C_L(d)$. Then H is quasisimple with $\tilde{H} = H/\langle d \rangle \cong U_6(2)$, $M(22)$, for $L = M(22)$, $M(23)$, respectively, as L is of type $\tilde{M}$. As H is transitive on A_d, $C_G(d) = HC_G(\langle a, d \rangle)$ for $a \in A_d$, by a Frattini argument. Let $X = \langle D_{a,d} \rangle$. Then by definition of the groups $M(22)$ and $M(23)$, and by 16.1 and 16.3,

$$X \cong PO_6^{+,\pi}(3), \qquad PO_7^{-\pi,\pi}(3),$$

respectively. Then by 15.10 and 16.5, $|N_{\text{Aut}(\tilde{H})}(\tilde{X}^{\tilde{H}}) : \tilde{H}| = 2, 1$, respectively. Further if $L = M(22)$ there is $L \le G_0 \le G$ with $|G_0 : L| = 2$ and $C_{G_0}(d)$ the extension of H by an involutory outer automorphism; for example this follows from 34.13. Thus $C_G(z) = C_{G_0}(z)C_G(\langle H, a \rangle)$. Finally by 9.5.5, $L = \langle H, a \rangle$, so $C_G(z) = C_{G_0}(z)$ and then $G = LC_G(z) = G_0$, completing the proof.

(37.3)   *Each involution in* $\text{Aut}(M(22))$ *centralizes a 3-transposition in* $M(22)$.

Proof: Let $L = M(22)$, $G = \text{Aut}(L)$, and suppose t is an involution in G fixing no member of the set D of 3-transpositions of L that make L of type $M(22)$. Then by Exercise 8.4, $d^t \in D_d$ for each $d \in D$.

Let $t \in T \in \text{Syl}_2(G)$, $A = \langle T \cap D \rangle$ and $M = N_G(A)$. Then by 25.7, $A \cong E_{2^{10}}$ is the Todd module for $M/A \cong \text{Aut}(M_{22})$. As t fixes no member of $T \cap D$ of order 22, the class of tA in M/A is determined by Exercise 7.6.5. Moreover by Exercise 7.6.6, we can pick T so that there is $b \in T \cap D$ with $u = bb^t \in Z(T)$ and t induces a nontrivial inner automorphism on $C_M(u)^\infty O_2(C_M(u))/O_2(C_M(u)) \cong A_5$.

Let $H = C_L(u)$ and $H^* = H/O_2(H)$. Then $H^* \cong \text{Aut}(U_4(2))$. As $d^t \in D_t$ for each $d \in C_D(u)$, $[d^*, d^{*t}] = 1$ for each $d^* \in C_D(u)^*$. But as t induces a nontrivial inner automorphism on $C_M(u)^\infty O_2(C_M(u))/O_2(C_M(u))$, $1 \ne t^* \in E(H^*)$, so Exercise 7.7 supplies a contradiction.

(37.4) *Let* $G = M(22)$, $M(23)$, *or* $M(24)$. *Then*

(1) *G has k classes of involutions* $\mathcal{J}_m$, $1 \leq m \leq k$, *where* $k = 4$ *if* $G = M(24)$ *and* $k = 3$ *otherwise, and with* $\mathcal{J}_m$ *the set of products of m distinct commuting 3-transpositions.*

(2) *If* $m \leq 3$ *and* $j \in \mathcal{J}_m$ *then there is a unique set* $\{d_1, \ldots, d_m\}$ *of commuting 3-transpositions with* $j = d_1 \cdots d_m$, *except when* $G = M(22)$ *and* $m = 3$ *where there are exactly two such sets.*

Proof: Let D be the set of 3-transpositions of G making G of type M, $T \in \mathrm{Syl}_2(G)$, $A = \langle T \cap D \rangle$, and $M = N_G(A)$. By 25.7, if $G = M(n)$ then A is the Todd module for $M/A \cong M_n$ and $T \cap D$ is of order n. Then by 22.1, 22.3, and 22.4, M has k orbits $\mathcal{I}_m$, $1 \leq m \leq k$, on $A^\#$, where $\mathcal{I}_m$ is the set of products of m distinct members of $T \cap D$. In particular $\mathcal{I}_m \subseteq \mathcal{J}_m \cap A$.

Suppose $j = d_1 \cdots d_m = e_1 \cdots e_r$ can be written as the product of commuting members of D in two ways with $r < m \leq 4$. Then $r \leq 3$, so by 8.4 and 8.11, d_i centralizes e_j for all j and hence $\langle d_i, e_j : i, j \rangle$ is a 2-group. Thus we may take $d_i, e_j \in T$ for all i, j and then we have a contradiction to $\mathcal{I}_m \cap \mathcal{I}_r = \emptyset$. It follows that $\mathcal{I}_m = \mathcal{J}_m \cap A$ for each m, and that $\mathcal{J}_m$, $1 \leq m \leq k$, are distinct conjugacy classes of involutions. Therefore to prove (1) it remains to show each involution t in G is the product of commuting members of D.

We first show t centralizes some member of D. Assume not. Then $G \neq M(23)$ as in that case the members of D are 2-central. Further by 37.3, $G \neq M(22)$, so $G = M(24)$. Let $d \in D$. By Exercise 8.4, $d^t \in D_d$, so $u = dd^t$ is an involution centralizing t. Then t acts on $H = \langle C_D(u) \rangle$. Now by 34.13, $\langle d, d^t \rangle = Z(H)$ and $H^* = H/Z(H) \cong M(22)$ with D^*_{d,d^t} the set of 3-transpositions in H^*. As t centralizes no member of $C_D(u)$, t is faithful on H^*, so by 37.3, t centralizes some $a^* \in D^*_{d,d^t}$. But then t acts on $D \cap \langle a, d, d^t \rangle = \{a, d, d^t\}$ and hence on a, a contradiction.

Therefore $t \in K = C_G(a)$ for some $a \in D$. Now $K^* = K/\langle a \rangle \cong U_6(2)$, $M(22)$ or $M(23)$ with D^*_a the set of 3-transpositions, so if $G \neq M(22)$ then by induction on the order of G, $t^* = d^*_1 \cdots d^*_r$ for some commuting d^*_i. The same holds if $G = M(22)$ as each involution in $U_6(2)$ is the product of commuting transvections (cf. 23.2). But now $t = a^\epsilon d_1 \cdots d_r$ is the product of commuting 3-transpositions, as desired.

Therefore (1) is at last established. Further if $j = d_1 \cdots d_m = e_1 \cdots e_r$ with $r \leq m \leq 3$ then we saw previously that $r = m$ and we may take $e_i, d_j \in T$. We must show $\{e_1, \ldots, e_m\} = \{d_1, \ldots, d_m\}$ except when $G = M(22)$ and $m = 3$. This is trivial when $m = 1$, so $m = 2$ or 3, where the result follows from 22.1.2, 22.3, and 22.4.

(37.5) *Let* $G = M(24)$, $L = F^*(G) \cong F_{24}$, *and* $j = d_1 \cdots d_m$ *the product of m distinct commuting 3-transpositions of G. Then*

(1) $\mathcal{J}_2 = j_2^L$ *and* $\mathcal{J}_4 = j_4^L$ *are the conjugacy classes of involutions of L and* $\mathcal{J}_1 = j_1^L$ *and* $\mathcal{J}_3 = j_3^L$ *are the classes of involutions in* $G - L$.

(2) $C_L(j_1) \cong M(23)$.

(3) $F^*(C_L(j_2)) = C_L(\langle d_1, d_2 \rangle)$ *is quasisimple with*

$$C_L(j_2)/\langle j_2 \rangle \cong \mathrm{Aut}(M(22)).$$

(4) $F^*(C_L(j_3))$ *is quasisimple with center* $\langle d_1 d_2, d_1 d_3 \rangle$ *and*

$$C_L(j_3)/\langle d_1 d_2, d_1 d_3 \rangle \cong \mathrm{Aut}(U_6(2)).$$

(5) $F^*(C_L(j_4)) \cong 2^{1+12}$ *and* $C_L(j_4)/F^*(C_L(j_4)) \cong \mathbf{Z}_2/U_4(3)/\mathbf{Z}_3$.

Proof: The first part follows from 37.4, and (2) is part of the definition of $M(24)$. Let $K = C_L(d_1)$, so that $K \cong M(23)$.

By 37.4.2, $j_2 = d_1 d_2$ is the unique way to write j_2 as the product of commuting 3-transpositions, while there is $t \in L$ with cycle (d_1, d_2) so $|C_L(j_2) : C_L(\langle d_1, d_2 \rangle)| = 2$. Also $K_2 = C_L(\langle d_1, d_2 \rangle) = C_K(j_2)$ and j_2 is a 3-transposition of K, so K_2 is quasisimple with $K_2/\langle j_2 \rangle \cong M(22)$. Now by (1), L has two classes of involutions with representatives j_2 and j_4, while by 34.1 one class has its centralizer described in (5) and by 34.13 the other has its centralizer described in (3). We conclude (3) and (5) hold.

Similarly by 37.4.2, $j_3 = d_1 d_2 d_3$ is the unique way to write j_3 as the product of three commuting 3-transpositions, so $L_3 = C_L(\langle d_1, d_2, d_3 \rangle) \trianglelefteq C_L(j_3)$ with $C_L(j_3)/L_3 \cong S_3$. Now

$$E = \langle d_1 d_2, d_1 d_3 \rangle = L \cap \langle d_1, d_2, d_3 \rangle \trianglelefteq C_L(j_3)$$

with $L_3 \leq C_L(E)$. Also $C_L(E)$ is quasisimple with $C_L(E)/E \cong U_6(2)$ by 34.11. But $C_L(E)$ centralizes E and $\langle d_1, d_2, d_3 \rangle/E$, so $C_L(E) = C_L(E)^\infty \leq C_L(\langle d_1, d_2, d_3 \rangle) = L_3$. That is $L_3 = C_L(E)$, so (4) holds.

(37.6) *Let* $G = M(22)$ *or* $M(23)$ *and* $j_m = d_1 \cdots d_m$ *be the product of m distinct commuting 3-transpositions of G. Then*

(1) $\mathcal{J}_m$, $1 \leq m \leq 3$, *are the conjugacy classes of involutions of G.*

(2) $C_G(j_1)$ *is quasisimple with* $C_G(j_1)/\langle j_1 \rangle \cong U_6(2)$ *or* $M(22)$, *respectively.*

(3) $|C_G(j_2) : C_G(\langle d_1, d_2 \rangle)| = 2$ *with* $\langle d_1, d_2 \rangle = Z(C_G(\langle d_1, d_2 \rangle)$ *and*

(a) *If* $G \cong M(23)$ *then* $C_G(\langle d_1, d_2 \rangle)$ *is quasisimple with* $C_G(j_2)/\langle d_1, d_2 \rangle$

the extension of $U_6(2)$ by an involutory outer automorphism.
(b) *If $G \cong M(22)$ then $C_G(j_2)/\langle d_1, d_2 \rangle \cong \text{Aut}(U_4(2))/E_{2^8}$.*
(4) *If $G \cong M(23)$ then $C_G(j_3)/C_G(\langle d_1, d_2, d_3 \rangle) \cong S_3$ with*

$$C_G(j_3)/\langle d_1, d_2, d_3 \rangle \cong \Gamma U_4(2)/E_{2^8}.$$

(5) *If $G \cong M(22)$ then $|C_G(j_3)| = 2^{16} \cdot 3^3$.*

Proof: We argue as in the proof of the previous lemma. Let $d_i \in T \in \text{Syl}_2(G)$, D the set of 3-transpositions of G, $A = \langle T \cap D \rangle$, and $M = N_G(A)$. Observe that when $G \cong M(22)$, Exercise 7.6 says

$$C_M(j_2)/O_2(C_M(j_2)) \cong S_5$$

with involutions in $C_M(j_2)$ with cycle (d_1, d_2) projecting on transpositions. Thus each such involution induces an outer automorphism on

$$C_G(\langle d_1, d_2 \rangle)/O_2(C_G(\langle d_2, d_2 \rangle)) \cong U_4(2).$$

This can be used to complete the proof of (3b). Similarly if $G \cong M(23)$ then $C_M(j_3)/A$ is E_{16} extended by $\Gamma L_2(4)$, which can be used to complete the proof of (4).

(37.7) *Let $L = M(22)$, D the set of 3-transpositions of L, and $G = \text{Aut}(L)$. Then*

  (1) *There are three classes of involutions in $G - L$ with representatives s, sa, and sab.*
  (2) *$C_L(s) \cong \text{Aut}(\Omega_8^+(2))$, $a, b \in C_D(s)$, and $b \in D_a$.*
  (3) *$C_L(sa) = C_L(\langle s, a \rangle) \cong \mathbf{Z}_2 \times Sp_6(2)$.*
  (4) *$C_L(sab) \cong O_6^-(2)/E_{64}$, $C_L(sab)/O_2(C_L(sab))$ acts naturally on*

$$O_2(C_L(sab)),$$

  and ab is a nonsingular point of that space.

Proof: Let E be the set of 3-transpositions in $K = M(24)$ and $u, v, w \in E$ with $u, v \in A_w$ and $v \in D_u$. Then $S = \langle u, v, w \rangle \cong S_4$, and by 25.6,

$$C_K(S) = \langle C_D(S) \rangle \cong \text{Aut}(\Omega_8^+(2)).$$

Let $t \in O_2(S)$ with $u^t = v$ and $L_0 = \langle E_{u,v} \rangle$, $G_0 = L_0\langle t \rangle$, and $G_0^* = G_0/\langle uv \rangle$. Notice $t = a_1 b_1$ for commuting $a_1, a_2 \in E \cap S$. Also by 37.5 we may regard $G = G_0^*$, $L = L_0^*$, and $D = E_{u,v}^*$.

Let $s = t^*$. Then $S = \langle x, u \rangle \langle uv, t \rangle$, where x is of order 3 and inverted by u and $E_4 \cong \langle uv, t \rangle \trianglelefteq S$. Therefore $C_K(\langle uv, t \rangle)$ is $\langle x, u \rangle$-invariant, so, as $C_K(\langle uv, t \rangle) = \langle uv, t \rangle C_K(\langle t, u \rangle)$, $\langle x \rangle = [\langle x \rangle, u]$ centralizes

$$C_K(\langle uv, t \rangle) / \langle uv, t \rangle$$

and hence

$$C_K(\langle t, uv \rangle) = C_K(S)\langle t, uv \rangle; \quad \text{therefore } C_{L_0}(t) = C_K(S)\langle uv \rangle,$$

so $C_L(s) = C_K(S)^* \cong \mathrm{Aut}(\Omega_8^+(2))$, establishing (2).

Now let $a_0 \in E_{u,v}$ and $b_0 \in E_{a_0,u,v}$ centralize t and define $a = a_0^*$ and $b = b_0^*$. Also let $H = C_L(a)$, so that $H/\langle a \rangle \cong U_6(2)$. Then s induces an outer automorphism on $H/\langle a \rangle$ by 37.6.3, since $E(C_K(a_0)) \cong M(23)$ and $H = C_{E(C_K(a_0))}(uv)/\langle uv \rangle$. Further $C_L(\langle s, a \rangle)$ is the centralizer in $\mathrm{Aut}(\Omega_8^+(2))$ of the 3-transposition a, so $C_L(\langle s, a \rangle) \cong \mathbf{Z}_2 \times Sp_6(2)$. Moreover as $t = a_1 b_1$ we have $ta_0 = a_1 b_1 a_0$ is the product of three commuting 3-transpositions of K so $C_{G_0}(ta_0) = C_{G_0}(\langle t, a_0 \rangle)$ by 37.4.2, since uv acts as the transposition (a_1, b_1) on $\{a_1, b_1, a_0\}$. Therefore $C_L(sa) = C_L(\langle s, a \rangle)$, establishing (3).

By 37.3, each involution in G is conjugate to an involution in $C_G(a)$, so each involution in $G - L$ is conjugate to an involution in $C_G(a) - H$. But by Exercise 4.16.6, each involution in $C_G(a) - H$ is conjugate to s, sa, sb, or sab. Also a is conjugate to b in $C_L(s)$ so sa is conjugate to sb. Finally s is the image of t, $tuv \in \mathcal{J}_2$, sa is the image of ta_0, $ta_0uv \in \mathcal{J}_3$, and sab is the image of ta_0b_0, $ta_0b_0uv \in \mathcal{J}_4$, so s, sa, and sab are in distinct classes of G. Therefore (1) holds.

It remains to establish (4). We saw in the previous paragraph that each involution in $C_G(a) - H$ that is G-conjugate to sab is H-conjugate to sab, so we have

$$B = C_D(sab) = a^G \cap C_G(sab) = a^{C_G(sab)}$$

Let $Y = \langle C_B(a) \rangle$ and $X = \langle B \rangle$. By Exercise 4.16.6, $Y = \langle a \rangle \times Y_1 = C_L(\langle a, s, b \rangle)$, where Y_1 is the centralizer of a transvection in $Sp_6(2)$.

Next $\{a, b\} = Z(Y) \cap B$ so $V_a = \{a, b\}$ and hence by 9.2, $ab \in O_2(X)$ and $2 = |V_a| = |O_2(X) : C_{O_2(X)}(a)|$. Also $O_2(Y_1) \cong E_{32}$ with Y_1 irreducible on $O_2(Y_1)/\langle ab \rangle$, so $C_{O_2(X)}(a) = \langle ab \rangle$ or $O_2(Y_1)$. The former case is out as $|C_X(ab) : C_X(\langle a, b \rangle)| \leq 2$ and X is transitive on B. Thus $O_2(X) \cong E_{64}$.

Now $YO_2(X)/O_2(X) \cong \mathbf{Z}_2 \times Sp_4(2)$, so by Exercise 8.2, $X/O_2(X) \cong O_6^\epsilon(2)$. Then as $X/O_2(X)$ is nontrivial on $O_2(X) \cong E_{64}$, $O_2(X)$ is the natural module for $X/O_2(X)$ with $\langle ab \rangle = [O_2(X), a]$ a nonsingular point. By a Frattini argument, $C_G(sab) = XC_G(\langle sab, a \rangle) = X$. Thus it remains to show $\epsilon = -1$.

But this follows as $O_6^\epsilon(2)$ is a section of $C_K(ta_0b_0)$ with $ta_0b_0 \in \mathcal{J}_4$. So by 37.5, $O_6^\epsilon(2)$ is a section of $U_4(3)$ whereas $O_6^+(2) \cong S_8$ has no faithful 6-dimensional representation over a field of odd characteristic.

(37.8) *Let $L = M(22)$, $G = \mathrm{Aut}(L)$, D the set of 3-transpositions of L, and $j_m = d_1 \cdots d_m$, $m = 1, 2, 3$, be the product of distinct commuting 3-transpositions $d_i \in D$. Then*

(1) *$F^*(C_G(j_2)) = Q$ is a large extraspecial 2-subgroup of G of order 2^{11} and $C_G(j_2)/Q \cong \mathrm{Aut}(U_4(2))$.*
(2) *There exists $T \in \mathrm{Syl}_2(G)$ and $E_{32} \cong B \leq A = \langle T \cap D \rangle$ such that $d_i \in B = \langle B \cap D \rangle$, $|B \cap D| = 6$, $T \leq M = N_G(B)$, and $N_M(A)$ induces S_6 on $B \cap D$.*
(3) *$R = O_2(M)$ is special with center B and $M/R = M_1/R \times M_2/R$, where $M_2 = C_G(B)$, $M_1 = \langle Q^M \rangle$, $M_1/R \cong S_6$, and $M_2/R \cong S_3$.*
(4) *$R/B \cong W_1 \otimes W_2$ as an M/R-module, where W_i is a faithful M_i/R-module of dimension $4, 2$ for $i = 1, 2$, respectively, and $W_1 \cong [B, M]^\sigma$ as an M_1/R-module, where σ induces an outer automorphism on*

$$M_1/R \cong S_6.$$

(5) *$C_G(j_3) \leq M$.*
(6) *M has five classes of subgroups of order 3 with representatives $\langle x_{1i} \rangle$, $\langle x_2 \rangle$, $\langle x_{1i}x_2 \rangle$, $i = 1, 2$, where $\langle x_2 \rangle \in \mathrm{Syl}_3(M_2)$ and $x_{1i} \in M_1$ such that x_{11}, x_{12} fix $0, 3$, members of $B \cap D$, respectively.*
(7) *$x_{12} \in x_{11}^L$, $x_{11} \in N_L(A)$, and $w_D(C_L(x_{11})) = 4$.*
(8) *x_2 is inverted by a member of D and $w_D(C_G(x_2)) = 6$.*
(9) *$w_D(C_G(x_2x_{12})) = 3$.*

Proof: Let $a = d_1$, $b = d_2$, $c = d_3$, and $H = C_L(a)$, and $K = C_L(\langle a, b \rangle)$. By 37.6, H is quasisimple with $H^* = H/\langle a \rangle \cong U_6(2)$ and b^* is a transvection in H^*. Thus $K^* \cong U_4(2)/2^{1+8}$. By 37.6, K is of index 2 in $C_L(j_2)$ with $Q_K = O_2(K) = F^*(C_L(j_2))$ and $C_L(j_2)/Q_K \cong \mathrm{Aut}(U_4(2))$. By 31.5.2, a is not a square in L, so if $x \in Q_K$ is of order 4 then as $Q_K^* \cong 2^{1+8}$, we have $x^2 = j_2$. Therefore $\Phi(Q_K) = \langle j_2 \rangle$.

Let $s \in C_G(j_2) - L$. As $C_L(j_2)/Q_K \cong \mathrm{Aut}(U_4(2)) = \mathrm{Aut}(K/Q_K)$, we can choose s so that $[s, K] \leq Q_K$ and $s^2 \in Q_K$. As K is irreducible on $Q_K/\langle a, b \rangle$, $[s, Q_K] \leq \langle a, b \rangle$ and $s^2 \in \langle a, b \rangle$. Therefore $\langle s, a, b \rangle \trianglelefteq \langle s, Q_K \rangle = Q$ and $Q = O_2(C_G(j_2)) = F^*(C_G(j_2))$.

We saw in paragraph three of the proof of 37.7 that

$$s_0 \in C_G(\langle a, b \rangle) - L$$

induces an outer automorphism on $C_L(a)$ and hence on K/Q_K, so

$$s \notin C_G(\langle a, b \rangle).$$

Therefore $\langle s, a, b \rangle = \langle s, a \rangle \cong D_8$, and as $\langle s, a, b \rangle \trianglelefteq Q$,

$$Q = \langle s, a \rangle * C_Q(\langle s, a \rangle)$$

and

$$C_Q(\langle s, a \rangle) = C_{Q_K}(s) \cong Q_K^* \cong 2^{1+8}.$$

Therefore $Q \cong 2^{1+10}$ is extraspecial. Finally by 8.7 in [SG], Q is large in G, since $j_2^G \cap Q \neq \{j_2\}$. Therefore (1) is established.

Next let $T \in \mathrm{Syl}_2(G)$ with $a, b, c \in T$ and set $A = \langle T \cap D \rangle$. By 25.7, $A \cong E_{2^{10}}$ is the Todd module for $N_L(A)/A \cong M_{22}$, with $T \cap D$ of order 22. Indeed viewing G as $C_{G_0}(a_0 b_0)/\langle a_0, b_0 \rangle$ for $G_0 = M(24)$ and a_0 and b_0, commuting 3-transpositions of G_0, 25.7 says $N_G(A)/A \cong \mathrm{Aut}(M_{22})$ and A is the Todd module for M/A. Thus by Exercise 7.8, there is $E_{32} \cong B \leq A$ such that $B = \langle B \cap D \rangle$ with $|B \cap D| = 6$ and $N_G(A) \cap N_G(B)$ induces S_6 on $B \cap D$. Let $M = N_G(B)$. As $N_G(A)$ is 3-transitive on $A \cap D$, we may take $a, b, c \in B$ and as $N_M(A)$ contains a Sylow 2-subgroup of G, we may take $T \leq N_M(A)$. Therefore (2) holds. Indeed as $N_M(A)$ induces S_6 on $B \cap D$ we may take $j_2 \in Z(T)$. Then $Q = O_2(C_G(j_2)) \leq T$.

As $H^* \cong U_6(2)$ with b^* a transvection in H^*, b^* is the unique transvection in $O_2(C_{H^*}(b^*)) = Q_K^*$, so $Q \cap D = \{a, b\}$. Hence $c \notin Q$. But

$$[c, Q_K/\langle a, b \rangle] \leq (Q_K \cap B)/\langle a, b \rangle$$

is of order 4, so $Q \cap B = Q_K \cap B = [B, Q_K] = [B, Q]$ is a hyperplane of B. Then BQ/Q is a transvection in $KQ/Q \cong U_4(2)$ and $B \cap D = BQ \cap D$, so

$$B \trianglelefteq C_G(j_2) \cap N_G(BQ).$$

As $C_G(j_2) \cap N_G(BQ)$ is maximal in $C_G(j_2)$, $C_G(j_2) \cap N_G(BQ) = C_M(j_2)$.

Let $M_1 = \langle Q^{N_M(A)} \rangle$. Then $M_1 \trianglelefteq N_M(A)$ so as $N_M(A)/O_2(N_M(A)) \cong S_6$ with $QO_2(N_M(A))/O_2(N_M(A)) \not\leq O^2(N_M(A))O_2(N_M(A))/O_2(N_M(A))$, we conclude $N_M(A) = M_1 O_2(N_M(A))$. Therefore

$$O^2(N_M(A)) \leq M_1 O_2(N_M(A)),$$

so as $M_1 \trianglelefteq N_M(A)$, we have $O^2(N_M(A)) \leq M_1$. Also

$$O_2(N_M(A)) = A O_2(O^2(N_M(A)))$$

with $|A : A \cap O^2(N_M(A))| = 2$, so $N_M(A) = M_1 A$ with $|A : A \cap M_1| \leq 2$.

Next as $N_M(A)$ is 2-transitive on $B \cap D$ and $Q = O_2(C_G(j_2))$ with $j_2 = ab$ and $a, b \in B \cap D$, we have $Q^{N_M(A)} = Q^M$, so $M_1 = \langle Q^M \rangle \trianglelefteq M$. Also $A \not\leq O_2(C_M(j_2))$ so $A \not\leq O_2(M)$. Thus $|A : A \cap M_1| = 2$ and $O_2(M_1) = O_2(M_2) = R = C_{M_1}(B)$. Let $M_2 = C_M(B)$. As M_1 induces S_6 on $B \cap D$, $M = M_1 M_2$. Notice $M_1 \cap M_2 = C_{M_1}(B) = R$ and $M = M_1/R \times M_2/R$ with $M_1/R \cong S_6$. So to complete the proof of (3) it remains to show $M_2/R \cong S_3$ and to show that R is special with center B.

First $M_2 \leq C_M(j_2)$ with $C_M(j_2) = M_2(C_M(j_2) \cap N_M(A))$ as $N_M(A)$ induces S_6 on $B \cap D$. Hence as

$$3 = |C_M(j_2) : C_M(j_2) \cap N_M(A)|,$$

$|M_2 : N_{M_2}(A)| = 3$. Then as $N_M(A) = M_1 A$ with $|A : A \cap M_1| = 2$, $|M_2 : R| = 6$ and $M_2/R \cong S_3$.

Next $R \leq M_2 = C_G(B)$ so $B \leq Z(R)$. By Exercise 7.8, M_1/R has 4-dimensional irreducibles $(A \cap M_1)/B$ and $R/(A \cap M_1)$ with $(A \cap M_1)/B \cong W_1^\sigma$, where $W_1 = [B, M_1] = [B, M]$ as an M_1/R-module and σ induces an outer automorphism on M_1/R. From the structure of $C_M(j_2)$, $B = C_R(x)$ for x of order 3 in M_2. Thus $C_{R/B}(x) = 1$ so (4) holds. In particular M is irreducible on R/B, so as $B \leq Z(R)$, either R is special and the proof of (3) is complete, or R is abelian. The latter is impossible as $Q \cap R$ is nonabelian since $|Q : Q \cap R| = 8$.

From the structure of M described in (3), $|C_M(j_3)| = 2^{17} \cdot 3^3$, while by 37.6, $|C_G(j_3)| = 2^{17} \cdot 3^3$, so (5) holds.

Part (6) follows from the structure of M described in (3). As $M_1 \leq N_G(A)$ and $N_L(A)/A \cong M_{22}$ has one class of elements of order 3, $x_{12} \in x_{11}^L$. Thus it remains to determine the width of the centralizer of each element of order 3 and to prove x_2 is inverted by a 3-transposition.

From the structure of K, $C_G(\langle a, b, c \rangle) = I$ has Sylow 3-subgroup $X_3 \times X_3^g$, where $g \in I$ and $X_3 \cong \mathbf{Z}_3$, and I has 3-classes X_i^I, $1 \leq i \leq 3$, of order 3, where X_1 centralizes $[Q_K/\langle a, b \rangle, B]$ and hence also B, and is inverted by a member of $I \cap D$, while X_2 and X_3 are faithful on $[Q_K/\langle a, b \rangle, B]$ and hence regular on $c^{Q_K} - \{c\}$. Next $\bar{K} = K/O_2(K) \cong U_4(2)$ with $\bar{D}$ the set of transvections and $w_{\bar{D}}(C_{\bar{K}}(\bar{X}_i)) = 1$ for $i = 1, 2$ and equaling 2 if $i = 2$. Therefore $C_{D \cap K}(X_1) = c^{Q_K} \cup \{a, b\}$ is of order 6, so that $w_D(C_G(X_1)) = 6$, while $w_D(C_G(X_i)) = w_D(C_K(X_i)) = w_{\bar{D}}(C_{\bar{K}}(X_i)) + 2 = 4, 3$ for $i = 2, 3$, respectively. Finally by 23.9, H has three classes of subgroups of order 3, so these must be X_i^H, $1 \leq i \leq 3$. Therefore we have proved

(37.9) *Let $G = M(22)$ and D the set of 3-transpositions of G. Then G has three conjugacy classes X_i^G, $1 \leq i \leq 3$, of subgroups of order 3 centralizing a member of D. Further $w_D(C_G(X_i)) = 6, 4, 3$, for $i = 1, 2, 3$, respectively.*

Returning to the proof of 37.8, observe that as

$$B \cap D \subseteq C_M(x_2) \quad \text{and} \quad |B \cap D| = 6,$$

37.9 says that $\langle x_2 \rangle \in X_1^L$. Then as X_1 is inverted by a member of $I \cap D$, (8) holds. Next $\langle x_2, x_{12} \rangle \in \text{Syl}_3(I)$ with $x_{12} \in N_G(A)$, so $|C_{A \cap D}(x_{12})| = 4$. By previous remarks, a Sylow 3-subgroup of I contains a unique member of X_1^L, so, as $\langle x_2 \rangle$ is in X_1^L, $\langle x_{12} \rangle$ is not, and hence $w_D(C_G(x_{12})) \leq 4$. Therefore by 37.9, $w_D(C_G(x_{12})) = 4$ and $\langle x_{12} \rangle \in X_2^L$. Similarly $\langle x_2 x_{12} \rangle \in X_3^L$, and the proof is complete.

13

Elements of order 3 in orthogonal groups over GF(3)

In this chapter we develop an algorithm for describing classes of elements of order 3 in orthogonal groups $O(V)$ over the field of order 3. The description of elements of order 3 is in terms of their action on the n-dimensional orthogonal space V and in particular their Jordon block structure on V.

Namely write $2^a \cdot 3^b_\epsilon$ for the set of elements $g \in O(V)$ such that g has a Jordon blocks of size 2 and b blocks of size 3, and the orthogonal space $[V, g]/C_{[V,g]}(g)$ is of discriminant ϵ. Then by Lemma 38.10, the classes of $O(V)$ are $2^a \cdot 3^b_\epsilon$, where a is even, $\epsilon = \pm 1$, and $2a + 3b \le n$ with $\mu(V) = \epsilon$ in case of equality. There is also a discussion as to how these classes split in groups G between $\Omega(V)$ and $O(V)$.

38. Orthogonal groups

In this section V is an n-dimensional orthogonal space over a field F of characteristic other than 2 with bilinear form $(\ ,\ )$ and $1 \ne g \in O(V)$ is unipotent. Let $\theta = 1 - g \in \text{End}(V)$, so that θ is nilpotent. Observe $V\theta = [V, g]$ and $\ker(\theta) = C_V(g)$.

Let m be the nilpotence degree of θ; that is

$$m = \min\{k : \theta^k = 0\}.$$

As $g \ne 1, m > 1$, and indeed

(38.1)   m is the maximum size of a Jordon block of g.

(38.2) $V\theta^k = \ker(\theta^k)^\perp$ for $k \ge 0$.

Proof: Induct on k; the result is trivial if $k = 0$. Let $x \in V$ and $y \in \ker(\theta^{k+1})$. Then $y\theta \in \ker(\theta^k)$ and g acts on $V\theta^k$ so $x\theta^k g \in V\theta^k$. Hence by induction on

$k, 0 = (x\theta^k g, y\theta) = (x\theta^k g, y - yg)$. Therefore

$$(x\theta^k g, y) = (x\theta^k g, yg) = (x\theta^k, y),$$

so

$$(x\theta^{k+1}, y) = (x\theta^k - x\theta^k g, y) = 0.$$

That is $V\theta^{k+1} \leq \ker(\theta^{k+1})^\perp$. Then as

$$\dim(V\theta^{k+1}) = n - \dim(\ker(\theta^{k+1})) = \dim(\ker(\theta^{k+1})^\perp),$$

we have $V\theta^{k+1} = \ker(\theta^{k+1})^\perp$, completing the proof.

Define the *degree* of $x \in V$ by

$$\deg(x) = \min\{k : x \in \ker(\theta^k)\}.$$

(38.3) *There exists a decomposition $V = V_1 \oplus \cdots V_m$, where V_k is the sum of Jordon blocks $V_{k,i}$, $1 \leq i \leq n_k$, of size k for g, and for each such decomposition*

(1) $V_i \leq \ker(\theta^r)$ *for* $i \leq r$, *while* $\dim(V_k\theta^r) = n_k(k-r)$ *and* $\dim(\ker(\theta^r) \cap V_k) = rn_k$ *for* $r < k \leq m$.

(2) $V_k\theta^r = \ker(\theta^{k-r}) \cap V_k$ *for* $1 \leq r < k$.

(3) $V\theta^{m-1} = V_m\theta^{m-1} = C_{V_m}(g) = \ker(\theta) \cap V_m$ *and*

$$C_{V_m}(g)^\perp = \ker(\theta^{m-1}) = V_1 \oplus \cdots V_{m-1} \oplus [V_m, g].$$

Proof: Parts (1) and (2) follow from the definition of Jordon block. By (1), $V\theta^{m-1} = V_m\theta^{m-1}$ and by (2), $V_m\theta^{m-1} = \ker(\theta) \cap V_m$. Of course $\ker(\theta) = C_V(g)$. Next by 38.2, $C_{V_m}(g)^\perp = (V\theta^{m-1})^\perp = \ker(\theta^{m-1})$. But by (1),

$$\ker(\theta^{m-1}) = V_1 \oplus \cdots V_{m-1} \oplus (\ker(\theta^{m-1}) \cap V_m),$$

whereas by (2), $V_m\theta = \ker(\theta^{m-1}) \cap V_m$. So as $[V, g] = V\theta$, (3) is established.

(38.4) *There exists an orthogonal decomposition $V = V_1 \perp \cdots \perp V_m$ where V_k is the sum of Jordon blocks of g of size k.*

Proof: Let $E = V\theta^{m-1}$ and pick a basis $e_1, \ldots, e_r$ for E and $x_i \in V$ with $x_i\theta^{m-1} = e_i$. Therefore x_i is of degree m, so the $F[g]$-submodule U_i generated by x_i is a Jordon block for g of size m with $C_{U_i}(g) = U_i\theta^{m-1} = Fe_i$. Let $U = U_1 + \cdots + U_r$.

As $Fe_i = C_{U_i}(g)$ and $e_1, \ldots, e_r$ are independent, $U = U_1 \oplus \cdots \oplus U_r$ and $E = C_U(g)$. Let $A = \mathrm{Rad}(U)$. By 38.2, $E^\perp \cap U = \ker(\theta^{m-1}) \cap U$ and hence is of codimension r in U. So as $r = \dim(E)$, $E \cap A = 0$ and therefore U is nondegenerate, so $V = U \oplus U^{\perp}$. Now by 38.3, the number of Jordon blocks of g of size m is $\dim(E) = r$, so all the blocks of g on $U^{\perp}$ are of size less than m. Hence proceeding by induction on $\dim(V)$, $U^{\perp} = V_1 \perp \cdots \perp V_{m-1}$ and we let $U = V_m$.

(38.5) *If $x \in v$ is of degree 2 then $[x, g]$ is a singular vector orthogonal to x.*

Proof: Let $y = [x, g]$. By 38.2, y is singular so

$$(x, x) = (xg, xg) = (x + y, x + y) = (x, x) + 2(x, y),$$

and therefore $(x, y) = 0$.

(38.6) *Assume V is the sum of Jordon blocks for g of size 2. Then*

(1) $V = V_1 \perp \cdots \perp V_k$ where V_i is g-invariant of dimension 4 and sign $+1$.
(2) V_i has a basis $\{e_1, e_2, x_1, x_2\}$ of singular vectors such that

$$[x_i, g] = e_i, \qquad (e_1, x_2) = 1, \qquad (e_2, x_1) = -1,$$

and all other inner products of basis vectors are 0.

Proof: Let $V = U_1 \oplus \cdots \oplus U_r$ be the sum of Jordon blocks for g. By hypothesis each block U_i is of dimension 2. Now U_1 has a basis $\{e_1, x_1\}$ with $[x_1, g] = e_1$. By 38.5, e_1 is singular and $(e_1, x_1) = 0$.

Next let $x_2 \in V - e_1^{\perp}$. By 38.2, $x_2 \notin \ker(\theta)$ so as $\ker(\theta)$ is the set of vectors of degree at most 1, $\deg(x_2) = 2$. Thus $F[g]x_2$ is a Jordon block of size 2, say $U_2 = F[g]x_2$ and $e_2 = [x_2, g]$. By 38.2, $\langle e_1, e_2 \rangle$ is totally singular. Multiplying x_2 by a suitable scalar, we may assume $(e_1, x_2) = 1$. Now

$$(x_1, x_2) = (x_1 g, x_2 g) = (x_1 + e_1, x_2 + e_2) = (x_1, x_2) + (x_1, e_2) + 1,$$

so $(x_1, e_2) = -1$.

Suppose $(x_1, x_1) \neq 0$ and let $a = (x_1, x_1)/2$ and $x_1' = x_1 + ae_2$. Replacing x_1 by x_1', we may assume x_1 is singular. Similarly we may take x_2 singular. Let $U = U_1 + U_2$, so that U is 4-dimensional and U_1 and U_2 are totally singular. Also $U_1^{\perp} \cap U_2 = 0 = U_1 \cap U_2^{\perp}$ from the preceding relations, so $U_i = U_i^{\perp} \cap U$ and therefore U is nondegenerate of Witt index 2. Therefore proceeding by induction on the dimension of V, we may take $V = U$.

Finally $x_1^{\perp} \cap U_2 = Fy_2$ where $y_2 = x_2 + be_2$ for some $b \in F$. As U_2 is totally singular, y_2 is singular. Also $[g, y_2] = [g, x_2] = e_2$, so replacing x_2 by y_2 we may assume $(x_1, x_2) = 0$. This completes the proof.

(38.7)   Assume F is finite and V is the sum of Jordon blocks of size 3. Let $E = V\theta^2$ and $Y = V\theta$. Then

(1) $V = V_1 \perp \cdots V_k$ where V_i is a nondegenerate Jordon block of g of size 3.

(2) Y/E is a k-dimensional orthogonal space under the induced form

$$(x + E, y + E) = (x, y),$$

and if $(y_i + E : 1 \le i \le k)$ is an orthogonal basis for Y/E then we can choose V_i with $y_i \in V_i$ for each i.

(3) The isomorphism type of the representation of g on the orthogonal space V is determined by the isometry type of the orthogonal space Y/E.

Proof: Observe first that by 38.3, $\dim(E) = \dim(Y/E) = k$ with E totally singular and $Y = E^\perp$. Thus $\bar{Y} = Y/E$ is an orthogonal space via $(\bar{x}, \bar{y}) = (x, y)$.

Let V_1 be a Jordon block of g, $x_1 \in V_1$ of degree 3, $y_1 = [x_1, g]$, and $e_1 = [y_1, g]$. Suppose first $(x_1, e_1) \ne 0$. Then $e_1 \notin \text{Rad}(V_1)$, so as $Fe_1 = C_{V_1}(g)$, V_1 is nondegenerate. Then $V = V_1 \oplus V_1^\perp$ and by induction on the dimension of V, $V_1^\perp = V_2 \perp \cdots \perp V_k$ is the orthogonal direct sum of Jordon blocks for g. Further $\bar{Y} = \bar{Y}_1 \perp \cdots \perp \bar{Y}_k$, where $Y_i = V_i\theta = Y \cap V_i$, so $y_i \in Y_i - E$ is nonsingular. Conversely given an orthogonal basis $(\bar{z}_i : 1 \le i \le k)$ for $\bar{Y}$, let $e_i = [z_i, g]$, $Z_i = \langle z_i, e_i \rangle$, and $W_i = \langle Z_j : j \ne i \rangle$. Then

$$A_i = \langle e_j : j \ne i \rangle = \text{Rad}(W_i^\perp)$$

with $W_i^\perp/A_i$ a Jordon block of size 3 for g and $Z_i = W_i^\perp\theta$. So each $x_i \in W_i^\perp - (Z_i + A_i)$ is of degree 3 with $X_i = \langle x_i, Z_i \rangle$ a Jordon block of size 3. Then X_1 is nondegenerate so $V = X_1 \oplus X_1^\perp$ with $W_1 = X_1^\perp\theta$, so by induction on $\dim(V)$, (2) holds.

Further replacing x_i by $x_i + be_i$ for suitable $b_i \in F$, we may take x_i singular. Let $c_i = (z_i, z_i)$. Then

$$(x_i, z_i) = (x_ig, z_ig) = (x_i + z_i, z_i + e_i) = (x_i, z_i) + c_i + (x_i, e_i),$$

so $(x_i, e_i) = -c_i$. Also

$$(x_i, x_i) = (x_i + z_i, x_i + z_i) = (x_i, x_i) + c_i + 2(x_i, z_i),$$

so $(x_i, z_i) = -c_i/2$. Hence the inner product of basis vectors is determined by the parameters c_i, $1 \le i \le k$. This implies (3).

Thus we have reduced to the case where for all $x \in V - Y$, $x\theta^2 \in x^\perp$. Then $e = x\theta^2$ is in the radical of $U = F[g]x = \langle x, y, e \rangle$, where $y = x\theta$. Passing to $e^\perp/Fe$ we have the Jordon block U/Fe of size 2, so by 38.5, y is singular and $\langle y, e \rangle \leq \text{Rad}(U)$. But now choosing $y_0 \in Y$ nonsingular and $x_0 \in V$ with $x_0 \theta = y_0$, $x_0 \theta^2 \notin x_0^\perp$, contrary to our assumption. Thus the proof is complete.

In the remainder of this section assume $\text{char}(F) = 3$ and g is of order 3. Then all Jordon blocks of g on V are of size 1, 2, or 3, so we can appeal to the lemmas in this section to conclude

(38.8)

(1) $V = V_1 \perp V_2 \perp V_3$, where V_k is the sum of Jordon blocks for g of size k.

(2) $V_3 = V_{3,1} \perp \cdots \perp V_{3,m_3}$, where $V_{3,i}$ is a nondegenerate Jordon block for g of size 3.

(3) $V_2 = V_{2,1} \perp \cdots \perp V_{2,m_2/2}$ where $V_{2,i}$ is a nondegenerate 4-subspace of V of sign $+1$ and $V_{2,i} = V_{2,i,1} \oplus V_{2,i,2}$ with $V_{2,i,j}$ a totally singular Jordon block for g of size 2.

(4) The conjugacy class of g in $O(V)$ is determined by the number m_k of Jordon blocks of g of size k, $1 \leq k \leq 3$, and the isometry type of $V\theta/C_{V\theta}(g) \cong V_3\theta/V_3\theta^2$, a nondegenerate orthogonal space of dimension m_3.

Proof: Part (1) follows from 38.4. Then (2) follows from 38.7 and (3) from 38.6.

Let $G = O(V)$. Certainly if $h \in g^G$ then h and g have the same Jordon decomposition and

$$V\theta_g/C_{V\theta_g}(g) \cong V\theta_h/C_{V\theta_h}(h).$$

Conversely assume $h \in G$ satisfies these constraints. Let U_i, U_{ij}, and so on, be the blocks in a decomposition for h. As

$$V\theta_g/C_{V\theta_g}(g) \cong V\theta_h/C_{V\theta_h}(h),$$

38.7.3 says there exists an isometry $\alpha : V_3 \to U_3$ inducing a quasiequivalence of the representations of g and h. By Witt's Lemma 10.9, α extends to $x \in G$. Then $V_3^\perp x = U_3^\perp$ and if $V_3 \neq 0$ then by induction on $\dim(V)$ there exists $y \in C_G(U_3)$ with $g_{|U_3^\perp}^{xy} = h_{|U_3^\perp}$. Now $g^{xy} = h$, establishing (4) in this case, so we may take $V_3 = 0$.

Next as h and g have the same number of Jordon blocks of size 2, by 38.6 there is an isometry $\beta : V_2 \to U_2$ inducing a quasiequivalence of the representations of g and h, and arguing as in the previous paragraph, β extends to $z \in G$ with $g^z = h$. Therefore (4) is established, completing the proof.

In the remainder of this section assume F is of order 3. Using 38.8 we can identify the conjugacy classes of subgroups of $O(V)$ of order 3. To do so we establish some notation.

NOTATION: Write $2^a \cdot 3^b_\epsilon$ for the set of all elements g of $O(V)$ of order 3 such that g has a Jordon blocks of size 2 and b Jordon blocks of size 3, and the orthogonal space $V\theta/C_{V\theta}(g)$ is of discriminant ϵ. In particular if $b = 0$ then g centralizes $V\theta$, so $\epsilon = +1$, since by convention the discriminant of the trivial space is 1. The following facts from Exercise 4.2 will be useful:

(38.9)

   (1) *If* $V = U \perp W$ *then* $\mu(V) = \mu(U)\mu(W)$.

   (2) *If* n *is even then* $\mu(V) = (-1)^{n/2} \, \mathrm{sgn}(V)$.

(38.10) *The conjugacy classes of $O(V)$ are $2^a 3^b_\epsilon$, where a is even, $\epsilon = \pm 1$, and $2a + 3b \leq n$ with $\epsilon = (-1)^n \mu(V)$ in case of equality.*

Proof: This is almost immediate from 38.8, although we do need to check that if $n = 2a + 3b$ then $\epsilon = (-1)^n \mu(V)$. This follows from parts (2), (3), and (5) of the following lemma:

(38.11) *Let* $V = V_1 \perp V_2 \perp V_3$ *with* V_k *the sum of Jordon blocks for g of size k and let* $g \in 2^a 3^b_\epsilon$. *Then*

   (1) $\mu(V_2) = 1$ *and* $\dim(V_2) \equiv 0 \bmod 4$.

   (2) $n + b \equiv \dim(V_1) \bmod 4$.

   (3) $\mu(V_1)\mu(V_2) = \mu(V_1 + V_3) = \mu(V)$.

   (4) *If U is a Jordon block of g of size 3 then*

$$\mu(U) = Q(y) = -\mu\,(U\theta/C_{U\theta}(g))$$

   for $y \in U\theta - C_{U\theta}(g)$.

   (5) $\mu(V_3) = (-1)^b \epsilon$.

Proof: V_2 is the orthogonal direct sum of 4-dimensional subspaces of sign $+1$, and by 38.9.2, each such subspace is of discriminant 1. Hence $\dim(V_2) \equiv 0 \bmod 4$ and $\mu(V_2) = 1$ by 38.9.1. That is (1) holds. Then part (2) follows as $\dim(V_2) \equiv 0 \bmod 4$ and $\dim(V_2) = 3b \equiv -b \bmod 4$. Similarly part (3) follows from 38.9.1 as $\mu(V_2) = +1$. Under the hypotheses of (4), $U = Fy \perp W$ with W a hyperbolic line, so by 38.9,

$$\mu(U) = \mu(W)\mu(Fy) = -\mu(Fy) = Q(y),$$

so (4) holds as Fy is isometric to $U\theta/C_{U\theta}(g)$. Now by 38.8,

$$V_3 = V_{3,1} \perp \cdots \perp V_{3,b}$$

with $V_{3,i}$ a Jordon block of size 3. Then setting $Y = V_3\theta$, $\bar{Y} = Y/C_Y(g)$, and $Y_i = V_{3,i} \cap Y$, we have $\bar{Y} = \bar{Y}_1 \perp \cdots \perp \bar{Y}_b$. Hence by (4) and 38.9,

$$\epsilon = \mu(\bar{Y}) = \prod_i \mu(\bar{Y}_i) = \prod_i -\mu(V_{3,i}) = (-1)^b \mu(V_3),$$

establishing (5).

Notice that by 38.10, if $b = 0$ or $2a + 3b = n$ then the sign ϵ for the class of g is forced, so we usually omit that sign.

(38.12) *Let $g \in 2^a \cdot 3_\epsilon^b$ and $V_1 = C_V(g)$. Then one of the following holds:*

(1) *$\Omega(V)$ is transitive on $2^a \cdot 3_\epsilon^b$.*

(2) *$n \equiv 0 \bmod 4$, $\mathrm{sgn}(V) = +1$, $n = 2a$, and $\Omega(V)$ has two or four orbits on 2^a. Further if r is a reflection then $\langle r \rangle \Omega(V)$ has one or two orbits on 2^a.*

(3) *$n \equiv 0 \bmod 4$, $\mathrm{sgn}(V) = +1$, $n = 2a + 4$, $b = 1$, $\Omega(V)$ has two orbits on $2^a \cdot 3_\epsilon$ and if $x \in V$ with $Q(x) = -\epsilon$ and r_x is the reflection through x, then $\langle r_x \rangle \Omega(V)$ is transitive on $2^a \cdot 3_\epsilon$.*

(4) *$n \equiv 1 \bmod 4$, $n = 2a + 1$, $\mu(V_1) = \mu(V)$, and $\Omega(V)$ has two orbits on 2^a.*

(5) *$n \equiv 3 \bmod 4$, $n = 2a + 3$, $b = 1$, $\mu(V) = -\epsilon$, and $\Omega(V)$ has two orbits on $2^a \cdot 3_\epsilon$.*

Proof: Let $L = \Omega(V)$, $G = O(V)$, and $h \in g^G$. Now L is transitive on decompositions $V = V_1 \perp V_{2,1} \perp \cdots \perp V_{2,a/2} \perp V_{3,1} \perp \cdots \perp V_{3,b}$ of the isometry type determining g^G, so without loss g and h have the same decomposition. For W a summand, let $L(W)$, $O(W)$ be the subgroup of L, G, trivial on all summands except W, respectively, and $g(W)$ the projection of g on $L(W)$. Therefore $g = \prod_W g(W)$ and for $W \neq V_1$, $L(W)$ has 2,4 orbits $\mathcal{O}_i(W)$ on elements of order 3 in $L(W)$ with the same orthogonal Jordan decomposition as $g(W)$, for $\dim(W) = 3, 4$, respectively. Further $O(W)$ is transitive on the orbits $\mathcal{O}_i(W)$, so there exists $r(W) \in O(W)$ with $g(W)^{r(W)} = h(W)$. Let $r = \prod_W r(W)$. Then $g^r = h$ and it remains to decide when $r \in LC_G(g)$.

Next, representatives for the cosets of L in G are $1, r_u, r_v, r_u r_v$, where r_x is the reflection with center x and $Q(u) = +1$ and $Q(v) = -1$. If $\dim(V_1) > 1$ we can choose $u, v \in V_1$ and hence $r_u, r_v \in C_G(g)$, so $G = LC_G(g)$. That is (1) holds unless $\dim(V_1) \leq 1$, so we assume $\dim(V_1) \leq 1$.

Next if $\dim(W) = 3$ let $x \in W$ with $Q(x) = -\mu(W)$ and $r_x \in G$ be the reflection through x. Then $-id_W \in C_G(g)$ and $-id_W r_x$ has a 2-dimensional commutator space of sign -1, so $-id_W r_x \in L$. That is $r_x \in C_G(g)L$ with $Q(x) = -\mu(W)$. But if $b > 1$ we can find $\bar{w}_\epsilon, \epsilon = \pm1$, in the orthogonal space $V\theta/C_{V\theta}(g)$ with $Q(\bar{w}_\epsilon) = \epsilon$, giving Jordon blocks W_ϵ of dimension 3 with $w_\epsilon \in W_\epsilon$, so that by 38.11.4, $\mu(W_\epsilon) = \epsilon$. Hence both reflections r_u and r_v are in $LC_G(g)$, so that (1) holds. Thus we may assume $b \leq 1$.

Suppose n is even. Then by 38.11.2, either $b = \dim(V_1) = 0$ or $b = \dim(V_1) = 1$. In the first case $V = V_2$ is of sign $+1$ by 38.11.1 and 38.9.2, so $n = 2a$ and $[V, g]$ is a maximal totally singular subspace of V. Now $\Omega(V)$ has two orbits on such subspaces fused by reflections, so as $G/L \cong E_4$, (2) holds.

So assume $b = 1 = \dim(V_1)$. Then $V_1 + V_3$ is a 4-dimensional space of sign $sgn(V) = \mu(V)$ by 38.11.3 and 38.9.2. Now $r_y \in C_G(g)$ for $y \in V_1$, and $Q(y) = -\mu(V_1)$, while $r_x \in LC_G(g)$ for $Q(x) = -\mu(V_3)$ by an earlier remark. Also by 38.11.5, $\epsilon = -\mu(V_3)$, so if $\epsilon = \mu(V_1)$, then (1) holds. Thus we may assume $\epsilon = -\mu(V_1)$, so by 38.9 and 38.11.3,

$$sgn(V) = \mu(V) = \mu(V_1)\mu(V_3) = (-\epsilon)(-\epsilon) = +1.$$

As $G/L \cong E_4$ and $r_y \in C(g)$ with $Q(y) = \epsilon$, L has two orbits on g^G and $\langle r_x \rangle L$ is transitive on g^G for $Q(x) = \mu(V_3) = -\epsilon$. Therefore (3) holds.

So assume n is odd. Then either $b = 0$ and $\dim(V_1) = 1$ or $b = 1$ and $\dim(V_1) = 0$. In the first case (4) holds and in the second case (5) holds.

If d is a reflection on V then $d^{\Omega(V)} = D$ is a conjugacy class of 3-transpositions of $\langle D \rangle \cong O_n^{\mu,\pi}(3)$ from Section 11. Indeed:

(38.13) *Let $n > 5$, $v \in V$ with $Q(v) = \pi = \pm1$, $d = t(v) \in O(V)$ the reflection through v, g an element of order 3 in $O(V)$ inverted by d, $S = V\theta^2$, $G = \langle d \rangle \Omega(V)$, $M = N_G(S)$, $r \in V - S^\perp$ singular, $W = \langle S, r \rangle$, and $\delta : S^\perp \to W^\perp$ the projection with respect to the decomposition $S^\perp = S \oplus W^\perp$. Then*

(1) $g \in 3_{-\pi}$.

(2) $M = K\langle t \rangle A$, *where* $A = O_3(M) \cong E_{3^{n-2}}$, $K = \langle C_{D_\pi}(W) \rangle \cong O_{n-2}^{-\mu,\pi}(3)$, $t \in G$ *is an involution with* $W \leq [V, t]$ *of dimension 4 and sign $+1$, and* $K\langle t \rangle \cong O(W^\perp)$.

(3) *The map* $h \mapsto [r, h]\delta$ *is an equivalence of the representations of K on A and $W^\perp$.*

(4) *The elements of $A^{\#}$ are in 2^2 or 3_ϵ with $h \in 2^2$ if $[r, h]$ is singular and $h \in 3_\epsilon$ if $\mu(F[r, h]) = \epsilon$. M is transitive on these three classes of elements of order 3.*

(5) $g \in A$ and for $h \in A \cap 3^\epsilon$, $A \leq N_G(\langle h \rangle) \leq M$ with $N_G(\langle h \rangle)/A \cong \mathbf{Z}_2 \times O(Z^\perp)$, where Z is a nondegenerate Jordon block of h, $\mu(Z) = -\epsilon$, and $\mu(Z^\perp) = -\epsilon\mu$.

Proof: g of order 3 is inverted if and only if $a = gd \in A_d$, in which case $[V, g] \leq [V, d] + [V, a]$ is of dimension at most 2 and contains the nonsingular vector v in case of equality. Therefore by 38.10, $g \in 3_{-\epsilon}$, where $\epsilon = Q(x)$ for $x \in V\theta - V\theta^2$, and $V\theta = [V, g] = [V, d] + [V, a]$. Then $v \in V\theta - V\theta^2$, so $\epsilon = \pi$. Therefore (1) holds.

Parts (2) and (3) are restatements of portions of Lemma 11.10. Also from 11.10, $A = C_G(S) \cap C_G(S^\perp/S)$, so $h \in A^\#$ if and only if $V\theta_h \leq S^\perp$ and $S^\perp\theta_h \leq S$. In particular $\dim(V\theta_h) \leq 1 + \dim(S^\perp\theta_h) \leq 2$ and $\dim(V\theta_h^2) \leq 1$, so we conclude from 38.10 that $h \in 2^2$ or 3_ϵ. In the first case $V\theta_h$ is totally singular so $[r, h]$ is singular. On the other hand if $h \in 3^\epsilon$ then $[r, h] \in V\theta_h - V\theta_h^2$, so $\mu(F[r, h]) = \epsilon$. Then using (3) and the fact that K has three orbits on points of $W^\perp$, we conclude that (4) holds. Similarly using the equivalence in (3),

$$N_{K\langle t \rangle}(\langle h \rangle) \cong N_{O(W^\perp)}(\langle [r, h]\delta \rangle) = O(\langle [r, h]\delta \rangle) \times O(Z^\perp) \cong \mathbf{Z}_2 \times O(Z^\perp),$$

where $W \leq Z$ is a Jordon block for h. By 38.9,

$$\mu(Z^\perp) = \mu(Z)\mu(V) = \mu(F[r, h])\mu(W)\mu = -\epsilon\mu.$$

(38.14) Let $g \in 2^2$, $G = O(V)$, and $X = \langle g \rangle$. Then $N_G(X)$ is the stabilizer in G of the totally singular line $[V, X]$ and is the split extension of $O_3(N_G(X)) \cong 3^{1+2(n-4)}$ by $GL_2(3) \times O(U^\perp)$, where U is a nondegenerate subspace of dimension $n - 4$ and discriminant $\mu(V)$.

Proof: As X is the unique subgroup X_0 of order 3 with $[V, X] = [V, X_0]$, $N_G(X)$ is the stabilizer in G of $[V, X]$. The result follows.

(38.15) Let $n = 7$, $\mu(V) = -1$, $G = \Omega(V) \cong \Omega_7(3)$, and D the set of involutions d in G with $[V, d]$ of dimension 6 and sign -1. Then

(1) G has seven classes of subgroups of order 3: 2^2, 3_+, 3_-, 3_+^2, 3_-^2, $2^2 \cdot 3_1$, and $2^2 \cdot 3_2$, with $2^2 \cdot 3_1 \cup 2^2 \cdot 3_2 = 2^2 \cdot 3_+$.

(2) 3_- is the set of subgroups of G of order 3 inverted by members of D.

(3) For $Z = \langle g \rangle \in 2^2$, $\langle C_D(Z) \rangle = K_1 * K_2 * K_3$ with $K_i \cong \mathbf{Z}_2/3^{1+2}$, $C_D(Z) = \bigcup_i K_i \cap D$, and

$$F^*(N_G(Z)) = Q = Q_1 * Q_2 * Q_3 \cong 3^{1+6},$$

where $Q_i = O_3(K_i)$.

(4) $3_- \cap Q = \left(\bigcup_i Q_i\right) - Z$,

$$(3_+ \cup 3_-^2) \cap Q = \bigcup_{i,j} (Q_i Q_j - (Q_i \cup Q_j)),$$

and $N_G(Z)$ has four orbits Δ_k, $1 \leq k \leq 4$, on $\Delta = Q - \bigcup_{i,j} Q_i Q_j$ where

$$\Delta_1 = Q \cap 2^2 - \{Z\}, \qquad \Delta_2 = 3_+^2 \cap Q,$$

$$\Delta_3 = 2^2 \cdot 3_1 \cap Q, \qquad \Delta_4 = 2^2 \cdot 3_2 \cap Q$$

are of length $3 \cdot 2^4$, $3^3 \cdot 2^4$, $3^2 \cdot 2^4$, and $3^2 \cdot 2^4$, respectively.

(5) For $X = \langle h \rangle \in 2^2 \cdot 3$, $V_0 = C_V(X)$ is a maximal totally singular subspace of V, $N_G(X) \leq N_G(V_0)$, $|O_3(N_G(X))| = 3^6$, and

$$N_G(X)/O_3(N_G(X)) \cong GL_2(3).$$

Proof: To prove (1) given 38.12, we must prove $2^2 \cdot 3_+$ splits into two classes under G. Let $h \in 2^2 \cdot 3_+$, $I = O(V)$, $G^+ = G\langle -1 \rangle$, $X = \langle h \rangle$, and $M = N_I(X)$. It suffices to show $M \leq G^+$. Let $V_0 = C_V(h)$ and $U = V\theta^2$. Then V_0 is a totally singular 3-subspace and U is a point in V_0. Therefore

$$M \leq N_I(V_0) \cap N_I(U) = P$$

a parabolic of I with $O_3(P)$ of order 3^8 and $P/O_3(P) \cong \mathbf{Z}_2 \times GL_2(3)$.

Next $V = V_2 \perp V_3$ with V_k the sum of Jordon blocks for h of size k. Further $h = h_2 h_3$ with $h_2 \in 2^2$, $h_3 \in 3_+$, and V_k the sum of nontrivial Jordon blocks for h_k. Then

$$C_I(h) \cap N_I(V_2) = C_{G^+}(h) \cap N_{G^+}(V_2) = \langle -1 \rangle (C_G(h) \cap N_G(V_2))$$

with $C_G(h) \cap N_G(V_2) \cong SL_2(3) \times E_9$ and h is inverted by $s \in N_G(V_2)$. Therefore $C_G(h) \cap N_G(V_2)$ contains a unique involution t and as $V_2 = [V, t]$, $C_I(ht) \leq N_I(V_2)$, so that $N_I(\langle ht \rangle) = N_{G^+}(\langle ht \rangle)$ and

$$N_G(\langle ht \rangle)/O_3(N_G(\langle h \rangle)) \cong GL_2(3) \cong (P \cap G)/O_3(P),$$

so $P \cap G = N_G(\langle ht \rangle) O_3(P)$ and then by a Frattini argument,

$$N_I(X) \leq N_I(\langle ht \rangle) O_3(P) = N_{G^+}(\langle ht \rangle) O_3(P) \leq G^+,$$

completing the proof of (1). Further we have also established (5) once we prove $|C_G(h)|_3 = 3^7$.

Notice $d = -r_x$ for $Fx = [V, d]^\perp$, so the subgroups inverted by d and r_x are the same. By 38.9, $Q(x) = -\mu(Fx) = -\mu(V)\mu([V, d]) = 1$, so $Q(x) = 1$. Now 38.13 implies (2).

Let $Z = \langle g \rangle \in 2^2$. By 38.14, $N_G(Z)$ is the stabilizer of the singular line $[V, Z]$. Now $C_V(Z) = [V, Z] \perp U_1 \perp U_2 \perp U_3$, where the union of the sets $[V, Z] + U_i - [V, Z]$, is the set of vectors $u \in C_V(Z)$ with $Q(u) = 1$. Let $I_i = C_I([V, Z] + U_j + U_k)$ for $\{i, j, k\} = \{1, 2, 3\}$. Then $I_i \leq M_i = C_I(U_j + U_k) \cong O_5(3)$, so $I_i = C_{M_i}([V, Z]) = Q_i \langle r_i \rangle$, where r_i is the reflection with center U_i and $Q_i \cong 3^{1+2}$ is the unipotent radical of $N_{M_i}([V, Z])$. Now $Q = Q_1 Q_2 Q_3 \cong 3^{1+6}$, so by 38.14, $Q = F^*(N_G(Z))$. Also $d_i = -r_i \in D$ and as Z centralizes $C_V(d)$ for each $d \in C_D(Z)$, $C_D(Z) = \bigcup_i d_i^Q$, and $\langle C_D(Z) \rangle = K_1 * K_2 * K_3$, where $K_i = \langle d_i \rangle Q_i$ and $d_i^Q = d_i^{Q_i}$. Therefore (3) is established.

Next for f of order 3 in G, $f \in Q$ if and only if f centralizes the flag $[V, Z] < C_V(Z) < V$, with $C_V(Z) = [V, Z]^{\perp}$. For example if $f \in 3_-$ then $V = V_3 \perp V_1$ where V_3 is a Jordon block of f of size 3 and V_1 is of dimension 4 and sign -1. Thus if l is a totally singular line centralized by f then $l = V\theta_f^2 + U$ where U is a singular point of V_1 and $C_{l^{\perp}}(f) = l \perp W$ where W is a complement to U in $U^{\perp} \cap V_1$, and is 2-dimensional of sign -1. It follows that $f \in Q$ if and only if $[V, Z]$ is such an l, in which case f centralizes two 3-spaces $l + Fw_1$ and $l + Fw_2$ with $w_i \in W$ and $Q(w_i) = 1$. Therefore $f \in Q_i$ for some i, so that $\bigcup_i Q_i - Z = Q \cap 3_-$.

Similarly we calculate that if $f \in 3_+$ or 3_+^2 and l is a totally singular line of $C_V(f)$ then f centralizes at most one 3-subspace $l + Fw$ of $C_V(f) \cap l^{\perp}$ with $Q(w) = 1$, and such a 3-subspace exists for suitable l. Also for f in the remaining classes, no such 3-subspace exists. Therefore $\bigcup_{i,j} Q_i Q_j - (Q_i \cup Q_j) = (3_+ \cup 3_+^2) \cap Q$. Also $C_G(f)$ is transitive on singular lines l such that f centralizes the flag $l < l^{\perp} < V$ (excluding $l = [V, f]$, when $f \in 2^2$) so $N_G(Z)$ has four orbits Δ_k, $1 \leq k \leq 4$, on the set Δ of part (4), with

$$\Delta_1 = 2^2 \cap Q - \{Z\}, \qquad \Delta_2 = 3_-^2 \cap Q,$$

$$\Delta_3 = 2^2 \cdot 3_1 \cap Q, \qquad \Delta_4 = 2^2 \cdot 3_2 \cap Q.$$

So it remains to calculate the lengths of these orbits.

First $f \in 2^2 \cap Q$ if and only if $[V, f] \leq [V, Z]^{\perp}$, so Δ_1 is of length $3 \cdot 2^4$ since this is the number of such lines $[V, f] \neq [V, Z]$. Second if $f \in 3_-^2$ then $V\theta_f^2$ is the unique singular line of $C_V(f)$, so $f \in Q$ if and only if $[V, Z] = V\theta_f^2$ in which case $N_G(\langle f \rangle) \leq N_G(Z)$. We calculate that if s is an involution centralizing f then as f acts on the eigenspaces of s, $[V, s]$ is 4-dimensional of sign $+1$ with $C_V(s)$ a Jordon block of size 3 and discriminant -1 for f. We then conclude $\langle s \rangle \in \text{Syl}_2(C_G(f))$ and $|N_G(\langle f \rangle)|_2 = 4$. Also $C_G(f)$ acts on the orthogonal spaces $C_V(f)/[V, Z]$ and $V\theta_f/[V, Z]$, whose isometry groups are $3'$-groups, and hence a Sylow 3-group T of $C_G(f)$ centralizes $[V, Z]^{\perp}/[V, Z]$.

Then T centralizes $[v, f]$ for each $v \in V\theta_f$, so, as $[V, Z]$ is generated by such commutators, T also centralizes $[V, Z]$. Therefore $T \leq Q$, so $T = C_Q(f)$ is of order 3^6 and hence

$$|\langle f \rangle^{N_G(Z)}| = |N_G(Z) : N_G(\langle f \rangle)| = 3^3 \cdot 2^4.$$

This leaves $\Delta_3 \cup \Delta_4$ of order $3^2 \cdot 2^5$ and as $2^2 \cdot 3_1$ and $2^2 \cdot 3_2$ are fused by a reflection in $N_I(Z) - G$, these orbits are each of length $3^2 \cdot 2^4$. In particular if $Y \in \Delta_3$ then a Sylow 3-subgroup of $N_G(Z) \cap N_G(Y)$ is of order 3^7 and Sylow in $N_G(Y)$, completing the proof of (5).

(38.16) *Let* $L \cong P\Omega_8^+(3)$ *and* G *a group with* $F^*(G) = L$ *and* $G/L \cong S_3$. *Let* V *be the natural (projective) module for* L. *Then* G *is determined up to isomorphism and*

(1) G *has a subgroup* $G_0 = \langle d \rangle L$ *of index 3 such that* d *induces a reflection on* V, *and conjugating in* $\mathrm{Aut}(L)$ *we may take* $[d, V]$ *of discriminant* -1. *Let* $D = d^G$.

(2) L *has 13 classes of subgroups of order 3:*

$$2^2, \ 3_\epsilon, \ 3_\epsilon^2, \ 2_{\epsilon,i}^4, \quad and \quad 2^2 \cdot 3_{\epsilon,i}, \qquad i = 1, 2, \epsilon = \pm 1.$$

(3) *The orbits of* G *on subgroups of* L *of order 3 are*

$$2^2, \ 3_- \cup 2_{-,1}^4 \cup 2_{-,2}^4, \ 3_+ \cup 2_{+,1}^4 \cup 2_{+,2}^4, \ 3_+^2, \ 3_-^2, \ 2^2 \cdot 3_{+,1},$$

and

$$2^2 \cdot 3_{+,2} \cup 2^2 \cdot 3_-.$$

(4) *For* $X \in 2^2 \cdot 3_{+,2}$, $N_G(X) = \langle d \rangle N_L(X)$ *where* $d \in C_D(X)$. *Further* $N_G(X) \leq N_{G_0}(V_0)$ *for a totally singular 3-subspace* V_0 *of* V,

$$|O_3(N_G(X))| = 3^9, \qquad N_G(X)/O_3(N_G(X)) \cong \mathbf{Z}_2 \times GL_2(3),$$

$N_G(X)$ *has* D-width 1, *and* $|C_D(X)| = 27$.

(5) *For* $Z \in 2^2$, $F^*(N_G(Z)) = Q \cong 3^{1+8}$, $\langle C_D(Z) \rangle / Q \cong \mathbf{Z}_2/3^{1+2}/Q_8^3$,

$$w_D(C_G(Z)) = 4, \qquad N_L(Z)/Q \cong \mathbf{Z}_2/SL_2(3)^3$$

with Q/Z *the tensor product of the natural modules for* $SL_2(3)$ *as an* $O^2(N_L(Z))/Q$-module, *and* $N_G(Z)/N_L(Z)$ *induces* S_3 *on the three* $SL_2(3)$ *factors.*

Proof: First $\mathrm{Out}(L) \cong S_4$, so G/L is determined up to conjugation in $\mathrm{Out}(L)$ and in particular (1) holds. Indeed $G_0 = \hat{G}_1/\langle -1 \rangle$, where $\hat{G}_1$ is $\hat{L} = \Omega(V)$

extended by a reflection $\hat{d}$ with image d in G_0 such that $[V, \hat{d}]$ has discriminant -1.

To prove (2) given 38.12 we must show that 2^4 breaks up into four orbits under L and $2^2 \cdot 3_\epsilon$ breaks up into two orbits under L. First d centralizes a member $g \in 2^2 \cdot 3_+$ and $C_L(d) \cong \Omega_7(3)$, so $2^2 \cdot 3_+ \cap C_L(d)$ breaks up into two orbits in $C_L(d)$ by 38.15. Also $C_{D_d}(g) = \emptyset$, so $C_G(g)$ is of D-width 1, so if $g, h \in 2^2 \cdot 3_+$ are fused in G, they are fused in $C_G(d)$, and therefore indeed $2^2 \cdot 3_+$ splits into two orbits under L. As $2^2 \cdot 3_-$ is conjugate to $2^2 \cdot 3_+$ under $\mathrm{Aut}(L)$, the same holds for that class. However by 38.12.3, d fuses the two orbits $2^2 \cdot 3_{-,1}$ and $2^2 \cdot 3_{-,2}$.

Similarly if $h \in 2^4$ then $[V, h]$ is a maximal totally singular subspace of V and L has two orbits $\mathcal{U}_1$ and $\mathcal{U}_2$ on such subspaces. Let 2^4_i consist of those subgroups $\langle h \rangle$ with $[V, h] \in \mathcal{U}_i$. Then L acts on 2^4_i and as d interchanges $\mathcal{U}_1$ and $\mathcal{U}_2$, d interchanges 2^4_1 and 2^4_2.

Next d inverts Y of order 3 in $G - L$ and Y is transitive on the three sets $\mathcal{P}_i$, $1 \leq i \leq 3$, of maximal parabolics of L with members of $\mathcal{P}_i$ stabilizing members of $\mathcal{U}_i$ for $i = 1, 2$, and $\mathcal{P}_3$ consisting of the stabilizers of singular points of V. Now if $P_i \in \mathcal{P}_i$ and A_i is the unipotent radical of P_i then subgroups of order 3 in A_3 are in 2^2, 3_-, and 3_+, and those in A_i are in 2^2 and 2^4_i for $i = 1, 2$. We conclude from the transitivity of G on $\{\mathcal{P}_1, \mathcal{P}_2, \mathcal{P}_3\}$ that 2^4_i decomposes into two orbits $2^4_{i,\epsilon}$ under L and under G, 2^2 and $3_\epsilon \cup 2^4_{1,\epsilon} \cup 2^4_{2,\epsilon}$, $\epsilon = \pm 1$ are orbits. Notice this completes the proof of (2).

Next $X \in 3^2_\epsilon$ is contained in a subgroup $K(\epsilon) \cong \Omega_6^{-\epsilon}(3)$ centralizing a 2-subspace $V(\epsilon)$ of discriminant ϵ. Now the image $K(\epsilon)^y$ of $K(\epsilon)$ under $1 \neq y \in Y$ acts on V with two irreducibles that are totally singular 4-subspaces if $\epsilon = -1$ and irreducibly with V the natural projective module for $U_4(3)$ if $\epsilon = +1$. In particular in the first case $K(-1)^y$ contains subgroups of type 2^2, 2^4, and 3^2_-, and we have seen the first two types are not fused to 3^2_+ in G, so 3^2_- is G-invariant. A similar argument shows 3^2_+ is G-invariant.

This leaves $2^2 \cdot 3_{\epsilon,i}$. Now d interchanges $2^2 \cdot 3_{-,1}$ and $2^2 \cdot 3_{-,2}$ and fixes $2^2 \cdot 3_{+,i}$ whereas a reflection r with $[V, r]$ of discriminant $+1$ fixes $2^2 \cdot 3_{-,i}$ and interchanges the remaining two classes. Thus dr has two cycles of length 2, so $\mathrm{Out}(L) \cong S_4$ acts faithfully as S_4 on the four classes, completing the proof of (3).

Let $X \in 2^2 \cdot 3_{+,2}$ and $M = N_G(X)$. Then X^L is moved by Y so $M = \langle d \rangle N_L(X)$, where $d \in C_D(X)$. Also if $X = \langle g \rangle$ then $C_{V\theta_g}(g) = V_0$ is a totally singular 3-subspace of V invariant under $N_{G_0}(X) = M$. We observed previously that M is of D-width 1, so by 8.6,

$$M = C_M(d)O_3(M).$$

Also $a \in C_D(X)$ if and only if $[V, a] \leq C_V(X) = V_0 \perp [V, d]$ if and only if $[V, a]$ is one of the 27 complements to V_0 in $C_V(X)$. Hence $|C_D(X)| = 27 = |M : C_M(d)|$. Finally by 38.15.5, $|O_3(C_M(d))| = 3^6$ and

$$C_M(d)/O_3(C_M(d)) \cong \mathbf{Z}_2 \times GL_2(3),$$

so the proof of (4) is complete.

Let $Z \in 2^2$. By 38.14, $F^*(N_L(Z)) = Q \cong 3^{1+8}$ and $O^2(N_L(Z))/Q \cong SL_2(3)^3$ with Q/Z the tensor product of the natural modules for the $SL_2(3)$ factors.

Let t be an involution in $O^2(N_L(Z))$ such that t inverts Q/Z. Then $N_G(Z) = Q(N_G(Z) \cap C_G(t))$ by a Frattini argument. Now $O^2(C_L(t)) = K_1 * \cdots * K_4$ with $K_i \cong SL_2(3)$ and $Z \leq K_4$, and $C_G(t)$ induces S_4 on $\{K_1, \ldots, K_4\}$ with

$$\left| \bigcap_i N_G(K_i) : K_1 \cdots K_4 \right| = 2.$$

We conclude $N_G(Z)/N_L(Z) \cong (N_G(Z) \cap C_G(t))/(N_L(Z) \cap C_L(t))$, $N_G(Z)$ induces S_3 on the three $SL_2(3)$ factors of $O^2(N_L(Z))/Q$, and

$$N_L(Z)/O^2(N_L(Z)) \cong \mathbf{Z}_2.$$

Finally if $d \in C_D(Z\langle t \rangle)$ then $[K_3, d] = 1$ and $K_1^d = K_2$, so

$$w_D(N_G(Z)) = w_D(\langle d \rangle K_1 K_2) = |d^{O_2(K_1 K_2)}| = 4$$

and $\langle C_D(Z) \rangle/Q \cong \langle C_D(Z\langle t \rangle) \rangle \cong \mathbf{Z}_2/3^{1+2}/Q_8^3$.

(38.17) *Let $3 \leq n \leq 7$, $G = O(V)$, S be a singular point of V, and $A = C_G(S) \cap C_G(S^\perp)$. Then $A \cong E_{3^{n-2}}$ and $A = J(P)$ for $A \leq P \in \mathrm{Syl}_3(G)$.*

Proof: If $n = 3$ or 4 then $P = A \cong E_3$ or E_9, respectively, so we may take $n \geq 5$. Then by 38.13 (or 11.10 when $n = 5$), $A \cong E_{3^{n-2}}$ and $O^2(N_G(A)) = AH$, where $H = O^2(O(W^\perp))$ and the map $\gamma : h \mapsto [r, h]\delta$ of 38.13.3 is an equivalence of the actions of H on A and $W^\perp$.

Let $HA^* = HA/A$ and regard H^* as $O^2(O(W^\perp))$ via γ. Suppose $E_{3^m} \cong B \leq HA$ with $m \geq n - 2$ and $B \neq A$. Then $B^* \neq 1$ and

$$m(B^*) \geq m(A/(A \cap B)) \geq m(A/C_A(B)) = \dim(W^\perp/C_{W^\perp}(B)).$$

By 38.10 applied to the action of $B^* \leq H^* = O(W^\perp)$, $\dim(W^\perp/C_{W^\perp}(B)) \geq 2$, so $m(B^*) \geq 2$. Thus $n - 2 \geq 4$, so $n \geq 6$ and hence $n = 6$ or 7. Moreover if $n = 6$ then $m(B^*) = 2 = m(H^*)$, so $B^* \in \mathrm{Syl}_3(H^*)$ and hence $C_{W^\perp}(B)$ is a point, so $\dim(W^\perp/C_{W^\perp}(B)) = 3 > m(B^*)$.

Therefore $n = 7$. By induction on n, $m(B^*) \leq m(H^*) = 3$ with B^* the unipotent radical of the stabilizer of a singular point R of $W^\perp$ in case of equality. In the latter case, $R = C_{W^\perp}(B)$, so $4 = m(W^\perp/C_{W^\perp}(B)) > m(B^*)$. Therefore $m(B^*) = 2$ and $\dim(C_{W^\perp}(B)) = 3$. In particular $C_{W^\perp}(B)$ contains a nonsingular point X, so $B^* \leq O((W + X)^\perp) \cong O_4^\epsilon(3)$, so as $m(B^*) = 2$, $B^* \in \mathrm{Syl}_3(O((W + X)^\perp))$ and then $C_{W^\perp}(B) = 2$, a contradiction.

Exercises

1. Let $G = PO_6^{+,\pi}(3)$, $L = E(G)$, D the set of reflections in G, and $g \in G$ of order 3 with $C_D(g) = \emptyset$. Prove
   (1) $g \in 3_+^2$.
   (2) If V is the natural projective module for G then $\mathbb{Z}_3 \times 3^{1+2} \cong C_L(g) \leq C_G(V\theta^2) \cap C_G(V\theta/V\theta^2)$.
   [*Hint:* To prove (1) show $h \in G$ of order 3 centralizes an element of D if and only if there is $v \in C_V(h)$ with $Q(v) = \pi$. In (2) show $C_L(g)$ is a 3-group by observing L has one class of involutions t^L and t centralizes only elements in 2^2 and 3_ϵ.]

14

Odd locals in Fischer groups

In this chapter we determine the subgroups of odd prime order and their normalizers in the Fischer groups. For primes $p > 3$, the p-local structure of the Fischer groups is fairly easy to determine, so almost all our time is spent on the 3-locals. When G is $M(22)$ or $M(23)$ our starting point is the D-subgroup L of G isomorphic to $\Omega_7(3)$ or $S_3/P\Omega_8^+(3)$, respectively. As L contains a Sylow 3-subgroup of G, each subgroup of order 3 in G is fused into L and it turns out that a large part of the normalizer of each such subgroup is visible in L. The enumeration in Chapter 13 of the classes of subgroups of L of order 3 is crucial to our analysis. When G is $M(24)$ we work primarily with the centralizer of a 3- transposition and with the 3-local of G isomorphic to $O_7(3)/E_{3^7}$.

At the end of Section 40, we produce the 3-local subgroup M of $M(23)$ remarked upon in Lemma 19.14. By Lemmas 19.13 and 19.14, $O_{3,2}(M)$ is the largest 4-generator 3-transposition group of width 1.

39. Subgroups of odd prime order in $M(22)$

In this section we determine the conjugacy classes of subgroups of odd prime order and the normalizers of such subgroups in $M(22)$. Thus in this section $G = M(22)$ and D is the class of 3-transpositions of G. We first observe

(39.1) There is a D-subgroup L of G with $L \cong \Omega_7(3)$.

Proof: Let $M = M(23)$, E the set of 3-transpositions of M, $d \in E$, and $H = C_M(d)$. Then as M is of type $\tilde{M}(23)$, $\bar{H} = H/\langle d \rangle \cong M(22)$ and if $a \in A_d$ then $K = \langle E_{a,d} \rangle \cong P\Omega_7^{-\pi,\pi}(3) \cong \Omega_7(3)$. Thus we can identify G with $\bar{H}$, D with $\bar{E}_d$, and let $L = \bar{K} \cong K$.

In the remainder of this section L is the D-subgroup of G supplied by 39.1. Let V be the natural $GF(3)L$-module; thus V is a 7-dimensional orthogonal space over $GF(3)$ with form (,). We normalize and take V of discriminant -1. Then $O_7^{-1,1}(3) = \langle D_- \rangle = \langle -I \rangle \times \Omega(V)$ by Exercise 4.12, so we view L as $\Omega(V)$ or $PO_7^{-1,1}(3) \cong \langle D_- \rangle / \langle -I \rangle$. Let $P \in \mathrm{Syl}_3(L)$ and $A = J(P)$.

(39.2)

(1) L has seven classes 3_+, 3_-, 2^2, 3_-^2, 3_+^2, $2^2 \cdot 3_1$, and $2^2 \cdot 3_2$ of subgroups of order 3, in the notation of Section 38. $2^2 \cdot 3_1 \cup 2^2 \cdot 3_2 = 2^2 \cdot 3_+$.

(2) $L \cap D$ is the class of involutions d of L with $[V, d]$ a subspace of V of dimension 6 and sign -1.

(3) Subgroups of L inverted by members of $L \cap D$ are in 3_-.

Proof: Part (1) is Lemma 38.15, since we have normalized so that $\mu(V) = -1$. As $\Omega_7^{-1,1}(3) = \langle -I \rangle \times L$, the set $L \cap D$ of 3-transpositions of L is the set of involutions $-r_x$, where r_x is the reflection through $x \in V$ with $Q(x) = 1$, and $[V, -r_x] = C_V(r_x)$ is of dimension 6 and discriminant $\mu(V)\mu(Fx) = 1$. Therefore (2) holds, and (2) and 38.15.2 imply (3).

(39.3)

(1) $A \cong E_{3^5}$ and $N_L(A)/A \cong PO_5(3) \cong O_5^{+,+}(3)$ with A the natural module for $N_L(A)/A$.

(2) $N_L(A)$ has three orbits on subgroups of A of order 3 with representatives in 3_+, 3_- and 2^2, corresponding to the classes of nonsingular points of the natural module A for $N_L(A)/A$ with stabilizer $O_4^+(3)$ and $O_4^-(3)$ and the singular points with stabilizer $O_3(3)/E_{27}$, respectively.

(3) For X of order 3 in A of type 3_ϵ, $N_L(X) \le N_L(A)$.

(4) $P \in \mathrm{Syl}_3(G)$.

Proof: As $|L|_3 = 3^9 = |G|_3$, part (4) holds. By 38.17, $A = J(P) \cong E_{3^5}$ is the unipotent radical of the stabilizer of a singular point of V. Then (1), (2), and (3) follow from 38.13, keeping in mind that $O_7^{-,+}(3) = L \times \langle -I \rangle$, so we must factor out $-I$ from all the subgroups recorded in 38.13.

(39.4)

(1) $N_G(A) = N_L(A)$ controls fusion in A.

(2) For $d \in D$, $a \in A_d$, and $X = \langle ad \rangle$, $N_G(X) = \langle a, d \rangle \times \langle D_{a,d} \rangle$ with $\langle a, d \rangle \cong S_3$, $\langle D_{a,d} \rangle = C_G(\langle a, d \rangle) \cong PO_6^{+,+}(3)$, $w_D(C_G(X)) = 6$, and $X^G \cap C_G(\langle a, d \rangle) = X^{C_G(\langle a,d \rangle)}$.

Proof: Adopt the notation of (2) with $d \in N_L(A)$ and $a \in dA$. By 25.9, $N_G(X) = \langle a, d \rangle \times \langle D_{a,d} \rangle$ with $K = \langle D_{a,d} \rangle \cong PO_6^{+,+}(3)$. Then by 11.10 and 11.11,

$$N_K(A) \cong PO_4^-(3)/E_{81},$$

so by 39.3.2

$$N_G(A) \cap N_G(X) \cong N_L(A) \cap N_G(X),$$

and hence $N_G(A) \cap N_G(X) \leq N_L(A)$. Further $K \leq N_G(X)$ is transitive on pairs (b, c) with $b, c \in D_{a,d} = C_D(X)$ and $c \in A_b$, so $C_G(\langle a, d \rangle) = K$ is transitive on $X^G \cap K$. That is (2) holds. Next as $A = J(P)$, $N_G(A)$ controls fusion in A by 2.8, so $N_L(X) \cap N_L(A) = N_G(X) \cap N_G(A)$ is transitive on $X^G \cap A \cap C_G(\langle a, d \rangle) = X^G \cap C_A(d)$. But as A is the natural module for $N_L(A)/A$, each subgroup of A of order 3 is fused into $C_A(d)$, so $X^G \cap A = X^{N_L(A)}$. Now $X^G \cap A = X^{N_L(A)}$ and $N_G(A) \cap N_G(X) \leq N_L(A)$, so that $N_G(A) = N_L(A)$. Therefore (1) holds.

(39.5) *Let $X \in A \cap 3_+$. Then $N_G(X) \leq N_L(A)$ and $N_G(X)/A \cong O_4^+(3)$ with $w_D(C_G(X)) = 4$.*

Proof: By 39.4.1, $N_G(A) = N_L(A)$, whereas by 39.3.3, $N_L(X) \leq N_L(A)$, so $N_L(X) = N_L(X) \cap N_L(A)$ is described in 39.3.2, and in particular the lemma holds if A is normal in $N_G(X)$. But $A = J(P)$ so if $A \leq O_3(N_G(X))$ then $A \trianglelefteq N_G(X)$. Thus it remains to show $A \leq O_3(N_G(X))$.

Let $d \in C_{D \cap L}(X)$. Then from the description of $N_L(X)$ in 39.3.2, 11.11.2, and Exercise 11.11.2, $\langle D_d \cap C_L(X) \rangle \cong S_3 \times S_3 \times S_3$. On the other hand setting $H = C_G(d)$, $H/\langle d \rangle \cong U_6(2)$, so the classes of subgroups of order 3 are described in 23.9, and in particular there is a unique class whose centralizer contains a D-subgroup isomorphic to S_3^3. Hence by 23.9, X is determined up to conjugacy in $C_G(d)$ and $\langle C_{D_d}(X) \rangle \cong S_3^3$, so $C_{D_d}(X) \subseteq L$. Then as $A = \langle X, [A, d]^{C_L(X)} \rangle$, 25.1.3 says $A \leq O_3(N_G(X))$, completing the proof.

(39.6) *Let $Z \in 2^2$. Then*

  (1) $\langle C_{D \cap L}(Z) \rangle = K_1 * K_2 * K_3$ *with* $K_i \cong \mathbf{Z}_2/3^{1+2}$ *and* $N_L(Z)$ *transitive on* $\{K_1, K_2, K_3\}$.
  (2) $F^*(N_L(Z)) = Q = Q_1 * Q_2 * Q_3 \cong 3^{1+6}$, *where* $Q_i = O_3(K_i)$.
  (3) *The subgroups of order 3 in* $Q_i - Z$ *are in* 3_-.
  (4) *The subgroups of order 3 in* $Q_1 Q_2 - (Q_1 \cup Q_2)$ *are in* 3_+ *and* 3_-^2.
  (5) *Each subgroup of order 3 in* L *is* L-*conjugate to a subgroup of* Q.
  (6) $Q = F^*(N_G(Z))$, $C_D(Z) \subseteq L$, *and* $w_D(C_G(Z)) = 3$.

(7) $O_{3,2}(N_G(Z))$ *is of index* 4 *in* $SU_3(2) * SU_3(2) * SU_3(2)$ *and*

$$N_G(Z)/O_{3,2}(N_G(Z)) \cong \mathbf{Z}_2/E_9.$$

Proof: See 38.15 for (1)–(5). In particular if $d \in K_1 \cap D$ then

$$K_2 K_3 = \langle C_L(Z) \cap D_d \rangle,$$

so by 23.9, X is determined up to conjugacy in $C_G(d)$, $C_G(Z\langle d\rangle)/\langle d\rangle$ is of index 3 in $GU_3(2) * GU_3(2)$, and $\langle C_{D_d}(Z)\rangle \cong K_2 K_2$, so $C_{D_d}(Z) \subseteq L$. Hence by 25.1.3, $Q = \langle [d, Q]^{N_L(Z)} \rangle \leq O_3(N_G(Z)) = Q_0$. As $C_G(d)$ is transitive on $Z^G \cap C_G(d)$, $C_G(Z)$ is transitive on $C_D(Z)$, so by 25.1.2, $d^{Q_0} = d^Q$, and therefore $Q_0 = Q C_{Q_0}(d)$. Now

$$C_{Q_0}(d) \leq O_3(N_G(Z\langle d\rangle)) = Q_2 Q_3,$$

so $Q_0 = Q$. By 9.3, d centralizes $O^3(F^*(N_G(Z)))$, so as $F^*(N_G(Z\langle d\rangle)) = Q_1 Q_2\langle d\rangle$, we have $Q = F^*(N_G(Z))$.

Next let $d_i \in K_i \cap D$, $i = 2, 3$. Then $z = d d_2 d_3$ inverts Q/Z, so

$$\langle z \rangle Q \trianglelefteq N_G(Z) = H.$$

Let $E = H \cap D$ and $H^* = H/Q\langle z\rangle$. Then $d^{*\perp} = d_i^{*\perp} = \{d^*, d_2^*, d_2^*\}$, so as $d^* = d_2^* d_3^*$, we conclude from 8.11.1 that $E^* = \{d^*, d_2^*, d_3^*\}$. That is $E \subseteq L$ and (6) is established.

Further $|H : C_H(d)Q| = 3$ and $C_H(d)/\langle d\rangle \cong C_G(Z\langle d\rangle)/\langle d\rangle$ is of index 3 in $GU_3(2) * GU_3(2)$, so $|H| = 2^7 \cdot 3^9$. Next the subgroup U of $\mathrm{Out}(Q)$ stabilizing each Q_i and centralizing Z is $SL_2(3) * SL_2(3) * SL_2(3)$, so the subgroup of $C_G(Z)$ stabilizing each Q_i is contained in the split extension $K \cong GU_3(2) * GU_3(2) * GU_3(2)$ of Q by U. So as $|O_{3,2}(H)| = 2^7 \cdot 3^7$, and $O_{2,3}(K) = 2^9 \cdot 3^7$, $O_{3,2}(H)$ is of index 4 in $O_{3,2}(K) \cong SU_3(2) * SU_3(2) * SU_3(2)$. Also

$$H O_{3,2}(H) \cong (H \cap L)/O_{3,2}(H \cap L) \cong \mathbf{Z}_2/E_9,$$

so (7) is established and the proof of 39.6 is complete.

However we pursue the setup of 39.6 a little further. Let $T \in \mathrm{Syl}_2(O_{3,2}(H))$ and $\tilde{Q} = Q/Z$. By 39.6, each subgroup Y of L of order 3 not of type 3_ϵ or 3_-^2 is fused into the set Δ of subgroups Y of Q of order 3 with $\tilde{Y}$ projecting nontrivially on each $\tilde{Q}_i$. Now T acts on $\tilde{Q}_2 \tilde{Q}_3 - (\tilde{Q}_2 \cup \tilde{Q}_3) = \Gamma$ of order 2^6 and $C_{\tilde{Q}_2 \tilde{Q}_3}(t) \leq \tilde{Q}_2$ or $\tilde{Q}_3$ for each $t \in T - \langle d\rangle$, so $T/\langle d\rangle$ is regular on Γ and hence TQ is transitive on the subgroups of order 3 in $Q_2 Q_3 - (Q_2 \cup Q_3)$. Therefore each member of Δ is conjugate to $\langle u_i v\rangle$, $1 \leq i \leq 4$, where $\langle \tilde{u}_i \rangle$, $1 \leq i \leq 4$, are the subgroups of $\tilde{Q}_1$ of order 3 and $v \in Q_2 Q_3 - (Q_2 \cup Q_3)$.

Also as QT is transitive on subgroups of order 3 in $Q_2 Q_3 - (Q_2 \cup Q_2)$, 39.6.4 says that the L-classes 3_+ and 3^2_- are fused in G.

Next there is $w \in N_H(T) - Q$ of order 3 acting on each Q_i and centralizing v. Further as $T/C_T(Q_1) \cong Q_8$, $\langle w \rangle$ fixes $\tilde{u}_1$ and is transitive on $\langle \tilde{u}_i \rangle$, $2 \le i \le 4$, so either H is transitive on Δ or H has two orbits on Δ of length $3 \cdot 2^6$ and $9 \cdot 2^6$. Now there is $Z \ne Z_0 \in 2^2 \cap Q$ and by 39.4, 3_+, 3_-, and 2^2 are in distinct G-classes, so by 39.6.3 and 39.6.4, $Z_0 \in \Delta$. So as H has at most two orbits on Δ, as each subgroup of G of order 3 is fused into Q under G by 39.6.5, and as the subgroups of order 3 in $Q - \Delta$ are in 3_+, 3_-, or 2^2, it follows that G has three or four classes of subgroups of order 3, and in the latter case H has orbits of length $3 \cdot 2^6$ and $9 \cdot 2^6$ on Δ. We prove there are four classes, namely:

(39.7)   *G has four classes Y_i^G, $1 \le i \le 4$, of subgroups of order 3:*

(1) $w_D(C_G(Y_1)) = 6$ and $Y_1^G \cap L = 3_-$.

(2) $w_D(C_G(Y_2)) = 4$ and $Y_2^G \cap L = 3_+ \cup 3^2_-$.

(3) $w_D(C_G(Y_3)) = 3$ and $Y_3^G \cap L = 2^2 \cup 2^2 \cdot 3_1$.

(4) $w_D(C_G(Y_4)) = 0$ and $Y_4^G \cap L = 3^2_+ \cup 2^2 \cdot 3_2$.

Adopting the notation of 37.8, let $Y = \langle x_2 x_{11} \rangle$ and $U = \langle x_{11} \rangle$. Let $K = C_G(Y)$. By 37.9, G has three classes Y_i^G, $1 \le i \le 3$, of subgroups of order 3 centralizing 3-transpositions. Moreover $w_D(C_G(Y_i)) = 6, 4, 3$, respectively. From the lemmas in this section we conclude $3_1 \subseteq Y_1^G$, $3_+ \subseteq Y_2^G$, and $2^2 \subseteq Y_3^G$. We also saw that 3^2_- is fused to 3_+ in G, so $3^2_- \subseteq Y_2^G$.

Next by 37.8, $w_D(C_G(U)) = 4$ so we may take $U \in 3_+ \cap A$. By 37.8, $C_B(U) = C_B(Y) = \langle j \rangle$ with $j \in \mathcal{J}_3$, and $|C_R(U)| = 2^5$ with $C_R(U)/\langle j \rangle = [C_R(U)/\langle j \rangle, x_{12}]$. By 39.5, $C_G(U) \le N_L(A)$ with

$$O^2 (C_G(U)/A) \cong SL_2(3) * SL_2(3),$$

so $C_R(U) \cong Q_8 * Q_8$. Again by 37.8, $C_R(Y) \le C_R(U)$ with $|C_R(Y)| = 8$ and $C_R(Y) \in \mathrm{Syl}_2(C_M(Y))$, so as $O^2(C_G(U)/A) \cong SL_2(3) * SL_2(3)$, we conclude $C_R(Y) \cong Q_8$. Thus

$$Q_8 \cong C_R(Y) \in \mathrm{Syl}_2(C_M(Y))$$

and by 37.8.5, $C_G(j) \le M$, so $C_R(Y) \in \mathrm{Syl}_2(C_G(Y\langle j \rangle))$. Then as $\langle j \rangle = Z(C_R(Y))$, we have $C_R(Y) \in \mathrm{Syl}_2(K)$, so $m_2(K) = 1$ and in particular as $j \in \mathcal{J}_3$, $w_D(K) = 0$. Therefore as G has at most four classes of subgroups of order 3 and exactly three classes of subgroups centralizing 3-transpositions, G has exactly four classes and $Y = Y_4$ is a representative of the fourth class.

We saw that H has two orbits on Δ of order $3 \cdot 2^6$ and $9 \cdot 2^6$, and from 38.15.4, $H \cap L$ has four orbits Δ_i, $1 \le i \le 4$, on Δ where $\Delta_1 = 2^2 \cap \Delta$ is of length $3 \cdot 2^4$, $\Delta_2 = 3_+^2 \cap \Delta$ is of length $3^3 \cdot 2^4$, and $\Delta_3 = 2^2 \cdot 3_1 \cap \Delta$ and $\Delta_4 = 2^2 \cdot 3_2 \cap \Delta$ are of length $3^2 \cdot 2^4$. We conclude that the orbits of H on Δ are $\Delta_1 \cup \Delta_3$ of length $3 \cdot 2^6$ and $\Delta_2 \cup \Delta_4$ of length $9 \cdot 2^6$, completing the proof of 39.7.

We next prove

(39.8) *Let $Y \in 2^2 \cdot 3_2$. Then $w_D(C_G(Y)) = 0$, $V_0 = C_V(Y)$ is a maximal totally singular subspace of V, $N_G(Y) = N_L(Y) \le N_L(V_0)$ with*

$$|O_3(N_G(Y))| = 3^6$$

and $N_G(Y)/O_3(N_G(Y)) \cong GL_2(3)$.

Proof: Let $K = C_G(Y)$. From 39.7, $w_D(K) = 0$ and we saw during the proof of 39.7 that $Q_8 \cong T = C_R(Y) \in \mathrm{Syl}_2(K)$ with $\langle j \rangle = Z(T)$ and $C_G(j) \le M = N_G(B)$ in the notation of 37.8. In particular by the Brauer–Suzuki Theorem [BrS], $K = O(K)C_K(j)$. As $C_G(j) \le M$ and $C_{M \cap K}(j)/O_3(C_{M \cap K}(j)) \cong GL_2(3)$ it follows that $K/O(K) \cong GL_2(3)$. Also

$$C_{O(K)}(j) = O(M \cap K) \cong E_9.$$

Next by 38.15.5, $V_0 = C_V(Y)$ is a maximal totally singular subspace of V and $N_L(Y)$ is contained in $N_L(V_0)$ with $|O_3(N_L(Y))| = 3^6$ and

$$N_L(Y)/O_3(N_L(Y)) \cong GL_2(3).$$

So $K = N_L(Y)O(K)$ and $O_3(N_L(Y)) \le O(K)$. Let $P_L \in \mathrm{Syl}_3(N_L(Y))$, $P_L \le P_Y \in \mathrm{Syl}_3(K)$, and $P_Y \le P_G \in \mathrm{Syl}_3(G)$. By 39.6, $Z(P_G) = Z$ is of order 3 with $F^*(N_G(Z)) = Q \cong 3^{1+6}$ and from the structure of $N_G(Z)$ described in 39.6, $|C_G(ZI)|_3 \le 3^6$ for I of order 3 in $N_G(Z) - Q$, so as $|P_L| = 3^7$, $Y \le Q$. Then from the proof of 39.7, $|Y^{N_G(Z)}| = 9 \cdot 2^6$, so $|P_Y| = 3^7 = |P_L|$. That is $P_L = P_Y \in \mathrm{Syl}_3(K)$, so $O_3(N_L(Y)) \in \mathrm{Syl}_3(O(K))$. Hence if $O(K)$ is a 3-group then $K = C_L(Y)$ and the proof is complete.

We observed earlier that $C_{O(K)}(j) \cong E_9$, whereas as $K = O(K)C_K(j)$, we have $[O_3(N_L(Y)), j] \le O(K)$. Now $[O_3(N_L(Y)), j]$ contains $Z_0 \in 2^2$, and by 39.6, $C_G(Z_0)$ is a $\{2, 3\}$-group, so $C_{O(K)}([O_3(N_L(Y)), j]) \le C_{O(K)}(Z_0)$ is a 3-group. Now Exercise 1.6 completes the proof.

Notice we have now determined the classes of subgroups of G of order 3 and their normalizers. For primes p greater than 3, the problem of determining the classes of subgroups of order p and their normalizers is much easier.

(39.9)

    (1) There is a D-subgroup I of G isomorphic to $\mathrm{Aut}(\Omega_8^+(2)) \cong S_3/\Omega_8^+(2)$.

    (2) I is a strongly 5-embedded subgroup of G.

    (3) Let $E \in \mathrm{Syl}_5(I)$, *so that* $E \in \mathrm{Syl}_5(G)$. *Then* $E \cong E_{25}$ *and* $N_G(E)/E \cong$ $\mathbf{Z}_2/(\mathbf{Z}_4 * SL_2(3))$ *with* $E = C_G(E)$.

    (4) G has one class X^G *of subgroups of order 5 and for* $X \le I$, $N_G(X) =$ $N_I(X) = F \times \langle C_D(X) \rangle$ *with F a Frobenius group of order 20 and* $\langle C_D(X) \rangle \cong S_5$.

Proof: First I exists by 37.7.2, so (1) holds. Then I has a D-subgroup $J \cong$ $O_8^+(2)$ of index 3. Let U be the natural module for J and write $U = U_1 \perp U_2$ where U_i is a nondegenerate 4-dimensional subspace of sign -1. Then

$$N_J(U_1) = J_1 \times J_2$$

where $J_i = C_J(U_i)$ induces $O_4^-(2) \cong S_5$ on U_{3-i} with the elements of $D \cap J_i$ inducing transvections and corresponding to transpositions of J_i. Let $d \in J_2 \cap D$ and $E \in \mathrm{Syl}_5(J_1 J_2)$. Then $E = E_1 \times E_2 \cong E_{25}$ with $E_i = E \cap J_i$ of order 5. As $|G|_5 = 25$, $E \in \mathrm{Syl}_5(G)$. Let $X = E_1$. As $C_G(d)/\langle d \rangle \cong U_6(2)$,

$$N_G(X\langle d \rangle) = C_G(d) \cap N_G(X) = F \times \langle d \rangle \times \langle D_d \cap N_G(X) \rangle,$$

with F Frobenius of order 20 and $\langle D_d \cap N_G(X) \rangle \cong S_3$. Then $\langle D_d \cap N_G(X) \rangle \cong$ $\langle D_d \cap J_2 \rangle$ and $F \cong N_{J_1}(X)$, so $D_d \cap N_G(X) \subseteq J_2$ and $F = N_{J_1}(X)$. By 25.3, $J_2 = \langle C_D(X) \rangle$. Therefore $N_G(X) = J_2(N_G(X) \cap C_G(d)) = J_2 F \le J$.

    Also $N_I(E)$ has the structure described in (3), and hence in particular is transitive on $E^\#$. Therefore G has one class of subgroups of order 5 and $N_G(E) = N_I(E)(N_G(E) \cap N_G(X) = N_I(X)$, establishing (2)–(4).

(39.10)   *G is transitive on its subgroups of order 7 and if X is of order 7 in G, then* $N_G(X) = F \times \langle C_D(X) \rangle$ *with F a Frobenius group of order 42 and* $\langle C_D(X) \rangle \cong S_3$.

Proof: Let $d \in D$, $a \in A_d$, and $M = C_G(\langle a, d \rangle)$. Then $|G|_7 = 7 = |M|_7$, so by Sylow, G has one class X^G of subgroups of order 7 and we can choose $X \le M$. Then $M \cong PO_6^{+,\pi}(3)$ so $N_M(X) = F$, where F is Frobenius of order 42. Further $N_G(\langle d \rangle X) = F \times \langle d \rangle$, so $N_G(X) = F\langle C_D(X) \rangle$ with $\langle a, d \rangle \le \langle C_D(X) \rangle$ of width 1. In particular $\langle C_D(X) \rangle = \langle d \rangle R$ where R is a 3-group. Then as $\langle ad \rangle X$ is self-centralizing, $R = \langle ad \rangle$, completing the proof.

(39.11)   *G is transitive on subgroups of order 11 and if X is such a subgroup then* $N_G(X) = F \times \langle d \rangle$ *with* $d \in D$ *and F Frobenius of order 55.*

Proof: Let $d \in D$ and $H = C_G(d)$. Then $|G|_{11} = |H|_{11} = 11$, so by Sylow, G is transitive on its subgroups X of order 11 and we may take $X \leq H$. Now $N_H(X) = F \times \langle d \rangle$, so $N_G(X) = F \langle C_D(X) \rangle$ with $\langle C_D(X) \rangle$ of width 1. As $C_G(\langle a, d \rangle)$ is of order prime to 11 for $a \in A_d$, it follows that $\{d\} = C_D(X)$, completing the proof.

(39.12) *G is transitive on it subgroups X of order 13 and $N_G(X)$ is Frobenius of order 78.*

Proof: As $|G|_{13} = 13$, G is transitive on its subgroups X of order 13 and $X \in \mathrm{Syl}_{13}(G)$. From Section 37 and the earlier lemmas in this section, it follows that $C_G(Y)$ is of order prime to 13 for each Y of order p and each prime $p \neq 13$. Thus $X = C_G(X)$ and $N_G(X)$ is Frobenius of order dividing $13 \cdot 12$. By Sylow's Theorem, $|N_G(X)| = 78$.

40. Subgroups of odd prime order in $M(23)$

In this section we determine the conjugacy classes of subgroups of odd prime order and the normalizers of such subgroups in $M(23)$. Many arguments are similar to those in the previous section on $M(22)$, except that the $M(23)$ case is easier.

In this section $G = M(23)$ and D is the class of 3-transpositions of G. We first produce a subgroup of G analogous to the subgroup of Lemma 39.1:

(40.1) *There is a D-subgroup L of G with $F^*(L) \cong P\Omega_8^+(3)$ and $L/F^*(L) \cong S_3$, such that for $d \in D \cap L$,*

$$C_L(d) \cong \mathbf{Z}_2 \times \Omega_7(3).$$

Proof: The argument is the same as that in 39.1. Let $M = M(24)$, E the set of 3-transpositions of M, and $d \in E$. Then by 16.12, $C_{E(M)}(d) \cong M(23)$ and if $a \in A_d$ then as M is of type $M(24)$, $K = \langle E_{a,d} \rangle \cong S_3 / P\Omega_8^+(3)$. Thus we may identify G with $C_{E(M)}(d)$, D with $dE_d = \{de : e \in E_d\}$, and let $L = \langle d E_{a,d} \rangle \cong K$.

For the remainder of this section L is the D-subgroup of G supplied by 40.1. Then by 38.16, L is determined up to isomorphism and has a D-subgroup L_0 of index 3 isomorphic to $PO_8^{+,+}(3)$. Let V be the natural projective $GF(3)L_0$-module; thus V is an 8-dimensional orthogonal space over $GF(3)$ of sign $+1$ with form $(\ ,\ )$. As in 38.16 we can choose notation so that the members of $L_0 \cap D$ are reflections t on V with $[V, t]$ of discriminant -1. Let $P \in \mathrm{Syl}_3(L)$.

(40.2)

(1) L has seven orbits $3_+ \cup 2^4_{+,1} \cup 2^4_{+,2},\ 3_- \cup 2^4_{-,1} \cup 2^4_{-,2},\ 2^2,\ 3^2_-,\ 3^2_+,\ 2^2 \cdot 3_{+,1}$, and $2^2 \cdot 3_{+,2} \cup 2^2 \cdot 3_-$ on subgroups of order 3 in $E(L)$, in the notation of 38.16.

(2) $L_0 \cap D$ is the class of reflections in L_0 and for $d \in L_0 \cap D$, $[V, d]$ is of discriminant -1.

(3) Subgroups of L inverted by members of $L_0 \cap D$ are in 3_-.

Proof: Part (1) is Lemma 38.16.3, and we have already observed that (2) holds. Then (2) and 38.13.1 imply (3).

(40.3) For $d \in D$, $a \in A_d$, and $X = \langle ad \rangle$, $N_G(X) = \langle a, d \rangle \times \langle D_{a,d} \rangle$ with $\langle a, d \rangle \cong S_3$, $\langle D_{a,d} \rangle = C_G(\langle a, d \rangle) \cong \Omega_7(3)$, and $w_D(C_G(X)) = 7$.

Proof: This is contained in 25.9.

(40.4) Let $X \in 3_+$. Then $N_G(X) \leq N_{L_0}(V_0)$ for a suitable singular point V_0 of V with $O_3(N_G(X)) = O_3(N_{L_0}(V_0)) \cong E_{3^6}$ and $N_G(X)/O_3(N_G(X)) \cong O_5(3)$ with $w_D(C_G(X)) = 5$.

Proof: First by 40.1, $N_L(X) \leq L_0$ as elements in $L - L_0$ of order 3 do not act on $X^{E(L)}$. Then 38.13 shows that $N_L(X) \leq N_{L_0}(V_0)$ for a suitable singular point V_0 of V and that $O_3(N_{L_0}(X)) = O_3(N_{L_0}(V_0)) \cong E_{3^6}$ with $N_{L_0}(X)/O_3(N_{L_0}(V_0)) \cong O_5(3)$. Thus it remains to show $N_G(X) \leq L$.

Let $d \in C_{D \cap L}(X)$; then $C_L(d)/\langle d \rangle \cong \Omega_7(3)$ plays the role in $C_G(d)/\langle d \rangle \cong M(22)$ of the group of 39.1 with X in the class 3_+ of $C_L(d)$ defined in 39.2. Therefore by 39.5, $C_G(d) \cap N_G(X) \leq L$ and $\langle N_{D_d}(X) \rangle = H$ is transitive on $N_{D_d}(X)$, and $H/O_{3,Z}(H) \cong O_5^{\mu, \pi}(3)$. Now 25.11 completes the proof.

(40.5) Let $Z \in 2^2$. Then

(1) $F^*(N_G(Z)) = F^*(N_L(Z)) \cong 3^{1+8}$.

(2) $C_D(Z) \subseteq L$, $w_D(C_G(Z)) = 4$, $\langle C_D(Z) \rangle / Q \cong \mathbf{Z}_2/3^{1+2}/Q_8^3$, and $N_{E(L)}(Z)/Q \cong \mathbf{Z}_2/SL_2(3)^2$.

(3) $N_G(Z)/Q \cong GL_2(3)/3^{1+2}/Q_8^3$.

Proof: Let $d \in C_{L \cap D}(Z)$ and $M = N_G(Z)$. As in the proof of the previous lemma, $C_L(d)/\langle d \rangle \cong \Omega_7(3)$ plays the role in $C_G(d)/\langle d \rangle \cong M(22)$ of the group of 39.1 with Z in the class 2^2 of $C_L(d)$ defined in 39.2. Therefore by

39.6, $C_{D_d}(Z) \subseteq L$ and $O_3(C_G(Z\langle d \rangle)) \leq L$. Hence by 25.1.3, and using 38.16.5,

$$O_3(M) = O_3(L \cap M) = Q \cong 3^{1+8}.$$

Similarly

$$w_D(C_G(Z)) = w_D(C_G(Z\langle d \rangle)) = 4$$

by 39.6.6, and by 25.11, $C_D(Z)) \subseteq L$, so (2) holds by 38.16.5. Then by (2), $M = C_M(d)\langle C_D(Z) \rangle$, so that $|M : M \cap L| = |C_M(d) : C_{M \cap L}(d)| = 4$ by 39.6 and 38.14. Then as $d \in O^2(C_G(Z\langle d \rangle))$ while d is faithful on

$$O_{3,2,3}(\langle C_D(Z) \rangle)/O_{3,2}(\langle C_D(Z) \rangle) \cong 3^{1+2},$$

$$M/O_{3,2,3}(\langle C_D(Z) \rangle) \cong GL_2(3),$$

completing the proof of (3).

(40.6) *Let $Y \in 2^2 \cdot 3_{+,2}$. Then*

$$w_D(C_G(Y)) = 1, \qquad N_G(Y) = N_{L_0}(Y), \qquad |O_3(N_G(Y))| = 3^9,$$

and $N_G(Y)/O_3(N_G(Y)) \cong \mathbf{Z}_2 \times GL_2(3)$.

Proof: Let $Y \in 2^2 \cdot 3_{+,i}$ for $i = 1$ or 2 and $K = N_G(Y)$. By 38.16.4 it suffices to show that if $i = 2$ then $K \leq L$. Let $d \in D \cap K \cap L$. As in the proofs of the previous lemmas, $C_L(d)/\langle d \rangle \cong \Omega_7(3)$ plays the role in $C_G(d)/\langle d \rangle \cong M(22)$ of the group of 39.1 with Y in the class $2^2 \cdot 3_{j(i)}$ of $C_L(d)$ defined in 39.2 for $j(i) = 1$ or 2. Pick i so that $j(i) = 2$. Then by 39.8, $w_D(C_K(d)) = 1$ with $C_K(d) \leq L$, $|O_3(C_K(d))| = 3^6$, and $C_K(d)/O_3(C_K(d)) \cong \mathbf{Z}_2 \times GL_2(3)$. Then $w_D(K) = 1$ and hence by 8.6, $K = O_3(K)C_K(d)$ and $O_3(K)$ is transitive on $K \cap D$.

Let Δ be the set of D-subgroups of K isomorphic to S_3. As $O_3(K)$ is transitive on $K \cap D$, there is a bijection $\pi : \langle a, d \rangle^K \mapsto a^{C_K(d)}$ between the orbits of K on Δ and the orbits of $C_K(d)$ on $K \cap D - \{d\}$. We show K has exactly two orbits on Δ. We also show $C_K(d)$ has at least two orbits on $D \cap L_0 \cap K - \{d\}$, and hence as $C_K(d) \leq L$, we conclude from the existence of π that $O_3(K) \leq L_0$. Therefore $K \leq L_0$, so $i = 2$ and the proof is complete. It remains to establish the two claims.

First the number of orbits of K on Δ is equal to the number of orbits of $M = C_G(\langle a, d \rangle)$ on $Y^G \cap M$, for $a \in A_d$. Now $M \cong \Omega_7(3)$ and by 39.7, M has two orbits on $Y^G \cap M$, since $Y^G \cap M$ is the set of subgroups Y_0 of M of order 3 such that $C_D(\langle d \rangle Y_0) = \{d\}$. Hence the first claim is established.

Second suppose $C_K(d)$ is transitive on $D \cap L_0 \cap K - \{d\}$. Then $K \cap L_0$ is 2-transitive on $D \cap L_0 \cap K$, which is of order 27 by 38.16.4. This is impossible as 13 divides $27 - 1$ but not $|K \cap L_0|$.

(40.7) *G has four classes Y_i^G, $1 \le i \le 4$, of subgroups of order 3:*

(1) $w_D(C_G(Y_1)) = 7$ and $Y_1^G \cap E(L) = 3_- \cup 2_{-,1}^4 \cup 2_{-,2}^4$.

(2) $w_D(C_G(Y_2)) = 5$ and $Y_2^G \cap E(L) = 3_+ \cup 2_{+,1}^4 \cup 2_{+,2}^4 \cup 3_-^2$.

(3) $w_D(C_G(Y_3)) = 4$ and $Y_3^G \cap E(L) = 2^2 \cup 2^2 \cdot 3_{+,1}$.

(4) $w_D(C_G(Y_4)) = 1$ and $Y_4^G \cap E(L) = 3_+^2 \cup 2^2 \cdot 3_{+,2} \cup 2^2 \cdot 3_-$.

Proof: Let $d \in L \cap D$. By 39.7, $C_G(d)$ has four classes of subgroups $Y_i^{C_G(d)}$, $1 \le i \le 4$, of order 3 and $w_D(C_G(Y_i)) = w_D(C_G(Y_i) \cap C_G(d)) = 7, 5, 4, 1$, for $i = 1, 2, 3, 4$, respectively. Hence there is no further G-fusion of these classes. Moreover by 39.7, we can choose $Y_i \in E(L)$ and the fusion of Y_i in $C_L(t)$ is described in 39.7. Combining this description with the list of classes in 40.2.1, we conclude each subgroup of $E(L)$ of order 3 is conjugate to some Y_i and the fusion is as described in 40.7.

Thus we have determined the classes of subgroups Y of order 3 fused into $E(L)$ and it remains only to show each subgroup of G of order 3 is fused into $E(L)$. As $|L|_3 = 3^{13} = |G|_3$, it suffices to show each subgroup Y of order 3 in $L - E(L)$ is fused into $E(L)$. Let $Z \in 2^2$, $M = C_G(Z)$, $P \in \mathrm{Syl}_3(M \cap L)$, and $R = P \cap E(L)$. Then $P \in \mathrm{Syl}_3(M)$ and by 40.5, $M^* = M/O_{3,2}(M) \cong GL_2(3)/3^{1+2}$ with $R^* \cong E_{27}$. In particular all subgroups of order 3 in $P^* - R^*$ are contained in $O_3(M^*)$ and all are fused under M^* into R^*. Therefore Y is conjugate to a subgroup of $E(L)$, completing the proof.

Notice we have now determined the classes of subgroups of G of order 3 and their normalizers. Again for primes p greater than 3, the problem of determining the classes of subgroups of order p and their normalizers is much easier.

(40.8)

(1) Let $E \in \mathrm{Syl}_5(G)$. Then $E \cong E_{25}$ and $N_G(E)/C_G(E) \cong \mathbf{Z}_2/(\mathbf{Z}_4 * SL_2(3))$ with $C_G(E) = E \times \langle d \rangle$ for some $d \in D$.

(2) G has one class X^G of subgroups of order 5 and $N_G(X) = F \times \langle C_D(X) \rangle$ with F a Frobenius group of order 20 and $\langle C_D(X) \rangle \cong S_7$.

Proof: Let $d \in D$ and $H = C_G(d)$. Then $|G|_5 = 5^2 = |H|_5$ so H contains a Sylow 5-subgroup E of G. By 39.9, $E \cong E_{25}$ and $N_H(E)/E\langle d \rangle \cong \mathbf{Z}_2/(\mathbf{Z}_4 * SL_2(3))$ with $E\langle d \rangle = C_H(E)$. Also by 39.9, H has one class X^H of subgroups of order 5 and $N_H(X) = \langle d \rangle \times F \times \langle C_{D_d}(X) \rangle$, with F a Frobenius group of order 20 and $\langle C_{D_d}(X) \rangle \cong S_5$. Therefore G has one class of subgroups of order 5 and by Exercise 8.2.7, $N_G(X) = F \times \langle C_D(X) \rangle$ with $\langle C_D(X) \rangle \cong S_7$. In particular for $X \le E$, $C_G(E) = C_G(X) \cap C_G(E) = E \langle d \rangle$, completing the proof of (1).

(40.9) *G is transitive on its subgroups of order* 7 *and if X is of order 7 in G, then $N_G(X) = F \times \langle C_D(X) \rangle$ with F a Frobenius group of order 42 and $\langle C_D(X) \rangle \cong S_5$.*

Proof: Argue as in the previous lemma using 39.10 and 25.3. Notice that 25.3 says $\langle C_D(X) \rangle$ is S_5 or S_3^2 and it must be the former as $C_G(X)$ contains an element of order 5 by 40.8.

(40.10) *G is transitive on subgroups of order* 11 *and if X is such a subgroup then $C_D(X) = \{d, b\}$ is of order 2 with $N_G(X) \cap C_G(d) = F \times \langle d, b \rangle$ of index 2 in $N_G(X)$, where F is Frobenius of order 55.*

Proof: Let $d \in D$ and $H = C_G(d)$. Then $|G|_{11} = |H|_{11} = 11$, so by Sylow, G is transitive on its subgroups X of order 11, we may take $X \le H$, and $N_G(X)$ is transitive on $C_D(X)$. Now by 39.11, $C_D(X) = \{d, b\}$ is of order 2 and $N_H(X) = F \times \langle d, b \rangle$. Finally if $a \in A_d$, then $C_G(\langle a, d \rangle)$ has order prime to 11, so $\{d, b\} = C_D(X)$, completing the proof.

(40.11) *G is transitive on it subgroups X of order* 13 *and $N_G(X) = F \times \langle C_D(X) \rangle$, where F is Frobenius of order 78 and $\langle C_D(X) \rangle \cong S_3$.*

Proof: Let $d \in D$ and $H = C_G(d)$. As $|G|_{13} = 13 = |H|_{13}$, a Sylow 13-group X of G is of order 13 and we may take $X \le H$. By 39.12, $N_H(X) = F \times \langle d \rangle$ with F Frobenius of order 78. Then $w_D(C_G(X)) = 1$, so $N_G(X) = F \times K$ where $K = \langle C_D(X) \rangle = O_3(K) \langle d \rangle$ and $O_3(K) \cong E_{3^n}$ is inverted by d. Finally for $a \in A_d$, $N_G(\langle ad \rangle) = \langle a, d \rangle \times \langle D_{a,d} \rangle$ and $\langle D_{a,d} \rangle$ contains an element of order 13, so $n = 1$, completing the proof.

(40.12)

(1) *If X is of order* 17 *then $N_G(X)$ is a Frobenius group of order* $17 \cdot 16$.

(2) *If X is of order* 23 *then $N_G(X)$ is a Frobenius group of order* $23 \cdot 11$.

Proof: The prime divisors p of $|G|$ with $p > 13$ are 17 and 23, which divide G to the first power. So G is transitive on its subgroups X of order p and $X \in$ $\mathrm{Syl}_p(G)$. By previous lemmas, X centralizes no elements of prime order $q \le 13$, so either $X = C_G(X)$ or G has an element of order $17 \cdot 23$. Now by 25.7, G contains a subgroup $M_{23}/E_{2^{11}}$ that contains a Frobenius group F of order $23 \cdot 11$. Let A and B be subgroups of F of order 23 and 11, respectively. If G has an element of order $23 \cdot 17$ then B acts on a subgroup C of $C_G(A)$ of order 17 and hence centralizes C, contrary to 40.10. Thus $A = C_G(A)$ so $N_G(A)$ is Frobenius of order dividing $23 \cdot 22$ and then by Sylow's Theorem, $N_G(A) = AB$. Similarly if X is of order 17 then $N_G(X)$ is Frobenius of order dividing $17 \cdot 16$ and by Sylow's Theorem this is the order of $N_G(X)$.

Notice we have completed the determination of the subgroups of $M(23)$ of prime order and also determined the normalizer of each such subgroup.

We close this section by proving the existence of a 3-local M of G that is the split extension of $R = O_3(M)$ of order 3^{10} by $K \cong GL_3(3)$, with $d \in Z(K) \cap D$ inverting $R/\Phi(R) \cong E_{27}$. Thus $G_0 = R\langle d \rangle$ is a D-subgroup of G and if $x_i\Phi(R)$, $1 \le i \le 3$, is a basis for $R/\Phi(R)$ with d inverting x_i, and $S = \{d, dx_1, dx_2, dx_3\}$ then (G_0, S) is a Hall system of rank 4 in the sense of Section 19. Therefore by 19.13 and 19.14, G_0 is the largest 4-generator 3-transposition group of width 1.

It turns out that the 3-local M is the third of the three maximal subgroups of G containing the Sylow 3-subgroup P of G. The other two are L and $N_G(Z(P))$.

Theorem 40.13. *There exists a 3-local M of G such that*

(1) *$R = O_3(M)$ is of class 4 and order 3^{10}, and R possesses a complement $K \cong GL_3(3)$ in M.*
(2) *The ascending central series $1 = Z_0(R) < \cdots < Z_4(R) = R$ of R is a chief series for M with $m(R_{i+1}/R_i) = 3$ for $i = 0, 2, 3$, $Z_2(R) \cong E_{81}$, and $Z_3(R) = \Phi(R)$.*
(3) *$Z(K) = \langle d \rangle$ with $d \in D$ and $\langle M \cap D \rangle = \langle d \rangle R$ with $Z_3(R) = C_R(d) \times [Z_2(R), d]$, $[Z_2(R), d]$ is of order 3, and $C_R(d)$ is special of order 3^6.*

Proof: Let $Z = Z(P)$, $Q = O_3(N_G(Z))$, $d \in C_D(Z)$, and X of order 3 in Q inverted by d. Then by 40.3, $N_G(X) = \langle d \rangle X \times J$, where $J = \langle C_D(X) \rangle \cong \Omega_7(3)$. Then the stabilizer I of a totally singular 3-dimensional subspace of the natural module for J is a maximal parabolic with $R_I = O_3(I)$ special of order 3^6, possessing a complement $K_I \cong L_3(3)$ acting irreducibly on $Z(R_I) = U \cong E_{27}$ and on R_I/U. Choose $Z \le R$. Then $I = \langle Q_I, Q_I^g, Q_I^h \rangle$, where

$Q_I = C_Q(d) \cong 3^{1+6}$ and $g, h \in I$ with $U = Z \times Z^g \times Z^h = Q_I \cap Q_I^g \cap Q_I^h$. Therefore

$$XU = XQ_I \cap XQ_I^g \cap XQ_I^h \leq Q \cap Q^g \cap Q^h = W.$$

Now as $\Phi(Q) = Z$, $\Phi(W) = 1$, so $W \leq C_Q(X) = XQ_I$, and thus $W = XU$.

Let $M_0 = \langle Q, Q^g, Q^h \rangle$. Then $[Q, U] = Z \leq U$, so $Q \leq N_G(U)$. Similarly Q^g and Q^h act on U, so $U \trianglelefteq M_0$. Also the same argument shows $W \trianglelefteq M_0$. Notice $I = \langle Q_I, Q_I^g, Q_I^h \rangle \leq M_0$.

Let $R = C_Q(U)C_{Q^g}(U)C_{Q^h}(U)$. As $Z \leq U$, $C_Q(U) \trianglelefteq C_G(U)$, so R is a 3-subgroup of G. Further $[Q, C_G(U)] \leq C_Q(U) \leq R$, so $R \trianglelefteq M_0$. As $C_Q(U)$ induces the group of transvections on W with center Z and axis U, and as R and W are normal in M_0 with $I \leq M_0$ inducing $SL(U)$ on U, we conclude $R/C_R(W)$ is the group of transvections of W with axis U. Then $|R/C_R(W)| = 27$. Now $|R| \leq |G|_3/|K_I|_3 = 3^{10}$ with $XR_I \leq C_R(W)$, $|XR_I| = 3^7$, and $|R/C_R(W)| = 27$, so we conclude $|R| = 3^{10}$, $XR_I = C_R(W)$, and M_0 contains a Sylow 3-subgroup of G, which we take to be P.

Let $K = \langle d \rangle K_I$ and $M = \langle d \rangle M_0$. Then $K \cong GL_3(3)$ and $KR/C_R(W)$ is the stabilizer in $GL(W)$ of U. Further $N_G(U)$ acts on $\bigcap_{x \in N_G(W)} Q^x = W$, so $N_G(U) = MC_G(W)$. Finally $C_G(W) = C_G(X) \cap C_G(U) = XR_I$, so $M = N_G(U)$.

Therefore $M = RK$ with $R = O_3(M)$ of order 3^{10} and $K \cong GL_3(3)$. As $R/XR_I \cong E_{27}$, $\Phi(R) \leq XR_I$ and as R_I is special with center U, $U \leq \Phi(R)$. Let $H = N_G(Z)$ and $H^* = H/Q$. Then

$$(H \cap M)^*/O_3((H \cap M)^*) \cong (H \cap M)/O_3(H \cap M) \cong \mathbf{Z}_2 \times GL_2(3)$$

and $|O_3((H \cap M)^*)| = 27$. So by 40.5.3, $(H \cap M)^* \cong \mathbf{Z}_2 \times (GL_3(3)/3^{1+2}$. Further $H \cap M$ is irreducible on $O_3(H \cap M)/R \cong E_9$, so $R^* = O_3((H \cap M)^*) \cong 3^{1+2}$. Therefore as $W \leq Q \cap R$, R/W is nonabelian and then as M is irreducible on R/XR_I and XR_I/W, we conclude R/W is special. In particular $XR_I = \Phi(R)W$, so, as $U \leq \Phi(R)$, $\Phi(R) = XR_I$ or $XR_I = X \times \Phi(R)$. But in the latter case as $X = [XR_i, d]$, we have $\Phi(R) = C_{XR_i}(d) = R_I$ and d inverts $R/\Phi(R)$. Then $R = [R, d] \leq C(R_I)$, contradicting R_I nonabelian.

So $\Phi(R) = XR_I = C_R(W)$. We conclude $U = Z_1(R)$, $W = Z_2(R)$, and $\Phi(R) = Z_3(R)$. Hence the lemma is established.

41. Subgroups of odd prime order in $M(24)$

In this section we determine the conjugacy classes of subgroups of odd prime order and the normalizers of such subgroups in $M(24)$. Thus in this section

$G = M(24)$ and D is the class of 3-transpositions of G. Let $d \in D$, $a \in A_d$, $X = \langle ad \rangle$, and $L = \langle D_{a,d} \rangle$. Recall

(41.1)

 (1) $N_G(X) = \langle a, d \rangle \times L$ with $\langle a, d \rangle \cong S_3$ and

$$L = C_G(\langle a, d \rangle) \cong S_3 / P\Omega_8^+(3).$$

 (2) $w_D(C_G(X)) = 8$.

Proof: By 25.9, $N_G(X) = \langle a, d \rangle \times \langle D_{a,d} \rangle$ with $\langle D_{a,d} \rangle \cong S_3 / P\Omega_8^+(3)$ as G is of type $M(24)$ and by 16.11. In particular $w_D(C_G(X))) = 8$ by Exercise 4.8.

Let $b_0 \in D_{a,d}$ and let V be the natural projective module for $E(L)\langle b_0 \rangle \cong PO_8^{+,+}(3)$, so that V has the structure of an 8-dimensional orthogonal space over $GF(3)$ of sign $+1$ and b_0 induces a reflection on V. Choose notation so that $[V, b_0]$ is of discriminant -1. We adopt the notation of 38.16 to describe the classes of subgroups of $E(L)$ of order 3.

(41.2) $C_G(X)$ *has two orbits on its D-subgroups $\langle b, c \rangle$ isomorphic to S_3: those with $c \in bE(L)$, for which*

$$C_G(\langle a, b, c, d \rangle) = \langle D_{a,b,c,d} \rangle \cong O_5^{+,+}(3) / E_{3^5},$$

and those with $c \notin bE(L)$, for which $C_G(\langle a, b, c, d \rangle) \cong G_2(3)$.

Proof: Let $b, c \in D_{a,d}$ with $c \in A_b$. By 9.5, $\langle b \rangle E(L)$ is rank 3 on $D_{a,d} \cap bE(L)$, so L is transitive on D-subgroups $\langle b, c \rangle \cong S_3$ with $c \in bE(L)$. By 11.12.5 and 38.13.5, $C_G(\langle a, b, c, d \rangle) = \langle D_{a,b,c,d} \rangle \cong O_5^{+,+}(3) / E_{3^5}$ for such b, c.

On the other hand let $b = b_0$ and suppose $c \notin bE(L)$. Then $C_{E(L)}(b) \cong \Omega_7(3)$ is the stabilizer in $E(L)$ of $[V, b]$ and as $b^{cb} = c$, $C_{E(L)}(c) = C_{E(L)}(b)^{cb}$ acts irreducibly on V and transitively on points of V of discriminant -1, with the stabilizer in $C_{E(L)}(c)$ of $[V, b]$ isomorphic to $G_2(3)$. Of course that stabilizer is $C_L(\langle b, c \rangle)$ and as $C_{E(L)}(c)$ is transitive on points of V of discriminant -1, L is transitive on D-subgroups $\langle b, c \rangle$ with $c \notin bE(L)$.

Let V_0 be a singular point in V, $A_X = O_3(N_{E(L)}(V_0))$, and $A = XA_X$.

(41.3) $A \cong E_{3^7}$ *and $N_G(A)$ is a D-subgroup of G such that A is the natural module of discriminant $+1$ for $N_G(A)/A \cong O_7^{+,+}(3)$ with X and members of*

$A_X \cap 3_-$ *points of A of discriminant* -1, *members of* $A_X \cap 2^2$ *singular points of A, and members of* $A_X \cap 3_+$ *points of A of discriminant* $+1$.

Proof: Let M_X be the stabilizer in $\langle b \rangle E(L)$ of V_0. By 38.13,

$$A_X = O_3(M_X) \cong E_{36}$$

is the natural module for $M_X / A_X \cong PO_6^+(3) \cong O_6^{-,+}(3)$. Thus

$$A = XA_X \cong E_{37}.$$

Let $M = N_G(A)$, $M^* = M/A$, and $E = N_D(A)$. Then E^* is a set of 3-transpositions of $\langle E^* \rangle$ and $d^* \in E^*$ with $C_{M^*}(d^*) = \langle d^* \rangle \times \langle E^*_{d*} \rangle$, $\langle E^*_{d*} \rangle \cong O_6^{-,+}(3)$, d^* inducing a reflection on A with center X, and $C_A(d)$ is the natural module for $\langle E^*_{d*} \rangle$.

Let $b = b_0 \in E \cap D_d$, $c \in A_b \cap bA$, and $I = \langle a, b, c, d, D_{a,b,c,d} \rangle$. By 41.2, $I = \langle a, d \rangle \times \langle b, c \rangle \times \langle D_{a,b,c,d} \rangle$ with $O_3(I) = A$ and $I^* \cong E_4 \times O_5^{+,+}(3)$.

Now $C_{M^*}(d^*)$ preserves a quadratic form q_d on $C_A(d)$ with $q_d(bc) = 1 = q_d(y)$ for each $y \in B = C_A(\langle a, b, c, d \rangle)$ inverted by a member of $D_{a,b,c,d}$. Similarly $C_{M^*}(b^*)$ preserves a quadratic form q_b on $C_A(b)$ with $q_b(ad) = 1 = q_b(y)$ for each such y. Then as I preserves q_b and q_d on B, we can extend q_b and q_d to a quadratic form q on A preserved by $M_0 = \langle C_M(d), C_M(b) \rangle$. But by Exercise 8.2.2, $\langle E^* \rangle = M_0^* = \langle E^* \rangle \cong O_7^{+,+}(3)$, so M_0^* is the subgroup of $O(A, q)$ generated by the set E^* of reflections $t(y)$ with $Q(y) = 1$. Further by a Frattini argument, $M = M_0 C_M(d) = M_0$.

By construction, $X = [A, d^*]$ is a point of the orthogonal space A of discriminant $-Q(ad) = -1$. Further M has two more orbits on the points of A, consisting of the singular points and the points of discriminant $+1$. By 38.13.4, the members of $2^2 \cap A_X$ are singular with respect to $q_{|C_A(d)} = q_d$ and the members of $A_X \cap 3_\epsilon$ are points of discriminant ϵ. Thus the proof of the lemma is complete.

(41.4) Let $Y \in 3_+ \cap A_X$. Then $N_G(Y) \leq N_G(A)$, $N_G(Y)/A \cong O_6^-(3)$, and $w_D(C_G(Y)) = 6$.

Proof: Let $M = N_G(A)$. By 41.3, $N_M(Y)/A$ is the stabilizer in $O_7^{+,+}(3)$ of the point Y of discriminant $+1$ in the orthogonal space A of discriminant $+1$, so $N_M(Y)/A \cong O_6^-(3)$ and hence $N_M(Y)$ is of width 6. So it remains to show $N_G(Y) \leq M$.

Let $d \in C_{D \cap M}(Y)$. Then the group $\langle d D_{a,d} \rangle \cong S_3/P\Omega_8^+(3)$ plays the role in $C_{E(G)}(d) \cong M(23)$ of the group of 40.1 with Y in the class 3_+ of $C_{E(G)}(d)$

defined in 40.2. Therefore by 40.4, $C_G(d) \cap N_G(Y) \leq M$. Then $N_G(Y) \leq M$ by 25.11.2.

(41.5) *Let $Z \in 2^2$. Then*

    (1) $F^*(N_G(Z)) = Q \cong 3^{1+10}$

    (2) $w_D(C_G(Z)) = 5$, and $\langle C_D(Z)\rangle/Q \cong \mathbf{Z}_2 \times U_5(2)$.

    (3) $N_G(Z)/Q \cong \mathbf{Z}_2 \times \text{Aut}(U_5(2))$.

    (4) $N_G(Z) = Q(N_G(Z) \cap C_G(t))$ where $t \in \mathcal{J}_3$ inverts Q/Z.

Proof: Let $M = N_G(Z)$ and $Q = O_3(M)$. As in the proof of the previous lemma, $\langle d D_{a,d}\rangle$ plays the role in $C_{E(G)}(d)$ of the group of 40.1 with Z in the class 2^2 of $C_{E(L)}(d)$ defined in 40.2. Therefore by 40.5, $C_{D_d}(Z) \subseteq L$, $O_3(C_G(Z\langle d\rangle)) = Q_d \leq L$ with $3^{1+8} \cong Q_d = F^*(C_L(Z)) \leq \langle C_{D_{a,d}}(X)\rangle$, and $w_D(C_G(Z)) = 5$. Further by 40.7, $C_G(d)$ is transitive on subgroups Z of $C_G(d)$ of order 3 with $w_D(C_G(Z)) = 5$, so M is transitive on $C_D(Z)$. Therefore by Exercise 3.1,

$$d^Q = \{e \in C_D(Z) : D_e \cap C_D(Z) = D_d \cap C_D(Z)\}$$
$$\subseteq \{e \in C_D(Z) \cap D_b : D_{e,b} \cap C_D(Z) = D_{d,b} \cap C_D(Z)\} = d^{Q_b}.$$

Then $Q = [Q, d]C_Q(d)$ with $C_Q(d) \leq O_3(C_M(d)) = Q_d$. But from the structure of $M \cap L$ described in 40.5, $M \cap L$ is irreducible on Q_d/Z and for $b \in M \cap D_{a,d}$, $[Q, b] \cong 3^{1+2}$. Therefore $C_Q(d) = Q_d$ and in particular $[Q_d, b] \leq Q$, so $[Q, b] = [Q_d, b] \cong 3^{1+2}$. Therefore as $d \in b^M$, $[Q, d] \cong 3^{1+2}$ centralizes $\langle C_{D_d}(Z)\rangle \leq Q_d$ and then $Q = [Q, d] \cap Q_b \cong 3^{1+10}$ with $a \in d[Q, d]$.

Next from 40.5 there exists an involution $t_L \in L \cap M$ inverting Q_d/Z such that $t_L = b_1 \cdots b_4$ is the product of four commuting members of $D_{a,c} \cap M$ and $\langle C_{D_d}(\langle t\rangle Z)\rangle/\langle d\rangle \cong \mathbf{Z}_2/3^{1+2}/Q_8^3$. Then $t = dt_L$ inverts Q/Z, so $M = QC_M(t)$. Further $t = db_1 \cdots b_4$ is the product of five commuting members of D, so as 5 is odd, $t \notin E(L)$, and hence by 37.4, $t \in D$ or $t \in \mathcal{J}_3$. As M is transitive on $C_D(Z)$, $t = d_1d_2d_3 \in \mathcal{J}_3$. Then by Exercise 8.2.8, $\langle C_D(Z\langle t\rangle)\rangle \cong \mathbf{Z}_2 \times U_5(2)$. But by 37.5, $C_G(t)/\langle d_1, d_2, d_3\rangle \cong \text{Aut}(U_6(2))$, so Z is determined up to conjugacy in $C_G(t)$ and $N_G(Z\langle t\rangle)/Z \cong \mathbf{Z}_2 \times \text{Aut}(U_5(2))$. Hence the lemma is established.

(41.6)

    (1) Let $E \in \text{Syl}_5(G)$. Then $E \cong E_{25}$ and $N_G(E) = S \times (N_G(E) \cap C_G(S))$, where $S = \langle C_D(E)\rangle \cong S_4$ and $(N_G(E) \cap C_G(S))/E \cong \mathbf{Z}_2/(\mathbf{Z}_4 * SL_2(3))$ with $C_G(E) = E \times S$.

(2) G has one class B^G of subgroups of order 5 and $N_G(B) = F \times \langle C_D(B) \rangle$ with F a Frobenius group of order 20 and $\langle C_D(B) \rangle \cong S_9$.

(3) Let X_i be a subgroup of $\langle C_D(B) \rangle$ with i cycles of length 3 in the representation of $\langle C_D(B) \rangle$ on nine points. Then $w_D(C_G(X_i)) = 8, 6, 5$ for $i = 1, 2, 3$, respectively.

Proof: Let $d \in D$ and $H = C_G(d)$. Then $|G|_5 = 5^2 = |H|_5$ so H contains a Sylow 5-subgroup E of G. By 40.8, $E \cong E_{25}$, $C_H(E) = E \langle d, b \rangle$ for $b \in D_d$, and $N_H(E)/C_H(E) \cong \mathbf{Z}_2/(\mathbf{Z}_4 * SL_2(3))$ with $E \langle d \rangle = C_H(E)$. Also by 40.9, H has one class B^H of subgroups of order 5 and $N_H(B) = \langle d \rangle \times F \times \langle C_{D_d}(B) \rangle$, with F a Frobenius group of order 20 and $\langle C_{D_d}(B) \rangle \cong S_7$. Therefore G has one class of subgroups of order 5 and by Exercise 8.2.7, $N_G(B) = F \times \langle C_D(B) \rangle$ with $\langle C_D(B) \rangle \cong S_9$. In particular for $B \leq E$, $C_G(E) = C_G(B) \cap C_G(E) = E \times S$, where $S = \langle C_D(E) \rangle \cong S_4$, completing the proof of (1) and (2).

Adopt the notation of (3). We may choose d to invert X_1 and then take $X_1 = X$. So by 41.1, $w_D(C_G(X_1)) = 8$. Then $C_G(B \langle a, d \rangle) = B \times K$ where $K = \langle C_D(B \langle a, d \rangle) \rangle \cong S_6$. In particular we may take $b_0 \in K \leq E(L) \langle b_0 \rangle$ and then $V = C_V(B) \perp [V, B]$ with $K = C_{E(L) \langle b_0 \rangle}([V, B]) \cong PO_4^-(3)$. Thus the classes of subgroups of order 3 in K are in 3_- and 3_+ and taking $X_2 \leq K$, $X_2 \in 3_+$. Therefore $w_D(C_G(X_2)) = 6$ by 41.4. Indeed we may take $X_2 \leq A_X$, so in the orthogonal space A of 41.3, X and X_2 are orthogonal points of discriminant -1 and $+1$, so the other two subgroups of order 3 in $X_1 X_2$ are fused into 2^2 under $N_G(A)$. We may take X_3 to be one of these, so by 41.5, $w_D(C_G(X_3)) = 5$.

(41.7) Let $Y \in 2^2 \cdot 3_{+,2}$. Then $w_D(C_G(Y)) = 2$, $|O_3(N_G(Y))| = 3^{12}$,

$\langle C_D(Y) \rangle / O_3(\langle C_D(Y) \rangle) \cong S_4$,   and $N_G(Y)/O_3(N_G(Y)) \cong S_4 \times GL_2(3)$.

Proof: Let $Y \in 2^2 \cdot 3_{+,2}$ and $M = N_G(Y)$. As in the proof of previous lemmas, and using 40.6, $C_M(d) \leq \langle d \rangle L$ with $w_D(C_M(d)) = 2$, $|O_3(C_M(d))| = 3^9$,

$$|O_3(C_M(d)) : O_3(C_M(d)) \cap C_G(b)| = 27 \quad \text{for } b \in D_{a,d} \cap M,$$

and $C_M(d)/O_3(C_M(d)) \langle d \rangle \cong \mathbf{Z}_2 \times GL_2(3)$. Then $w_D(M) = w_D(C_M(d)) = 2$. By 40.7, $C_G(d)$ is transitive on its subgroups Y of order 3 with $w_D(C_G(Y)) = 2$, so M is transitive on $M \cap D$. Therefore by Exercise 8.5, setting $K = \langle C_D(X) \rangle$, we have $K/O_3(K) \cong E_4$, S_4, or S_5. As $w_D(C_G(Y)) = 2$, the last case is out by 41.6. Thus to prove the second holds, it suffices to show Y centralizes a D-subgroup isomorphic to S_4.

Let $b \in D_d \cap M$ and $I = E(C_G(bd))$. By 37.5, $I/\langle bd \rangle \cong M(22)$ and $C_G(bd) = \langle s \rangle (I \times \langle d \rangle)$, where from the proof of 37.7, $s \in (bd)^G$ induces an outer automorphism on I with $C_G(\langle bd, s \rangle) \cong S_3/\Omega_8^+(2) = C_G(S)$ for a D-subgroup $S \cong S_4$ containing s, b, d. So it suffices to show Y centralizes a member of s^I.

By 39.7, I is transitive on its subgroups Y of order 3 with $w_D(C_G(Y)) = 2$, so by a Frattini argument, Y centralizes an involution $r \in sI$. By 37.7, I has three orbits on 4-groups $\langle r, bd \rangle$ with $r \in sI$ with representatives chosen so that $r = s$, su or suv for $du, dv \in C_D(X)$. If $r = s$ we are done and similarly from 37.7, $C_I(su) \leq C_I(s)$, so we may take $r = suv$. In this case $C_I(r)$ is transitive on its subgroups Y of order 3 with $w_D(C_{I\langle d \rangle}(\langle r, Y \rangle)) = 2$, so $C_I(Y)$ is transitive on $r^I \cap rC_I(Y)$. However $C_{I\langle r \rangle}(Y)/O_3(C_I(Y))\langle bd \rangle \cong \mathbf{Z}_2 \times \mathbf{SL}_2(3)$ by 39.8, so r centralizes an involution $t \in C_I(Y) - \langle bd \rangle$ and $tr \notin r^{C_I(Y)}$, so $tr \notin r^I$, reducing us to an earlier case.

So $K/O_3(K) \cong S_4$. Therefore $O_3(C_M(d)) \leq O_3(M)$ and as

$$C_M(d)/O_3(C_M(d)) \cong E_4 \times GL_2(3), \qquad M/O_3(M) \cong S_4 \times GL_2(3).$$

Also $[O_3(M), b] \leq O_3(C_M(d))$ by 8.4, so $|O_3(M) : C_{O_3(M)}(d)| = |O_3(M) : C_{O_3(M)}(b)| = |O_3(C_M(d) : O_3(C_M(d)) \cap C_G(b)| = 27$, and then as

$$C_{O_3(M)}(d) = O_3(C_M(d))$$

is of order 3^9, $|O_3(M)| = 3^{12}$, completing the proof.

(41.8) *Let $b \in D_{a,d}$, $c \in D_{a,d} - bE(L)$, $T = \langle a, b, c, d \rangle$ and $K = C_G(T)$. Then*

(1) $K \cong G_2(3)$ and KT is of index 2 in $N_G(T) = N_G(K)$.

(2) If Y is a subgroup of T of order 3 distinct from X and $\langle bc \rangle$ then $N_G(Y) \leq N_G(T)$ and $w_D(C_G(Y)) = 0$.

(3) G is transitive on its subgroups B of order 13 and, for B of order 13 in K, $N_G(B) \leq N_G(K)$ and FT is of index 2 in $N_G(B)$, where $F = N_K(B)$ is Frobenius of order 78.

Proof: By 41.2, $K \cong G_2(3)$ and there exists $g \in G$ interchanging $\langle a, d \rangle$ and $\langle b, c \rangle$. Thus g acts on T and then on $K = C_G(T)$.

Let $H = C_G(d)$. As $|G|_{13} = 13 = |K|_{13}$, a Sylow 13-group B of K is of order 13 and Sylow in G. By 40.11, $N_H(B) = F \times \langle b, c \rangle \langle d \rangle$ with F Frobenius of order 78. Then $w_D(C_G(B)) = 2$, and as $N_K(B)$ is of order 78, $F = N_K(B)$. Let $S = \langle C_D(B) \rangle$. By 25.3, $S \cong S_3^2$ or S_5. The last case is out since by 41.6, elements of order 5 don't centralize elements of order 13. Therefore $S = T$.

Next let $Y \leq S$ be of order 3 and distinct from X and $\langle bc \rangle$. Claim $Y \notin X^G$. For if so, Y is inverted by $e \in D$ with $[N_G(Y), e] \leq Y$, so e acts on XY and $\langle e, d \rangle$ induces D_8 on XY, contradicting D a set of 3-transpositions. So indeed $Y \notin X^G$ and hence $N_G(XY)$ acts on $\{X, \langle bc \rangle\}$ and hence $|N_G(XY) : TK| = 2$. Further by a Frattini argument, $N_G(K) = K(N_G(K) \cap N_G(B)) \leq N_G(XY)$ establishing (1) and (3).

Suppose Y centralizes a 3-transposition. Then by 40.7, Y^G is determined by $w_D(C_G(Y)) = w$ and $w = 8, 6, 5$ or 2. We have shown $Y \notin X^G$ so $w \neq 8$. As 13 divides the order of $C_G(Y)$, $w \neq 6$ by 41.4. Finally as 7 divides $|C_G(Y)|$, $w \neq 5$ or 2 by 41.5 and 41.7.

So $w_D(C_G(Y)) = 0$ and it remains to show $N_G(Y) \leq N_G(T)$. Let $z \in K$ be an involution and $M = C_G(z)$. Then $z \in \mathcal{J}_4$ so by 37.5.5, $Q = F^*(M) \cong 2^{1+12}$ and $F^*(M/Q)$ is the perfect central extension of $\mathbf{Z}_3$ by $U_4(3)$. Let $M^* = M/O_{2,3}(M)$, $E = D \cap M$, and $M_0 = \langle E \rangle$. Then by 29.9, 29.10, and 28.17.2, $M^* \cong PO_6^-(3)$ and $M_0^* \cong PO_6^{+,+}(3)$. Now for $d \in E$, $[d, O_{2,3}(M)] \leq Q$ and $|D \cap dQ| = 4$, so $C_{E^*}(Y^*) = C_E(Y^*)$ and therefore as $w_D(C_G(Y)) = 0$, $w_{E^*}(C_{M^*}(Y^*)) = 0$. We conclude from Exercise 13.1 that $C_{E(M^*)}(Y^*) \cong \mathbf{Z}_3 \times 3^{1+2}$.

Next from the structure of L, $C_Q(X) \cong 2^{1+8}$ and $C_Q(XY) \cong Q_8^2$. In particular $[Q, X] \cong Q_8^2 \cong [C_Q(X), bc]$, so $Q = [C_Q(bc), X] * [C_Q(X), bc] * C_Q(T)$ with each factor isomorphic to Q_8^2 and $C_Q(Y) = C_Q(T)$. Thus letting $Y = \langle y \rangle$ and regarding $\tilde{Q} = Q/Z$ as a 6-dimensional space over $GF(4)$ for $E(M/Q)$, y has eigenvalues ω, ω^{-1}, and 1 of multiplicity 1, where $\omega \in GF(4)$ is of order 3. Therefore a Sylow 3-subgroup of $C_M(Y)$ is abelian, so as $C_{E(M^*)}(Y^*) \cong \mathbf{Z}_3 \times 3^{1+2}$, we conclude $C_{M^\infty}(Y)^*$ is of index 3 in $C_{E(M^*)}(Y^*)$ and is isomorphic to $E_9 \times SL_2(3)^2$. Then as $|M \cap E(G) : M^\infty| = 2$, $|C_{M \cap E(G)}(Y)| \leq 2^6 \cdot 3^4 = |TK \cap M|$, so we conclude $C_{E(G)}(Y\langle z \rangle) = C_{TK}(z) \leq XYK$.

Let $I = C_{E(G)}(Y)$. As K has one class of involutions, it follows that either $I = XYK$ or XYK is strongly embedded in I. In the latter case by 7.6 in [SG], K has a subgroup of odd order transitive on the involutions in K. This is impossible as K has $|K : C_K(z)| = 3^4 \cdot 7 \cdot 13$ involutions, but no subgroup of odd order divisible by that integer. Therefore $XYK = I$ and then by a Frattini argument, $N_G(Y) \leq XYKN_G(B) \leq N_G(T)$.

(41.9) $w_D(C_G(Y)) > 0$ *for each subgroup Y of $XE(L)$ of order 3.*

Proof: If $Y \leq E(L)$ then $d \in C_D(Y)$, so we may take $Y = \langle y \rangle$ where $y = xl$ with $X = \langle x \rangle$ and l of order 3 in $E(L)$. Let $b = b_0$. An element k of $E(L)$ centralizes an element of $bE(L) \cap D$ precisely when $C_V(k)$ contains a point of discriminant -1, so l centralizes such an element unless l is of type 2^4 or $2^2 \cdot 3_-$. Further by 38.16.3, elements of type 2^4 and $2^2 \cdot 3_-$ are fused under L to elements centralizing members of $bE(L) \cap D$, so l centralizes some $c \in D_{a,d}$. Then $c \in C_D(X\langle l \rangle) \subseteq C_D(Y)$.

(41.10) *G has five classes Y_i^G, $1 \le i \le 5$, of subgroups of order 3 with $w_D(C_G(Y_i)) = 8, 6, 5, 2, 0$ for $i = 1, 2, 3, 4, 5$, respectively.*

Proof: By 40.7, $C_G(d)$ has four classes of subgroups $Y_i^{C_G(d)}$, $1 \le i \le 4$, of order 3 and $w_D(C_G(Y_i)) = w_D(C_G(Y_i) \cap C_G(d)) = 8, 6, 5, 2$, for $i = 1, 2, 3, 4$, respectively. Hence there is no further G-fusion of these classes. Thus it remains to show G is transitive on the set $\mathcal{Y}$ of subgroups of order 3 that centralize no 3-transposition. Let $\mathcal{Y}'$ consist of those subgroups Y of order 3 described in 41.8.2. By 41.2, $\mathcal{Y}'$ is an orbit of G on $\mathcal{Y}$, so we must show $\mathcal{Y} = \mathcal{Y}'$.

Let $Z = \langle z \rangle$ be a 3-central subgroup of order 3 centralizing d, so that $N_G(Z)$ is described in 41.5. Let $M = \langle C_D(Z) \rangle$, $Q = O_3(M)$, and $M^* = M/Q$. By 41.5, $Q \cong 3^{1+10}$ and $M^* \cong \mathbf{Z}_2 \times U_5(2)$ with $M = C_M(t)Q$ where t is an involution with $C_Q(t) = Z$. As $d \notin O^2(G)$, $d = ts$ where s^* is a transvection in $E(M^*) \cong U_5(2)$. As $|M|_3 = |G|_3 = 3^{16}$, it suffices to show $M \cap \mathcal{Y} \subseteq \mathcal{Y}'$.

Let U be the natural module for $E(M^*)$. If h^* is of order 3 in $E(M^*)$ then U is the orthogonal direct sum of eigenspaces for h^* and some eigenspace has dimension at least 2, so h^* centralizes the transvections with centers in that eigenspace. Therefore each subgroup of M of order 3 is fused into $C_M(d^*)$ under M, so it suffices to show $\mathcal{Y} \cap C_M(d^*) \subseteq \mathcal{Y}'$.

We saw during the proof of 41.5 that $Q_d = [Q, d] \cong 3^{1+2}$,

$$\langle D_d \cap M \rangle \cong \mathbf{Z}_2/3^{1+2}/Q_8^3/3^{1+8},$$

and $D_d \cap M \subseteq C_M(Q_d)$. But by 40.5, $C_{M^*}(d^*) \cong \mathbf{Z}_2 \times GL_2(3)/3^{1+2}/Q_8^3$ and as $Z = C_Q(t)$, for $b \in D_d \cap M$ we have $tb \in O^2(M)$ with $Z = C_{Q_d}(tb)$, so $\langle D_d \cap M \rangle = C_M(Q_d)$ and $C_M(d)/C_M(Q_d) \cong GL_2(3)$.

Let $Y = \langle y \rangle$ be of order 3 in $C_M(d^*)$. Then $y = y_+ y_-$ with $y_+ \in C_M(d)$, $y_- \in Q_d$ inverted by d, and y_+ and y_- 3-elements. Take $Y \in \mathcal{Y}$. Then $C_D(Y) = \emptyset$, so $y_- \ne 1$. Claim $y_+ \in C_G(y_-)$ and $yz \in y^{Q_d}$. For if not, $y_+ \notin C_M(Q_d)$, so y_+, and hence also y, induces an outer automorphism on Q_d. Therefore $YQ_d \cong \mathbf{Z}_3 \, wr \, \mathbf{Z}_3$, so its maximal subgroups of exponent 3 are Q_d and $J(YQ_d) = J \cong E_{27}$. So as y induces an outer automorphism on Q_d, $y \in J$ and then as $y_+ \in C_{YQ_d}(d) \le J$, the claim holds.

As d inverts y_-, we may take $X = \langle y_- \rangle$. Hence by 41.9, $y_+ \notin E(L)$. But $K^* = O^2(E(L) \cap M)^* \cong SL_2(3) * SL_2(3) * SL_2(3)$, so y permutes the 3

$SL_2(3)$ factors transitively. Therefore K^* is transitive on the elements of order 3 in y^*K^*, so all such elements are conjugate to y^* and $C_{K^*}(y^*) \cong SL_2(3)$. In particular it suffices to show that for each y' of order 3 in $yC_Q(d)$, $\langle y' \rangle \in \mathcal{Y}'$.

Also by 38.16.5, $C_Q(d)/Z$ is the tensor product of natural modules for the 3 $SL_2(3)$ factors. Hence $C_Q(d) = Q_3 * Q_1$ where Q_3/Z is the sum of two Jordon blocks of size 3 for y, $[Q_1, y] \le Z$, and $C_{K^*}(y^*)$ induces $SL_2(3)$ on Q_1/Z. So Q_3/Z is transitive on elements of order 3 in yQ_3/Z and $C_{K^*}(y^*)$ is transitive on $(Q_1/Z)^{\#}$, so each element of order 3 in $yC_Q(d)$ is conjugate under $C_K(y^*)$ to y or yg for any given fixed $g \in Q_1Q_2 - Q_2$, where $Q_2/Z = C_{Q_3/Z}(y)$, so it suffices to show $\langle yg \rangle \in Y^G$. Also $Q_2 \cong E_{27}$ and $yz \in y^{Q_d}$.

Now pick $Y \in \mathcal{Y}'$ with $YX = O_3(C_G(Y))$. Then by 41.8,

$$E = E(C_G(Y)) \cong G_2(3)$$

and $O^2(C_G(Y)) = YX \times E$. Now $3^{1+2} \times E_9 \cong Q_1Q_2 = O_3(O^2(C_K(Y)) \le E$ so $Q_1Q_2 = O_3(C_E(Z))$ and we may take Z to be a long root subgroup of E. By the Chevalley commutator relations there is a long root element $g \in Q_1Q_2 - Q_2$, so as g and z are fused in E, $yg \in (yz)^E$. Hence as $yz \in y^Q$, $\langle yg \rangle \in Y^G$, completing the proof.

Notice we have now determined the classes of subgroups of G of order 3 and their normalizers.

(41.11)

(1) G has two conjugacy classes X_i^G, $i = 1, 2$, of subgroups of order 7.

(2) $w_D(C_G(X_1)) = 3$ and $N_G(X_1) = F \times \langle C_D(X_1) \rangle$ with F a Frobenius group of order 42 and $\langle C_D(X_1) \rangle \cong S_7$.

(3) $w_D(C_G(X_2)) = 0$ and $N_G(X_2)$ is the split extension of $P \cong 7^{1+2}$ by $\mathbf{Z}_6$ wr $\mathbf{Z}_2$, acting as the global stabilizer of two maximal subgroups E and E' of P. Further $X_1^G \cap P$ is the set of subgroups of order 7 distinct from X_2 and contained in E or E'.

(4) G has a subgroup isomorphic to Held's sporadic group He.

Proof: By 40.9, a Sylow 7-subgroup X_1 of $H = C_G(d)$ is of order 7 and $N_H(X_1) = F \times \langle C_{H \cap D}(X_1) \rangle$ with F Frobenius of order 42 and $\langle C_{H \cap D}(X_1) \rangle \cong \mathbf{Z}_2 \times S_5$. In particular $w_D(C_G(X_1)) = 3$ and H is transitive on $X_1^G \cap H$ so $C_G(X_1)$ is transitive on $C_D(X_1)$. Therefore by Exercise 8.2.7, $\langle C_D(X_1) \rangle \cong S_7$ and $N_G(X_1) = \langle C_D(X_1) \rangle N_H(X_1) = \langle C_D(X_1) \rangle \times F$.

Let $E \in \mathrm{Syl}_7(N_G(X_1))$. Then $E = X_1 \times X_2$ with $X_2 = E \cap \langle C_D(X_1) \rangle$ of order 7, and $N_G(X_1) \cap N_G(E) = F \times F_2$, where $F_2 = N_{\langle C_D(X_1) \rangle}(E)$ is Frobenius of order 42. Let f, s be the involutions in F, F_2, respectively, and

$t = fs$. Now $F \leq H \cap E(G)$, so $f \in E(G)$ and hence $f \in \mathcal{J}_2 \cup \mathcal{J}_4$ by 37.5. On the other hand as s is the product of three members of $C_D(X_1)$, $s \in \mathcal{J}_3$ and $t = fs \notin E(G) \cup D$ so $t \in \mathcal{J}_3$ by 37.5. Notice also that FF_2 has three orbits $\{X_1\}$, $\{X_2\}$, and X_0^F of order 1,1,6, respectively, on the subgroups of E of order 7.

Next $C_G(E) = C_G(E) \cap N_G(X_1) = E$ and t inverts E, so $R = N_G(E) \cap C_G(t)$ is a complement to E in $N_G(E)$ and is faithful on E, so $R \leq GL(E)$. Let $R_1 = F \cap R$ and $R_2 = F_2 \cap R$. Notice that $R_1 R_2$ is the stabilizer in $GL(E)$ of the pair of points X_1 and X_2.

Let $P \in \mathrm{Syl}_7(N_G(E))$. As $|G|_7 = 7^3$, $P \in \mathrm{Syl}_7(G)$. Also P is the split extension of E by $P \cap R$ of order 7 so $P \cong 7^{1+2}$. Further $P \cap R$ fuses X_1 into the orbit X_0^F of order 6 and $X_2 = Z(P)$, so as $E \in \mathrm{Syl}_7(N_G(X_1))$, $X_2 \notin X_1^G$. Thus R acts on X_2, $X_1^G \cap E = X_1^P = \{X_1\} \cup X_0^F$ is of order 7, and by an order argument $R = (P \cap R)FF_2$ is the stabilizer in $GL(E)$ of the point X_2. In particular $P \trianglelefteq N_G(E)$ and $N_G(E)$ is the split extension of P by $R_1 R_2$.

Next as $t \in \mathcal{J}_3$, $t = d_1 d_2 d_3$ is the product of three 3-transpositions and $X_3 = P \cap R \leq O^{7'}(C_G(t)) \leq C_G(d_i)$, so $X_3 \in X_1^G$. Now PF has four orbits $\{X_2\}$, X_1^P, X_3^P, and X_4^{FP} of order 1, 7, 7, 42, on subgroups of P of order 7. We see in a moment that X_2 is not weakly closed in P with respect to G, so $X_2^G = \{X_2\} \cup X_4^{FP}$ and $X_1^G \cap P = X_1^P \cup X_3^P$ consists of the subgroups of order 7 in E and $E' = X_2 X_3$ distinct from X_2. In particular G has two classes of subgroups of order 7. Also $N_G(P)$ acts on $\{E, E'\}$.

The proof of 32.5 in [SG] shows G has a subgroup G_0 isomorphic to He. As $|G_0|_7 = 7^3 = |G|_7$, we may take $P \in \mathrm{Syl}_7(G_0)$. Then the discussion at the end of Section 52 of [SG] shows that X_2 is not weakly closed in P with respect to G_0, as claimed.

As E, E' are Sylow in $N_G(X_1)$, $N_G(X_3)$, respectively, as $E, E' \trianglelefteq P$, and as $X_3 \in X_1^G$, X_1 and X_3 are fused in $N_G(P)$.

Next $C_G(P) = C_G(P) \cap C_G(E) = C_E(P) = X_2$ so $N_G(P)/P$ is faithful on P/X_2. Then as f inverts P/X_2, S is a complement to P in $N_G(P)$, where $S = N_G(P) \cap C_G(f)$. Then $S \leq GL(P/X_2)$ and S is contained in the global stabilizer $\mathbf{Z}_6 \ wr \ \mathbf{Z}_2$ of the pair $\{E/X_2, E'/X_2\}$ of points of P/X_2. As E and E' are fused in $N_G(P)$ and $R_1 R_2$ is the pointwise stabilizer of $\{E/X_2, E'/X_2\}$, we conclude $S \cong \mathbf{Z}_6 \ wr \ \mathbf{Z}_2$ is the global stabilizer.

Let $K = N_G(X_2)$. To complete the proof of 41.11, it remains only to show that $K = N_G(P)$. Recall $f \in \mathcal{J}_2 \cup \mathcal{J}_4$. If $f \in \mathcal{J}_2$ then $f = d_1 d_2$ is the product of two 3-transpositions centralized by X_2, contradicting $X_2 \notin X_1^G$. So $f \in \mathcal{J}_4$. Then from the proof of 28.20, all nontrivial orbits of $C_G(f)$ on $O_2(C_G(f))/\langle f \rangle$ have order divisible by 7, so $\langle f \rangle = O_2(C_G(f)) \cap C_G(X_2)$. Then by Exercise 14.1 and a Frattini argument, $N_G(X_2) \cap C_G(f)$ is of order

$7 \cdot 2^3 \cdot 3^2 = |SX_2|$, so $SX_2 = C_K(f)$. In particular a Sylow 2-subgroup of $C_G(X_2\langle f\rangle)$ is cyclic of order 4, so $K = O(K)S_2$, where $S_2 \in \text{Syl}_2(S)$. Then as $C_{O(C_G(X_2))}(f) \cong Z_{21}$, we conclude $P \trianglelefteq O(K)$, so indeed $K = N_G(P)$.

(41.12) *G is transitive on subgroups of order 11 and if B is such a subgroup then $w_D(C_G(B)) = 3$ with $\langle C_D(B)\rangle \cong E_8$, $N_G(B)$ induces S_3 on $C_D(B)$, and $N_G(B) \cap C_G(C_D(B)) = F \times \langle C_D(B)\rangle$, where F is Frobenius of order 55.*

Proof: Let $d \in D$ and $H = C_G(d)$. Then $|G|_{11} = |H|_{11} = 11$, so by Sylow, G is transitive on its subgroups B of order 11, we may take $B \leq H$, and $N_G(B)$ is transitive on $C_D(B)$. Now by 40.10, $C_{D\cap H}(X)$ is of order 3 and $N_H(B) \cap C_G(C_{D\cap H}(B)) = F \times \langle C_{H\cap D}(B)\rangle$. Finally if $a \in A_d$, then $C_G(\langle a, d\rangle)$ has order prime to 11, so $C_D(X) \subseteq H$, completing the proof.

(41.13)

    (1) If B is of order 17 then $C_D(B)$ is of order 1 and $N_{E(G)}(B)$ is a Frobenius group of order $17 \cdot 16$.

    (2) If B is of order 23 then $C_D(B)$ is of order 1 and $N_{E(G)}(X)$ is a Frobenius group of order $23 \cdot 11$.

Proof: This follows from 40.12 via a now familiar argument.

(41.14) *If B is of order 29 then $N_G(B)$ is Frobenius of order $29 \cdot 28$.*

Proof: Use Sylow's Theorem and the usual argument.

Exercises

1. Let $G \cong U_4(3)$ and X be of order 7 in G. Prove $N_G(X)$ is a Frobenius group of order 21.

15

Normalizers of subgroups of prime order in Fischer groups

In this chapter we summarize in Tables $M(22)$, $M(23)$, and $M(24)$ much of the information we generated in Part III about the classes of subgroups of prime order in the Fischer groups and their normalizers.

Column 1 lists the class. To be consistent with other references, we have chosen the same notation for the classes used in the *Atlas* [At] and in the tables in the Gorenstein and Lyons Memoir [GL]. Thus for example pX denotes the Xth class of subgroups of order p. For $M(22)$ and $M(24)$ we have included the classes of outer involutory automorphisms; to emphasize that these involutions do not live in the simple normal subgroup, we write (Out) after the class.

Column 2 describes the normalizer in G of type M of a subgroup in the class. We use our usual notational conventions. In particular recall from Section 2 that

$$H_n / H_{n-1} / \cdots / H_1$$

denotes a group K with a normal series $1 = K_0 \trianglelefteq K_1 \trianglelefteq \cdots \trianglelefteq K_n = K$ such that $K_i / K_{i-1} \cong H_i$ for $1 \le i \le n$.

Although we list the centralizers of involutory outer automorphisms of $G = M(22)$ in Table $M(22)$, these are centralizers in $M(22)$, not in $\mathrm{Aut}(M(22))$, and similarly the normalizers of subgroups of prime order are normalizers in $M(22)$, not in $\mathrm{Aut}(M(22))$. On the other hand in Table $M(24)$, our normalizers are normalizers in the 3-transposition group $M(24)$, *not in its simple derived subgroup*.

Most of the normalizers are listed explicitly in earlier sections in Part III. For example the classes of involutions and their centralizers are listed in Section 37, and the classes of subgroups of odd prime order and their normalizers are listed in Sections 39, 40, and 41, for $M(22)$, $M(23)$, and $M(24)$, respectively. A small amount of extra effort is necessary to retrieve the structure of the centralizers $C_G(t)$ of two of the inner involutions $t \in G = M(22)$. There is a good description of $C_A(t)$ in Lemma 37.8, where $A = \mathrm{Aut}(G)$, but we want

250

$C_G(t)$. Notice $2B = \mathcal{J}_1$, $2A = \mathcal{J}_2$, and $2C = \mathcal{J}_3$. If $j_2 \in \mathcal{J}_2 = 2A$, then by
37.8, $F^*(C_A(j_2)) = Q \cong D_8^5$ and from the proof of that lemma, $Q \nleq G$, with
$F^*(C_G(j_2)) = Q \cap G \cong \mathbf{Z}_2 \times D_8^4$. Thus $C_G(j_2)$ is as indicated in the table.

Similarly by 37.8.5, $C_A(j_3)$ is contained in the subgroup M of A described
in that lemma. Now by 37.8.3 and 37.8.4, $F^*(M) = R$ is special with center
$B \cong E_{32}$, $R/B \cong E_{2^8}$, and $M/R = M_1/R \times M_2/R$, where $M_1 = C_G(B)$ with
$M_1/B \cong S_3$ and $M_2 = \langle Q^M \rangle$ with $M_2/R \cong S_6$. Further from the proof of 37.8,
$R \leq M \cap G = M_1 N_2$ with $N_2/R \cong A_6$. Next from 37.8, B is the permutation
module for M_2 on $B \cap D$ of degree 6, modulo its center, with $j_3 = d_1 d_2 d_3$ the
product of three members of $B \cap D$, so $C_G(j_3) = M_1 C_{N_2}(j_3)$ with $C_{M_2}(j_3)/R$
the stabilizer in $N_2/R = A_6$ of the partition $B \cap D = \{\{d_1, d_2, d_3\}, \{d_4, d_5, d_6\}\}$,
so $C_{N_2}(j_3)/R \cong \mathbf{Z}_4/E_9$. Thus the centralizer of $j_3 \in 2C$ is as described in the
table.

Table $M(22)$

$G = M(22)$; $|G| = 2^{17} \cdot 3^9 \cdot 5^2 \cdot 7 \cdot 11 \cdot 13$

Normalizers of subgroups of prime order

$2A$	$O_6^-(2)/(\mathbf{Z}_2 \times D_8^4)$
$2B$	$U_6(2)/\mathbf{Z}_2$ quasisimple
$2C$	$((\mathbf{Z}_4/E_9) \times S_3)/\text{special}(2^8/2^5)$
$2D(\text{Out})$	$S_3/\Omega_8^+(2) = \text{Aut}(\Omega_8^+(2))$
$2E(\text{Out})$	$\mathbf{Z}_2 \times \text{Sp}_6(2)$
$2E(\text{Out})$	$O_6^-(2)/E_{64}$
$3A$	$\mathbf{Z}_2/E_9/Q_8^3/3^{1+6}$
$3B$	$S_3 \times (\mathbf{Z}_2/U_4(3))$
$3C$	$O_4^+(3)/E_{35}$
$3D$	$GL_2(3)/3^6$
$5A$	$(\mathbf{Z}_4/\mathbf{Z}_5) \times S_5$
$7A$	$(\mathbf{Z}_6/\mathbf{Z}_7) \times S_3$
$11A$	$(\mathbf{Z}_5/\mathbf{Z}_{11}) \times \mathbf{Z}_2$
$13A$	$\mathbf{Z}_6/\mathbf{Z}_{13}$

Table $M(23)$

$$G = M(23); \ |G| = 2^{18} \cdot 3^{13} \cdot 5^2 \cdot 7 \cdot 11 \cdot 13 \cdot 17 \cdot 23$$

Normalizers of subgroups of prime order

$2A$	$M(22)/\mathbf{Z}_2$ quasisimple
$2B$	$\mathbf{Z}_2/U_6(2)/E_4$
$2C$	$\mathbf{Z}_2/(\mathbf{Z}_3 \times U_4(2))/(E_4 \times D_8^4)$
$3A$	$GL_2(3)/3^{1+2}/Q_8^3/3^{1+8}$
$3B$	$S_3 \times \Omega_7(3)$
$3C$	$O_5(3)/E_{36}$
$3D$	$(\mathbf{Z}_2 \times GL_2(3))/3^9$
$5A$	$(\mathbf{Z}_4/\mathbf{Z}_5) \times S_7$
$7A$	$(\mathbf{Z}_6/\mathbf{Z}_7) \times S_5$
$11A$	$\mathbf{Z}_2/((\mathbf{Z}_5/\mathbf{Z}_{11}) \times E_4)$
$13A$	$(\mathbf{Z}_6/\mathbf{Z}_{13}) \times S_3$
$17A$	$\mathbf{Z}_{16}/\mathbf{Z}_{17}$
$23A$	$\mathbf{Z}_{11}/\mathbf{Z}_{23}$

Table $M(24)$

$$G = M(24); \ |G| = 2^{22} \cdot 3^{16} \cdot 5^2 \cdot 7^3 \cdot 11 \cdot 13 \cdot 17 \cdot 23 \cdot 29$$

Normalizers of subgroups of prime order

$2A$	$E_4/U_4(3)/\mathbf{Z}_3/D_8^6$
$2B$	$\mathrm{Aut}(M(22))/E_4$
$2C(\mathrm{Out})$	$\mathbf{Z}_2 \times M(23)$
$2D(\mathrm{Out})$	$\mathrm{Aut}(U_6(2))/E_8$
$3A$	$(\mathbf{Z}_2 \times \mathrm{Aut}(U_5(2)))/3^{1+10}$
$3B$	$S_3 \times (S_3/P\Omega_8^+(3))$
$3C$	$O_6^-(3)/E_{3^7}$
$3D$	$(S_4 \times GL_2(3))/3^{12}$
$3E$	$\mathbf{Z}_2/((\mathbf{Z}_2/E_9) \times G_2(3))$
$5A$	$(\mathbf{Z}_4/\mathbf{Z}_5) \times S_9$
$7A$	$(\mathbf{Z}_6/\mathbf{Z}_7) \times S_7$
$7B$	$(\mathbf{Z}_6 \ wr \ \mathbf{Z}_2)/7^{1+2}$
$11A$	$S_3/((\mathbf{Z}_5/\mathbf{Z}_{11}) \times E_8)$
$13A$	$\mathbf{Z}_2/(S_3 \times S_3 \times (\mathbf{Z}_6/\mathbf{Z}_{13}))$
$17A$	$(\mathbf{Z}_{16}/\mathbf{Z}_{17}) \times \mathbf{Z}_2$
$23A$	$(\mathbf{Z}_{11}/\mathbf{Z}_{23}) \times \mathbf{Z}_2$
$29A$	$\mathbf{Z}_{28}/\mathbf{Z}_{29}$

References

[A1] M. Aschbacher, On finite groups generated by odd transpositions, I–IV, *J. Alg.* **26** (1973), 479–491.

[A2] M. Aschbacher, A homomorphism theorem for finite graphs, *Proc. AMS* **54** (1976), 468–470.

[A3] M. Aschbacher, On the maximal subgroups of the finite classical groups, *Invent. Math.* **76** (1984), 469–514.

[AH] M. Aschbacher and M. Hall, Groups generated by a class of elements of order 3, *J. Alg.* **24** (1973), 591–612.

[AS] M. Aschbacher and Y. Segev, Extending morphisms of groups and graphs, *Ann. Math.* **135** (1992), 297–323.

[At] J. Conway, R. Curtis, S. Norton, R. Parker, and R. Wilson, *An Atlas of Finite Groups*, Clarendon Press, Oxford (1985.

[B] N. Bourbaki, *Groupes et algèbres de Lie,* 4, 5, 6, Hermann, Paris, 1968.

[BrS] R. Brauer and M. Suzuki, On finite groups of even order whose 2-Sylow subgroup is a quaternion group, *Proc. Nat. Acad. Sci. USA* **45** (1959), 1757–1759.

[BS] F. Buekenhout and E. Shult, Foundations of polar geometry, *Geom. Ded.* **3** (1974), 155–170.

[Bu] F. Buekenhout, La géométrie des groupes des Fischer, Free University of Brussels, Lecture notes, 1974.

[CH1] H. Cuypers and J. Hall, The classification of 3-transposition groups with trivial center, *Groups, Combinatorics and Geometry*, Cambridge University Press, Cambridge, 1992, pp. 121–138.

[CH2] H. Cuypers and J. Hall, The 3-tranposition groups with trivial center, *J. Alg.* **178** (1995), 149–193.

[DGW] S. Danielson, M. Guterman, and R. Weiss, On Fischer's characterization of Σ_5 and Σ_n, *Comm. Alg.* **11** (1983), 1501–1510.

[F1] B. Fischer , Finite groups generated by 3-transpositions, University of Warwick, Lecture notes, 1969.

[F2] B. Fischer, Finite groups generated by 3-transpositions, *Invent. Math.* **13** (1971), 232–246.

[FGT] M. Aschbacher, *Finite Group Theory*, Cambridge University Press, Cambridge, 1986.

[G] G. Glauberman, Central elements of core-free groups, *J. Alg.* **4** (1966), 403–420.

[GL] D. Gorenstein and R. Lyons, The local structure of finite groups of characteristic 2 type, *Memoirs AMS* **276** (1983), 1–731.

[H1] J. Hall, On the order of Hall triple systems, *J. Comb. Theory, Series A* **29** (1980), 261–262.

[H2] J. Hall, Graphs, geometry, 3-transpositions, and symplectic F_2-transvection groups, *Proc. London Math. Soc.* **58** (1989), 89–111.

[H3] J. Hall, Some 3-transposition groups with normal 2-subgroups, *Proc. London Math. Soc.* **58** (1989), 112–136.

[H4] M. Hall, Jr., Automorphisms of Steiner triple systems, *Proc. Symp. Pure Math.* **6** (1962), 47–66.

[Hi] D. Higman, Finite permutation groups of rank 3, *M. Zeit.* **86** (1964), 145–156.

[Hu1] D. Hunt, Character tables of certain finite simple groups, *Bull. Australian Math. Soc.* **5** (1971), 1–42.

[Hu2] D. Hunt, A characterization of the finite simple group $M(22)$, *J. Alg.* **21** (1972), 103–112.

[Hu3] D. Hunt, A characterization of the finite simple group $M(23)$, *J. Alg.* **26** (1973), 431–439.

[P] D. Parrott, Characterizations of the Fischer groups, I, II, III, *Trans. AMS* **265** (1981), 303–347.

[SG] M. Aschbacher, *Sporadic Groups*, Cambridge University Press, Cambridge, 1994.

[SW] U. Semmler and R. Weiss, On Fischer's characterization of Σ_4 and Σ_5, *Arch. Math.* **36** (1981), 120–124.

[T1] F. Timmesfeld, A characterization of the Chevalley and Steinberg groups over F_2, *Geom. Ded.* **1** (1973), 269–323.

[T2] F. Timmesfeld, Groups generated by root-involutions, I, II, *J. Alg.* **35** (1975), 367–441.

[vBW] J. van Bon and R. Weiss, An existence lemma for groups generated by 3-transpositions, *Invent. Math.* **109** (1992), 519–534.

[W1] R. Weiss, A uniqueness lemma for groups generated by 3-transpositions, *Math. Proc. Camb. Phil. Soc.* **97** (1985), 421–431.

[W2] R. Weiss, On Fischer's characterization of $Sp_{2n}(2)$ and $U_n(2)$, *Comm. Alg.* **11** (1983), 2527–2554.

[W3] R. Weiss, 3-transpositions in infinite groups, *Math. Proc. Camb. Phil. Soc.* **96** (1984), 371–377.

[Z1] F. Zara, Sur les couples Fischeriens de largeur 1, *Europ. J. Comb.* **4** (1983), 185–199.

[Z2] F. Zara, Classification des couples Fischeriens, Thèse Université de Picardie, 1984.

[Z3] F. Zara, A first step toward a classification of Fischer groups, *Geom. Ded.* **25** (1988), 503–512.

Symbols

Index